Springer Series in Reliability Engineering

Series Editor

Hoang Pham, Department of Industrial and Systems Engineering
Rutgers University, Piscataway, NJ, USA

More information about this series at http://www.springer.com/series/6917

Jan-Erik Vinnem · Willy Røed

Offshore Risk Assessment Vol. 2

Principles, Modelling and Applications of QRA Studies

Fourth Edition

Springer

Jan-Erik Vinnem
Faculty of Engineering
Norwegian University of Science
and Technology
Trondheim, Norway

Willy Røed
Faculty of Science and Technology
University of Stavanger
Stavanger, Norway

ISSN 1614-7839   ISSN 2196-999X   (electronic)
Springer Series in Reliability Engineering
ISBN 978-1-4471-7450-9   ISBN 978-1-4471-7448-6   (eBook)
https://doi.org/10.1007/978-1-4471-7448-6

This Springer imprint is published by the registered company Springer-Verlag London Ltd. part of Springer Nature.
The registered company address is: The Campus, 4 Crinan Street, London, N1 9XW, United Kingdom

Preface to Fourth Edition

This is the fourth edition of the book; the first edition was published in 1999, the second edition in 2007 and the third edition in 2014. This version represents a new development where we have been two authors co-operating on the updating of the manuscript. My earlier M.Sc. and Ph.D. student and good friend, Willy Røed, Adjunct Professor at the University of Stavanger, Department of Safety, Economics and Planning, and consultant in Proactima in Stavanger, has shared the workload with me regarding the updating of the chapters and the new text.

The original author has since 2014 been professor at NTNU, Department of Marine Technology, Trondheim, in marine operational risk. The authors have been involved in several significant research projects during the last few years. Some few major accidents worldwide have had a considerable effect on the HES performance and awareness, in addition to the substantial fall of oil price in 2013. This prompted a need for a further update of the book.

Norwegian offshore installations have for more than 40 years been large and complex and manned installations with extensive maintenance needs. Manning levels have been reduced from a few hundred to around one hundred and considerably less for recent installations. There is now an interest in unmanned, simple wellhead installations, as they are less expensive to install than subsea production equipment. The first normally unmanned wellhead installation in the Norwegian sector will start to operate in stand-alone mode from the first half of 2019. It is envisaged that within a few years, also unmanned production installations, fixed and floating, will be developed. Unmanned installations have been used in other offshore sectors (UK, the Netherlands, USA, etc.), but these have been small, simple, shallow water installations. In future, we may find quite complex, unmanned production installations also in deep water if technology optimists are to be believed. An extensive new chapter (22) is devoted to discussion of some of the challenges that unmanned installations are faced with.

Gratitude is expressed to the Springer Nature Publishers (London), in particular Executive Editor Anthony Doyle, Project Coordinator Arun Kumar Anbalagan and their staff, for agreeing to publish the fourth edition of this book and for providing inspiring and valuable advice and assistance throughout the process.

Appendix A has been updated with the latest overview of some of the important software tools that are commonly used in offshore risk assessment. Thanks to all the consultancies and software suppliers who have provided the information required for the update of this appendix.

There are also several people who have kindly contributed to relevant information on various aspects or prepared small text sections for us to use; Dr. Haibo Chen, consultant to Lloyd's Register Consulting, China; Silje Frost Budde, Safetec Nordic; Sandra Hogenboom, DNV GL (Ph.D. student at NTNU); Petter Johnsen, Presight Solutions; Kenneth Titlestad, Sopra Steria; Dr. Xingwei Zhen, Dalian University of Technology, China; Odd W. Brude, DNV GL; Kaia Stødle, Ph.D. student at UiS; and Eldbjørg Holmaas, Proactima. Many thanks to all of you for valuable assistance.

Karoline Lilleås Skretting, M.Sc. student at the Department of Marine Technology, NTNU, has been our assistant during the final stages of the revision work. She has mainly been assisting with the updating of Appendix A, in addition to some other editorial tasks, and this has been very helpful to finishing the revision work in a timely manner. We are grateful for the excellent work done by Karoline. The Department of Marine Technology, NTNU, has supported the publishing of the book by paying the salary for Karoline.

Safetec Nordic has allowed us to use their QRAToolkit software, whereas ConocoPhillips, Equinor and Vår Energi have allowed us to use the results from a Joint Industry Project on collision risk modelling. We are also grateful to the RISP project and its contributors for providing access to relevant information and results before the RISP project is finalised.

May 2019   Jan-Erik Vinnem
                                                      Professor, Norwegian University
                                                           of Science and Technology
                                                                Trondheim, Norway

                                                                      Willy Røed
                                             Adjunct Professor, University of Stavanger
                                                                Stavanger, Norway

Preface to Third Edition

This is the third edition of the book; the first edition was published in 1999 and the second edition in 2007. The author has since then returned to an adjunct professorship at University of Stavanger, Norway, teaching a course in applied offshore risk assessment. Starting from January 2013, the author is also adjunct professor at NTNU, Trondheim in marine operational risk. The author has been involved in several significant research projects during the last few years. Several major accidents worldwide have had considerable effect on the HES performance and awareness, not the least the Macondo accident in 2010. This prompted a need for a further update of the book.

Norwegian offshore regulations were profoundly revised around the beginning of the new century, with new regulation becoming law from 2002. A limited revision was implemented from 2011, mainly limited to the integration of onshore petroleum facilities. This edition of the book captures some of the experience and challenges from the application of the new regulations. The important aspects of the new regulations are also briefly discussed, see Chapter 1.

About 30 major accidents and incidents are discussed at some length in Chapters 4 and 5 (Macondo accident), in order to demonstrate what problems have been experienced in the past. I have increased the emphasis on this subject in both the second and third editions, because it is essential that also new generations may learn from what occurred in the past. Where available, observations about barrier performance are discussed in addition to the sequence of events and lessons learned.

It is often claimed "what is measured will be focused upon". This implies that even if QRA studies have several weaknesses and limitations, quantification is the best way to focus the attention in major hazard risk management. This is also one of the lessons from the Macondo accident, in the author's view. It has therefore been surprising to realise how strong the opposition to QRA studies still is at the end of 2012 from many professionals in major international oil companies. This has to some extent given further inspiration to update this book, about a topic I consider crucial for improvement of major hazard risk management in the offshore petroleum industry.

Thanks are expressed to Springer London publishers, in particular Senior Editor Anthony Doyle and his staff, for agreeing to publish the third edition of this book, and for providing inspiring and valuable advice and assistance throughout the process.

Appendix A presents an overview of some of the important software tools that are commonly used in offshore risk assessment. Thanks to all the consultancies and software suppliers who have provided the information required for the update of this appendix.

There are also several people who have kindly contributed with relevant information on various aspects; Torleif Husebø, PSA; Prof. Stein Haugen, NTNU; Celma Regina Hellebust, Hellebust International Consultants, Prof. Bernt Aadnøy, UiS and Dr. Haibo Chen, Scandpower Inc. China. Many thanks for valuable assistance to all of you.

Meihua Fang has been my assistant during the final stages of the revision work, during her stay in Norway as the wife of a student at UiS in an international MSc. program in offshore risk management. Meihua has a M.Sc. degree in safety technology and engineering from China University of Geosciences in Beijing and HES management experience from SINOPEC in China and Latin America, and has been an ideal assistant. The main task has been the updating of Appendix A, in addition to several other editorial tasks, which has been very helpful to finish the revision work in a timely manner. I am very grateful for the excellent assistance provided by Meihua.

Last, but not least, I am very grateful to those companies that responded positively when asked for a modest support in order to cover the expenses involved in the production of this third revision. My warmest and most sincere thanks go to these companies:

- Faroe Petroleum Norway
- Norwegian oil and gas association
- Total E&P Norway
- VNG Norway

Bryne, May 2013   Jan-Erik Vinnem
                                                                    Adjunct Professor
                                            University of Stavanger & NTNU

Preface to Second Edition

This is the second edition of the book; the first edition was published in 1999. The author has since then taken up a full professorship at University of Stavanger, Norway, teaching courses in offshore risk analysis and management. This prompted a need for an update of the book. The fact that several important developments have occurred since 1999 also implied that a major revision was required.

The oil price has reached its peak in 2006, at the highest level ever (nominally). But the economic climate is at the same time such that every effort is made to scrutinise how costs may be curtailed and profit maximised. This will in many circumstances call for careful consideration of risks, not just an 'off the shelf risk analysis', but a carefully planned and broad-ranging assessment of options and possibilities to reduce risk.

Norwegian offshore regulations were profoundly revised around the beginning of the new century, with new regulation becoming law from 2002. This second edition of the book captures some of the experience and challenges from the first 4–5 years of application of the new regulations. The important aspects of the new regulations are also briefly discussed, see Chapter 1.

The first Norwegian White Paper on HES management in the offshore industry was published in 2001, and the second in 2006. One of the needs identified in this paper was the need to perform more extensive R&D work in this field, and a significant programme has been running in the period 2002–06. Some of the new results included in the second edition of the book result from that R&D initiative.

About 20 major accidents, mainly from the North Sea, are discussed at some length in Chapter 4, in order to demonstrate what problems have been experienced in the past. I have put more emphasis on this subject in the second edition, because it is essential that also new generations may learn from what occurred in the past. Where available, observations about barrier performance are discussed in addition to the sequence of events and lessons learned.

When it comes to management of risk and decision-making based upon results from risk analyses, this is discussed separately in a book published in parallel with my colleague at University of Stavanger, Professor Terje Aven, also published by

Springer in 2007. Interested readers are referred to this work, 'Risk Management, with Applications from the Offshore Petroleum Industry'.

Thanks are also expressed to Springer London publishers, in particular Professor Pham and Senior Editor Anthony Doyle, for agreeing to publish the second edition of this book, and for providing inspiring and valuable advice throughout the process. Simon Rees has given valuable assistance and support during production of the camera-ready manuscript.

Appendix A presents an overview of some of the important software tools that are commonly used. Thanks to all the consultancies and software suppliers who have provided the information required for this appendix.

In preparing the second edition of the book, I have been fortunate to have kind assistance from many colleagues and friends, who have provided invaluable support and assistance. First of all I want to express sincere thanks and gratitude to my friend David R. Bayly, Crandon Consultants, who has also this time assisted with improvement of the English language, as well as providing technical comments and suggestions. I don't know how I could have reached the same result without David's kind assistance.

My colleague at UiS, Professor Terje Aven has contributed significantly to the discussion of statistical treatment of risk and uncertainty. I am very pleased that this important improvement has been made. Dr. Haibo Chen, Scandpower Risk Management Beijing Inc has contributed valuable text regarding the analysis of DP systems on mobile installations.

Safetec Nordic AS has allowed use of several of their tools as input to the descriptions and cases. I want to express my gratitude for allowing this, and in particular express thanks to the following; Thomas Eriksen, Stein Haugen and Arnstein Skogset.

There are also several people who have contributed with relevant information on various technical details; Finn Wickstrøm, Aker Kvaerner and Graham Dalzell, TBS[3]. My daughter, Margrete, has assisted in the editing of the manuscript. Many thanks to all of you.

Bryne, January 2007   Professor, Jan-Erik Vinnem
                                                                   University of Stavanger

Preface to First Edition

From a modest start in Norway as a research tool in the late 1970s, Quantified Risk Assessment (QRA) for offshore installations has become a key issue in the management of Safety, Health and Environment in the oil and gas industries throughout the entire North Sea. While the initiatives in the early stages often came from the authorities, the use is now mainly driven by the industry itself. The QRA is seen as a vehicle to gain extended flexibility with respect to achievement of an acceptable safety standard in offshore operations. The models may be weak in some areas and the knowledge is sometimes limited, but studies are nevertheless used effectively in the search for concept improvement and optimisation of design and operation.

This book results from working with offshore QRAs for more than 20 years. The author has, during this period, had the opportunity to practice and evaluate the use of such studies from different perspectives; the consultancy's, the operating company's, the researcher's and the educator's point of view.

The author has for several years taught a course in risk analysis of marine structures at the Faculty of Marine Technology, NTNU, Trondheim, Norway. The starting point for the manuscript was the need to update the lecture notes.

It is hoped that this book in the future also may be a useful reference source for a wider audience. There has been for some years a rapid expansion of the use of risk assessments for the offshore oil and gas activities. It is expected that the expansion is going to continue for some time, as the offshore petroleum industry expands into new regions and meets new challenges in old regions.

The oil price reached its lowest level for many years, during the first quarter of 1999. One might be tempted to think that the economic climate may prohibit further attention to risk assessment and safety improvement. The opposite is probably more correct. As a friend in Statoil expressed not so long ago: 'Whenever the margins are getting tighter, the need for risk assessments increases, as new and more optimised solutions are sought, each needing an assessment of risk'.

In Norway, the beginning of 1999 is also the time when the Norwegian Petroleum Directorate is preparing a major revision of the regulations for offshore installations and operations, anticipated to come into effect in 2001. It has obviously not been possible to capture the final requirements of the new regulations, but an attempt has been made to capture the new trends in the regulations, to the extent they are known.

There have over the last 10–15 years been published a few textbooks on risk assessment, most of them are devoted to relatively generic topics. Some are also focused on the risk management aspects, in general and with offshore applicability. None are known to address the needs and topics of the use of QRA studies by the offshore industry in particular. The present work is trying to bridge this gap.

The use of QRA studies is somewhat special in Northern Europe, and particularly in Norway. The use of these techniques is dominated by offshore applications, with the main emphasis on quantification of risk to personnel. Furthermore, the risk to personnel is virtually never concerned with exposure of the public to hazards. Thus, the studies are rarely challenged from a methodology point of view. Most people will probably see this as an advantage, but it also has some drawbacks. Such challenges may namely also lead to improvements in the methodology. It may not be quite coincidental that the interest in modelling improvement and development sometimes has been rather low between the risk analysts working with North Sea applications.

This book attempts to describe the state-of-the-art with respect to modelling in QRA studies for offshore installations and operations. It also identifies some of the weaknesses and areas where further development should be made. I hope that further improvement may be inspired through these descriptions.

About the Contents

A Quantified Risk Assessment of an offshore installation has the following main steps:

1. Hazard identification
2. Cause and probability analysis
3. Accidental scenarios analysis
4. Consequence, damage and impairment analysis
5. Escape, evacuation and rescue analysis
6. Fatality risk assessment
7. Analysis of risk reducing measures

This book is structured in much the same way. There is at least one chapter (sometimes more) devoted to each of the different steps, in mainly the same order as mentioned above. Quite a few additional chapters are included in the text, on risk analysis methodology, analytical approaches for escalation, escape, evacuation and rescue analysis of safety and emergency systems, as well as risk control.

It is important to learn from past experience, particularly from previous accidents. A dozen major accidents, mainly from the North Sea, are discussed at the end of Chapter 4, in order to demonstrate what problems that have been experienced in the past.

The main hazards to offshore structures are fire, explosion, collision and falling objects. These hazards and the analysis of them are discussed in separate chapters. Risk mitigation and control are discussed in two chapters, followed by an outline of an alternative approach to risk modelling, specially focused on risk relating to short duration activities. Applications to shipping are finally discussed, mainly relating to production and storage tankers, but also with a view to applications to shipping in general.

Acknowledgements

Parts of the material used in developing these chapters were initially prepared for a course conducted for PETRAD (Program for Petroleum Management and Administration), Stavanger, Norway. Many thanks to PETRAD for allowing the material to be used in other contexts.

Some of the studies that have formed the main input to the statistical overview sections were financed by Statoil, Norsk Hydro, Saga Petroleum, Elf Petroleum Norge and the Norwegian Petroleum Directorate. The author is grateful that these companies have allowed these studies to be made publically available.

Direct financial support was received from Faculty of Marine Technology, NTNU, this is gratefully acknowledged. My part time position as Professor at Faculty of Marine Technology, NTNU, has also given the opportunity to devote time to prepare lecture notes and illustrations over several years. The consultancy work in Preventor AS has nevertheless financed the majority of the work, including the external services.

Thanks are also expressed to Kluwer Academic Publishers, Dordrecht, The Netherlands, for agreeing to publish this book, and for providing inspiring and valuable advice throughout the process.

Appendix A presents an overview of some of the important software tools that are commonly used. Thanks to all the consultancies and software suppliers who have provided the information required for this appendix. Appendix B is a direct copy of the normative text in the NORSOK Guideline for Risk and Emergency Preparedness analysis, reproduced with kind permission from the NORSOK secretariat.

Some of the consultancies have kindly given permission to use some of their material, their kind assistance is hereby being gratefully acknowledged. DNV shall be thanked for allowing their database Worldwide Offshore Accident Databank (WOAD) to be used free of charge, as input to the statistics in the book. The Fire Research Laboratory at SINTEF has given kind permission to use illustrations from their fire on sea research, and Scandpower has granted permission to use an illustration of the risk assessment methodology. Safetec Nordic has given kind permission to use results and illustrations from their software Collide.

I am particularly indebted to several persons who have offered very valuable help in turning this into a final manuscript. My colleague Dr. David Bayly, Crandon

Consultants, has reviewed the raw manuscript and contributed with many valuable comments of both a technical and linguistic nature. The importance of providing clear and concise text can never be overestimated, the efforts made in this regard are therefore of utmost importance. This unique contribution has combined extensive linguistic improvements with pointed comments and additional thoughts on the technical subjects. I am very grateful to you, David, for your extensive efforts directed at improvement of the raw manuscript.

My oldest son, John Erling, has helped me with several of the case studies that are used in the text, plus quite a few of the illustrations. My part time secretary, Mrs. Annbjørg Krogedal, has had to devote a lot effort to decipher a challenging handwriting, thank you for enthusiasm and patience. Assistance with the proof-reading has been provided by Ms. Kjersti G. Petersen, thanks also to Kjersti for enthusiastic and valuable assistance. Finally, M.Sc. Haibo Chen has also helped with the proof reading and checking of consistency in the text, your kind assistance is gratefully acknowledged.

Bryne, May 1999   Jan-Erik Vinnem

Contents

Abbreviations

AIBN	Accident Investigation Board Norway
AIR	Average Individual Risk
AIS	Automatic Identification System
ALARP	As Low As Reasonably Practicable
ALK	Alexander L. Kielland [accident]
ALS	Accidental limit state
ANP	National Petroleum Agency [Brazil]
AR	Accident Rate
ARCS	Admiralty Raster Chart Services
ARPA	Automated Radar Plotting Aid
ASCV	Annulus safety check valve
ASEA	Agency of Safety, Energy and Environmental enforcement [Mexico]
ASV	Annular safety valve
ATM	Air Traffic Management
bara	Bar absolute
barg	Bar gauge (overpressure)
BAST	Best Available and Safety Technology
BBD	Barrier Block Diagram
bbls	Barrels
BBN	Bayesian belief network
BD	Blowdown
BDV	Blowdown Valves
BF	Barrier Function
BFETS	Blast and Fire Engineering for Topside Systems
BHP	Broken Hill Proprietary Company Limited
BLEVE	Boiling Liquid Expanding Vapour Explosion
BOE	Barrels of Oil Equivalent
BOEMRE	Bureau of Offshore Energy Management, Regulation and Enforcement [USA]

BOP	Blowout Preventer
BORA	Barrier and Operational Risk Analysis
BP	British Petroleum (formerly)
BSEE	Bureau of Safety and Environmental Enforcement [USA]
CAA	Civil Aviation Authority
CAD	Computer Aided Design
CAPEX	Capital expenditure
CBA	Cost Benefit Analysis
CCA	Cause-Consequence Analysis
CCPS	Center for Chemical Process Safety
CCR	Central Control Room
CDSM	Cidade de São Mateus
CFD	Computational Fluid Dynamics
CNLOPB	Canada-Newfoundland and Labrador Offshore Petroleum Board
CNOOC	China National Offshore Oil Corporation
CNSOPB	Canada-Nova Scotia Offshore Petroleum Board
CO	Carbon monoxide
CO_2	Carbon dioxide
CPA	Closest Point of Approach
CPP	Controllable Pitch Propeller
CREAM	Cognitive reliability and error analysis method
CRIOP	Crisis intervention and operability analysis
CSE	Concept Safety Evaluation
DAE	Design Accidental Events
DAL	Design Accidental Load
DEA	Danish Energy Agency
DeAL	Design Accidental Load
DFU	Defined situations of hazard and accident (Definert fare- og ulykkessituasjon)
DGPS	Differential Global Positioning Systems
DHJIT	Deepwater Horizon Joint Investigation Team
DHSG	Deepwater Horizon Study Group
DHSV	DownHole Safety Valve
DiAL	Dimensioning Accidental Load
DNV GL	Det Norske Veritas—Germanischer Lloyd
DOL	Department of Labor [New Zealand]
DP	Dynamic Positioning
DSB	Directorate for Civil Protection and Emergency Planning (Direktoratet for samfunnssikkerhet og beredskap)
DSHA	Defined situations of hazard and accident (same as DFU)
DWT	Dead Weight Tonnes
E&P	Exploration and Production
E&P Forum	Previous name of organisation now called IOGP
EASA	European Aviation Safety Authority

EER	Escape, Evacuation and Rescue
EERS	Evacuation, Escape and Rescue Strategy
EESLR	Risk due to explosion escalation by small leaks
EFSLR	Risk due to fire escalation by small leaks
EGPWS	Enhanced Group Proximity Warning System
EIA	Environmental Impact Assessment
EIF	Environmental Impact Factor
EPIM	E&P Information Management [association]
EQD	Emergency Quick Disconnection [system]
EQDC	Emergency Quick Disconnect
ERA	Environmental risk analysis
ESD	Emergency Shutdown
ESREL	European Safety and Reliability
ESV	Emergency Shutdown Valve
ETA	Event Tree Analysis
Ex	Explosion [protected]
FAHTS	Fire And Heat Transfer Simulations
FAR	Fatal Accident Rate
FCC	Frigg Central Complex
FEM	Finite Element Method
FES	Fire and Explosion Strategy
Fi-Fi	Fire Fighting
FLACS	Flame Accelerator Software
FLAR	Flight Accident Rate
FMEA	Failure Mode and Effect Analysis
FMECA	Failure Mode, Effect and Criticality Analysis
f-N	Cumulative distribution of number of fatalities
FPPY	Fatalities per platform year
FPS	Floating Production System
FPSO	Floating Production, Storage and Off-Loading Unit
FPU	Floating Production Unit
FRC	Fast Rescue Craft
FSU	Floating Storage Unit
FTA	Fault Tree Analysis
GBS	Gravity Base Structure
GIR	Group Individual Risk
GIS	Geographical Information System
GoM	Gulf of Mexico
GPS	Global Positioning System
GR	Group Risk
GRP	Glass fibre Reinforced Plastic
GRT	Gross Register Tons
HAZAN	Hazard Analysis
HAZID	Hazard Identification
HAZOP	Hazard and Operability Study

HC	Hydrocarbon
HCL	Hybrid Causal Logic; Hydrocarbon Leak
HCLIP	Hydrocarbon Leak and Inventory Project
HCR	Hydrocarbon Release
HEP	Human error probability
HES	Health, Environment and Safety
HF	Human factors
HIPPS	High Integrity Pressure Protection System
HMI	Human–machine interface
HOF	Human and Organisational Factors
HOFO	Helicopter offshore operations
HP	High pressure
HR	Human Reliability
HRA	Human Reliability Analysis
HRO	High reliability organisation
HSE	Health and Safety Executive [UK]
HSS	Helicopter safety studies
IAEA	International Atomic Energy Agency
ICT	Information and communications technology
IEC	International Electrotechnical Commission
IMEMS	International Marine Environmental Modeling Seminar
IMO	International Maritime Organization
IO	Integrated operations
IOGP	International Oil and Gas Producers Association
IR	Individual Risk
IRF	International Regulators' Forum
IRIS	International Research Institute of Stavanger
IRPA	Individual Risk per Annum
ISO	International Organisation for Standardisation
JIP	Joint Industry Project
JU	Jack-up
KFX	Kameleon Fire Ex
kN	Kilonewton (10^3 N)
KNM	Royal Norwegian Navy
KPI	Key Performance Indicator
kW	Kilowatt (10^3 W)
LCC	Life Cycle Cost
LEL	Lower Explosion Level
LFL	Lower Flammability Level
LNG	Liquefied Natural Gas
LOPA	Layers of protection analysis
LP	Low pressure
M&O	Management and Operation
MGB	Main Gearbox
MIRA	Environmental risk analysis (Miljørettet risikoanalyse)

MIRMAP	Modelling of Instantaneous Risk for Major Accident Prevention
MISOF	Modelling of ignition sources on offshore oil and gas facilities
MJ	Megajoule (10^6 J)
MMI	Man–Machine Interface (see also HMI)
MMS	Minerals Management Service (now BEMRE)
MNOK	Million Norwegian kroner
MO	Human and organisational (or HO)
MOB	Man overboard
MOC	Management of Change
MODU	Mobile Offshore Drilling Unit
MOEX	Mitsui Oil Exploration
MP	Main (propulsion) Power
MSF	Module Support Frame; Main Safety Function
MTBF	Mean time between failures
MTO	Man, Technology and Organisation
MUSD	Million US Dollar
NCS	Norwegian Continental Shelf
NEA	National Environment Agency
NGOs	Non-governmental organizations
nm	Nautical Mile
NMD	Norwegian Maritime Authority
NOPSEMA	National Offshore Petroleum Safety and Environmental Management Authority [Australia]
NOROG	Norwegian Oil and Gas [association]
NORSOK	Norwegian Offshore standardisation organisation (Norsk Sokkels Konkurranseposisjon)
NOU	Norwegian Offshore Report (Norges Offentlige Utredninger)
NPD	Norwegian Petroleum Directorate
NPPs	Nuclear Power Plants
NPV	Net Present Value
NTNU	Norwegian University of Science and Technology
NUI	Normally unmanned installation
OCS	Outer Continental Shelf
OIM	Offshore Installation Manager
OMT	Organisational, Human and Technology
ONGC	Oil and Natural Gas Corporation
OPEX	Operational expenditure
OR	Overall risk
OREDA	Offshore and onshore reliability data
OSD	Offshore Division
OSPAR	Oslo and Paris Convention

OTS	Operational Condition Safety ('Operasjonell Tilstand Sikkerhet')
P&ID	Piping and Instrumentation Drawing
PCCC	Pressure-containing anti-corrosion cap
PDO	Plan for Development and Operation
PDQ	Production, Drilling and Quarters
PDS	Reliability of computer-based safety systems
PETRAD	Program for Petroleum Management and Administration
Petro-HRA	Human reliability analysis for petroleum industry
PFD	Process flow diagram; Probability of failure on demand
PFEER	Prevention of Fire and explosion and Emergency Response
PFP	Passive Fire Protection
PGS	Petroleum Geo-Services ASA
PHA	Preliminary Hazard Analysis
PLATO	Software for dynamic event tree analysis
PLL	Potential Loss of Life
PLOFAM	Process leak for offshore installations frequency assessment model
PLS	Progressive Limit State
PM	Position mooring
POB	Personnel On Board
PR	Performance Requirements
PRA	Probabilistic risk analysis
PRS	Position Reference System
PS	Performance Standards
PSA	Petroleum Safety Authority Norway; Probabilistic Safety Assessment
PSAM	Probabilistic Safety Assessment and Management
PSD	Process Shut Down
PSF	Performance shaping factor
PSV	Pressure Safety Valves
PWV	Production wing valve
QA	Quality Assurance
QC	Quality Control
QM	Quality Management
QP	[Frigg] Quarters Platform
QRA	Quantified Risk Assessment; Quantified Risk Analysis
R&D	Research and Development
RABL	Risk Assessment of Buoyancy Loss
RAC	Risk Acceptance Criteria
RACON	Radar signal amplification
RAE	Residual Accidental Events
RAMS	Reliability, Availability, Maintainability, Safety

RIDDOR	Reporting of Injuries, Diseases and Dangerous Occurrences Regulations [UK]
RIF	Risk Influencing Factor
RISP	Risk informed decision support in development projects
RNNP	Risk level in the Norwegian petroleum activity (Risikonivå i norsk petroleumsvirksomhet)
ROS	Risk and vulnerability study (Risiko- og sårbarhetsstudie)
ROV	Remote Operated Vehicle
RPM	Rotations per minute
RRM	Risk Reducing Measure
RTC	Risk tolerance criteria
SA	Shelter area
SAFOP	Safety and Operability Study
SAR	Search and Rescue
SBM	Single buoy mooring
SBV	Standby Vessel
SCR	Safety Case Regulations
SDS	Safe disconnection system
SERA	Safetec Explosion Risk Assessor
SIL	Safety Integrity Level
SIS	Safety instrumented systems
SJA	Safe Job Analysis
SLR	Risk due to small leaks
SNA	Snorre Alpha [installation]
SOLAS	Safety of Life at Sea
SOV	Service operations vessel
SPAR-H	Standardized plant analysis risk—human reliability analysis
SRA	Society of risk analysis
SSIV	Subsea Isolation Valve
SSM	State Supervision of Mines (Netherland)
ST	Shuttle tanker
STAMP	Systems-Theoretic Accident Model and Processes
STEP	Sequential time event plotting
SUPER-TEMPCALC	Software for 2D temperature analysis
TASEF-2	Software for 2D temperature analysis of structures exposed to fire
TCAS	Traffic Collision Avoidance System
TCP2	[Frigg] Treatment Platform 2
TCPA	Time to closest point of approach
TH	Thruster
THERP	Technique for human error-rate prediction
TLP	Tension Leg Platform
TP1	[Frigg] Treatment Platform 1
TR	Temporary Refuge

TRA	Total Risk Analysis
TST	Technical Safety Condition ('Teknisk Sikkerhets Tilstand')
TTS	Technical Condition Safety ('Teknisk Tilstand Sikkerhet')
UEL	Upper Explosive Limit
UFL	Upper Flammability Limit
UKCS	UK Continental Shelf
ULS	Ultimate limit state
UPP	Unmanned production platform
UPS	Underwater Production System; Uninterruptible Power Supply
US CSB	United States Chemical Safety Board
US GoM	United States Gulf of Mexico
US OCS	United States Outer Continental Shelf
USD	US Dollar
USFOS	Software for nonlinear and dynamic analysis of structures
VEC	Valued Ecological Component
VHF	Very High Frequency
VOC	Volatile Organic Compounds
VTS	Vessel Traffic System
W2W	Walk-to-walk
WCPF	Worst credible process fire
WHP	Wellhead platform
WOAD	Worldwide Offshore Accident Database (ref. DNV GL)
WP	Work Permit
X-MAS TREE	Christmas Tree (safety valve assembly)

Part III
Risk Analysis, Presentation and Evaluation Process

Chapter 14
Methodology for Quantified Risk Assessment

14.1 Analytical Steps and Elements

14.1.1 Analytical Elements

'Risk analysis' is defined in the NORSOK standard [1], as a structured use of available information to identify hazards and describe risk. QRA therefore has to be focused on:

- Identification of applicable hazards.
- Description (including quantification) of applicable risks to personnel, environment and assets.

The analytical elements of risk assessment are those that are required to identify the relevant hazards and to assess the risk arising from them. These elements may include all, or some, of the following:

- Identification of initiating events
- Cause analysis

  - qualitative evaluation of possible causes
  - probability analysis in order to determine the probability of certain scenarios

- Consequence analysis

  - consequence loads, related to physical effects of accidents
  - response analysis, related to response of the facilities, when exposed to accidental loads
  - probability analysis, related to the probability that these loads and responses occur
  - quantification of consequences in terms of injury to personnel, damage to environment and/or to assets.

© Springer-Verlag London Ltd., part of Springer Nature 2020
J.-E. Vinnem and W. Røed, *Offshore Risk Assessment Vol. 2*, Springer Series
in Reliability Engineering, https://doi.org/10.1007/978-1-4471-7448-6_14

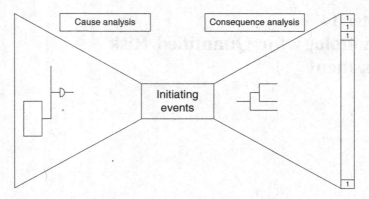

Fig. 14.1 Model for representation of the process risk assessment

The elements of QRA are shown diagrammatically in Fig. 14.1. This diagram is sometimes called the 'Bow-tie diagram' for obvious reasons. Figure 14.1 presents the main elements of risk assessment. The starting point is the identification of initiating Events, followed by cause analysis and consequence analysis. The practical execution of a risk assessment is often described as:

1. Identification of critical events
2. Coarse consequence analysis (order of magnitude)
3. Cause analysis (qualitative)
4. Quantitative cause analysis
5. Detailed consequence analysis
6. Risk calculation.

When an offshore or marine structure is considered, the consequence loads are mainly related to the following:

- Fire loads from ignited hydrocarbon releases
- Explosion loads from ignition of hydrocarbon gas clouds
- Structural impact from collisions, falling objects, etc.
- Environmental loads.

The consequence analysis is an extensive effort involving many different disciplines, and covering a series of steps including:

- Accident scenario analysis of possible event sequences
- Analysis of accidental loads, related to fire, explosion, impact
- Analysis of the response of systems and equipment to accidental loads
- Analysis of final consequences to personnel, environment, and assets
- Escalation analysis, relating to how accidents may spread from the initial equipment to other equipment and areas.

Each of these steps may include extensive studies and modelling. An overview is presented in this chapter and further details are discussed in later chapters.

14.1.2 Identification of Initiating Events

The identification of initiating events is often called hazard identification, usually abbreviated to HAZID. Hazard identification should include:

- A broad review of possible hazards and sources of accidents, with particular emphasis on ensuring that no relevant hazards are overlooked
- A coarse classification into critical hazards (as opposed to noncritical) for subsequent analysis
- Explicit statement of the criteria used in the screening of the hazards
- Clear documentation of the evaluations made in the categorisation of hazards as non-critical.

There are very few formalised tools available for hazard identification, and thus it is most important that the work is undertaken in a structured and systematic manner. The following approaches may be used to assist in hazard identification:

- Check lists
- Accident and failure statistics
- Hazard and Operability (HAZOP) Studies
- Comparison with detailed studies
- Experience from previous similar projects, concepts, systems, equipment and operations.

It will usually be essential that the reasons why hazards are classified as 'non-critical' are well documented, in order to demonstrate that the events in question could be safely disregarded, without distorting the completeness of the risk picture. This may be expressed as follows: The only requirements for extensive documentation of the hazard identification should apply to those hazards that are disregarded.

14.1.3 Cause Analysis

It is important to identify the causes of hazards or initiating events, as these will be the starting point of potential accident sequences. The cause analysis has three objectives:

- Identification of the combination of causes that may lead to initiating events
- Assessment of probability of initiating events
- Identify the possibilities for risk reducing measures.

The first objective is mainly qualitative, while the latter is quantitative. These two aspects are discussed separately below.

14.1.3.1   Qualitative Cause Analysis Techniques

Qualitative analysis of causes is sometimes the initial step, to be followed by subsequent quantification. In other cases the qualitative analysis of causes is the only step implemented, because data is unavailable for quantification, or quantification is not required for some reason. Qualitative analyses are intended to:

- Identify causes and conditions that may lead to the occurrence of initiating events
- Identify combinations that will result in such an occurrence
- Establish the basis for possible later quantitative analysis.

The cause analysis has a lot in common with traditional reliability analysis, and the applicable tools are also taken from the reliability analysis field. The following analysis techniques are the most common:

- Hazard and Operability Analysis (HAZOP)
- Fault Tree Analysis (FTA)
- Preliminary Hazard Analysis (PHA)
- Failure Mode and Effect Analysis (FMEA)
- Human error analysis techniques, such as Task analysis and Error mode analysis [2, 3].

The focus on operational causes has increased considerably in the last few years. Human error analysis is therefore increasingly more important (see also the following subsection).

Identification of causes gives the basis for prevention of accidents, if potential causes can be eliminated or controlled. Qualitative cause analysis may therefore be important, even though no subsequent quantification is performed.

14.1.3.2   Quantitative Cause Analysis Techniques

Quantitative analysis of possible causes of hazards is intended to establish the probability of occurrence of initiating events. The following analysis techniques are the most common:

- Fault Tree Analysis (FTA)
- Event Tree Analysis (ETA)
- Synthesis models
- Monte Carlo simulation
- Human error quantification techniques, such as THERP or similar [3]
- Calculation of frequency of initiating events from historical statistical data
- BORA methodology for analysis of hydrocarbon leaks and consequence barriers.

Fault Tree Analysis is described in several text books (see for instance [4]). A brief summary is presented in Sect. 6.2.1. Event Tree Analysis is introduced in Sect. 15.3. An example of a synthesis model is one which predicts collision risk (Chap. 9). Monte Carlo simulation is very specialised, and the interested reader is referred to [4].

Calculation of the frequency of initiating events from historical statistics is actually the most commonly used approach, although it is questionable whether it should be regarded as a type of cause analysis within the methodology of Fig. 14.1. It is currently being criticised quite heavily, because its only focus is to provide a number, without providing information that may be used to prevent accidents from occurring. It is therefore recommended that it should not be used as extensively as in the past. This aspect is discussed later in this chapter. The following requirements should apply when a frequency analysis has to be used:

- The robustness of the data used is to be considered.
- Both the data and the models in which the data is used need to be suitable in relation to the context of the study.
- The extent of the data basis has to be sufficiently broad to produce robust conclusions.
- The use of data should take account of possible trends if they can be substantiated.

Data that are used must be consistent with relevant operations and phases.

14.1.4 Modelling of Accident Sequences

The 'consequence' part of the Bow-tie diagram (Fig. 14.1) actually comprises three parts, the modelling of accident sequences, the analysis of physical consequences (effects) and quantification of consequences for personnel, environment and assets. The second of these parts is discussed in the following section, whereas the third part is discussed in Sect. 12.3.3, limited to personnel. The most important challenge in the use of QRA is the modelling of accident sequences, both with respect to the steps that these sequences may take, and the timing aspects of the sequences. It has been recognised for some time that the current tools are far from adequate in these regards, but they are still being used. Alternative approaches that have been suggested have never really been recognised as being superior, and therefore have not been broadly adopted.

Offshore installations usually offer a number of barriers functions and elements to contain hazards or to mitigate the effects should a hazard be realised. These barriers are usually quite sufficient in order to prevent serious accidents. They have their limitations related to the design basis, and they may not be sufficient in the most severe accident conditions. The accident sequences are made up of a series of steps which define the various escalation possibilities. Each step is usually related to the possible function, or failure of, the barriers involved.

Figure 14.2 presents a schematic representation of the barrier concept. There are different types of barriers shown in the diagram including the following:

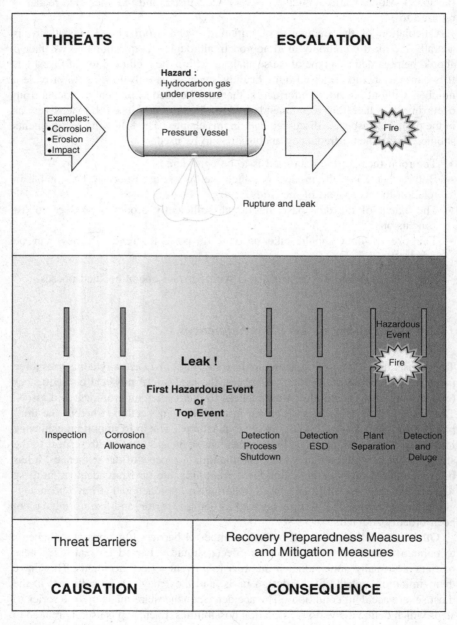

Fig. 14.2 Barrier diagram

- Causation/threat barriers
- Consequence/mitigation barriers
- Technical (hardware) barriers
- Procedural barriers.

The causation barriers in this example are only of the procedural type, whereas the consequence barriers are a mix of technical and procedural barriers.

The tools usually employed in the accident sequence analysis, are as listed below. It may be noted that the first approach is by far the most widely used:

- Event tree analysis
- Cause consequence diagrams
- Influence diagrams.

One of the main drawbacks of the event tree is that it is a static structure. The event tree analysis has in some instances been further developed, in the sense that sometimes 'dynamic trees' are developed, whereby the creation of the tree is dependent on the consequence calculations. 'Directional trees' have also been developed in other instances, whereby the trees are developed in six degrees of freedom, in order to reflect better the different capabilities with respect to resistance against escalation in different directions.

It is important to underline the fact that barriers can also be of the procedural type. Often only technical barriers are considered in the barrier analysis. It is important to underline that also non-technical aspects need to be considered in the accident sequences. Figure 14.3 presents an idealised sketch of Human and Organisational Factors (HOF).

MTO analysis is another technique for modelling accident sequences, with particular application in investigation of accidents and incidents. The MTO approach is outlined in Sect. 6.5.4. Integration of HOF and QRA is also mentioned in Sect. 15.6.5.

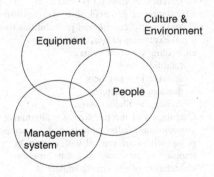

Fig. 14.3 Conceptual model
for HOF aspects

14.1.5   Consequence Analysis

Fire and explosion consequences are often assessed by using a series of calculation steps, which mainly are the same for all scenarios which may involve fire and/or explosion. The following is a brief overview of the main calculation steps for:

- Fire and explosion following leak from process system
- Fire and explosion following uncontrolled blowout.

An overview of elements of the consequence calculations for fire and explosion is presented in Table 14.1.

14.1.6   Risk Calculation, Analysis and Assessment

Figure 14.4 shows an iterative loop, which indicates the importance of risk evaluation (i.e. the consideration of results from the risk analysis in relation to risk tolerance criteria) as an integral part of the safety management process. If the risk results are unacceptable, a new loop is created through implementation of risk reducing measures, and an updating of the risk analysis to reflect these changes.

Table 14.1 Overview of fire and explosion calculations

Fire and explosion relating to process leaks	Fire and explosion relating to blowout

• Release rate as a function of time for small, medium and large releases without ESD, with ESD and with ESD and blowdown
• Calculation of external flammable cloud size for the large releases to serve as a basis for calculation of ignition probability and explosion overpressure
• Calculation of ignition probabilities
• Calculation of gas explosion overpressure
• Calculation of conditional probabilities that a gas explosion causes overpressure exceeding specified values:
  – fatalities
  – damage to equipment
  – damage to structures
  – damage to Shelter Area
• Calculation of the fire sizes for alternative scenarios and radiation impact at prescribed points which are critical with respect to trapping of personnel or safety functions
• Calculation of the smoke impact at prescribed points which are critical with respect to trapping of personnel or safety functions
	• Release rate as a function of time for small, medium and large releases without ESD, with ESD and with ESD and blowdown
	• Calculation of external flammable cloud size for the large releases to serve as a basis for calculation of ignition probability and explosion overpressure
	• Calculation of ignition probabilities
	• Calculation of gas explosion overpressure
	• Calculation of conditional probabilities that a gas explosion causes overpressure exceeding specified values:
	  – fatalities
	  – damage to equipment
	  – damage to structures
	  – damage to Shelter Area
	• Calculation of the fire sizes for alternative scenarios and radiation impact at prescribed points which are critical with respect to trapping of personnel or safety functions
	• Calculation of the smoke impact at prescribed points which are critical with respect to trapping of personnel or safety functions

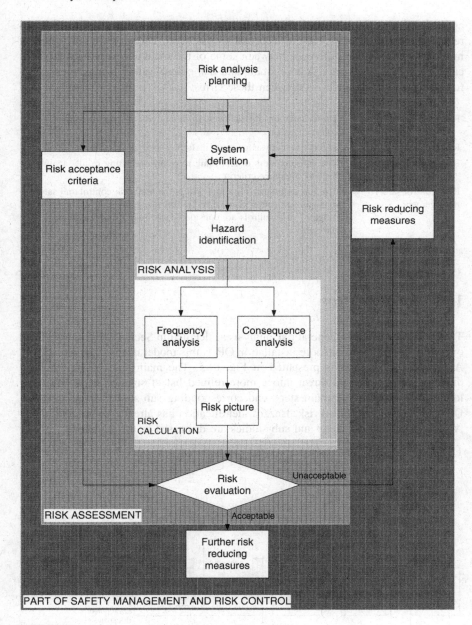

Fig. 14.4 Iterative loop for risk assessment and risk control

It should be noted that the two outcomes of the 'Risk evaluation' box are 'unacceptable' and 'acceptable'. This is based on an assumed approach whereby adherence to explicit risk tolerance criteria is ensured as the first point, followed by an ALARP evaluation. This is the reason why the diagram refers to 'further risk reducing measures', in the case of 'acceptable' risk level. The diagram may be used in order to explain the difference between some of the main terms in the field; risk calculation; risk analysis; risk assessment and risk control/safety management. The following are the distinctions between these terms:

Risk calculation	The calculation of risk levels from analysis of frequencies and consequences
Risk analysis	The entire analysis process, including the risk calculation
Risk assessment	The entire process of analysing risk and evaluating results against the risk tolerance criteria
Risk control/safety management	The risk assessment is a tool within the total risk control and safety management (often also called HES management), in which a wide range of risk controls are considered, including risk reduction measures

14.2   Analysis Steps

The principal steps of a general risk analysis as detailed in Sect. 14.1 give a general overview only. For an offshore installation QRA, this model needs to be expanded. A practical illustration is presented in Fig. 14.5. The main steps shown in this diagram may be broken down into a more refined list of smaller steps. The following list presents the main steps and corresponding sub-studies in a complete QRA evaluating personnel risk. Hazard identification has already been mentioned. All the remaining elements and sub-studies are discussed in this chapter.

1. Hazard Identification

    1.1. Systematic hazard review
    1.2. Top event spectrum

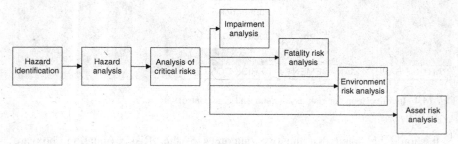

Fig. 14.5 Main elements of the offshore QRA

2. Hazard Analysis

   2.1. Blowout hazard study
   2.2. Riser/pipeline hazard study
   2.3. Process hazard study
   2.4. Fire and smoke analysis
   2.5. Explosion analysis
   2.6. Dropped object hazard study
   2.7. Collision hazard study
   2.8. Structural failure study
   2.9. Overall event tree study

3. Analysis of critical risks

   3.1. Barrier study
   3.2. Detailed probability study
   3.3. Detailed consequence study
   3.4. Revised event tree study

4. Impairment analysis

   4.1. Escape ways impairment study
   4.2. Shelter area impairment study
   4.3. Evacuation impairment study
   4.4. Impairment study of command and control safety function

5. Fatality risk analysis

   5.1. Immediate fatality risk study
   5.2. Escape ways risk study
   5.3. Shelter area risk study
   5.4. Evacuation means availability study
   5.5. Evacuation risk study
   5.6. Pick-up and rescue risk study
   5.7. Overall fatality risk summation.

14.2.1 Requirements for Analytical Approach

A risk assessment should always start with a system description. A precise definition of the system being analysed assists both the analyst, and more importantly those who will review, or evaluate, the study and results. The system description should include:

- Description of technical system, relevant activities and operational phases
- Statement of the time period to which the analysis relates

- Statement of the personnel groups, the external environment and the assets to which the risk assessment relates
- Capabilities of the system in relation to its ability to tolerate failures and its vulnerability to accidental effects.

The purpose of the system description is to make the analysis sufficiently transparent, so that other personnel are able to review and judge it. Additional requirements for the planning of QRA studies are as follows:

- When the QRA is carried out, the data needs to be adapted as far as possible to the purpose of the study. Databases (local, national, and international) need to be considered in this context, as well as the use of relevant company and industry experience.
- Prior to the decision to start a QRA, careful consideration should be given to whether the available data are adequate to allow reliable conclusions to be reached.
- Simple comparative studies based upon limited data, may still, in some instances, provide useful conclusions to assist decision making.

Another important aspect is to ensure a comprehensive documentation of assumptions and premises for the analysis. Assumptions and premises need to be documented where they belong in relation to calculations. In addition, there should be a summary of all assumptions and premises as a reference source.

14.3   Hazard Modelling and Cause Analysis

This section considers important aspects relating to the modelling of hazards, and important causes of initiating events.

14.3.1   Blowout Hazard Study

A blowout hazard study is carried out to determine the blowout probabilities in different phases, and with different well-related operations. Such a study also considers the outflow conditions under various scenarios, in order to determine both the fire potential and the pollution potential.

The input to a blowout hazard study is the drilling, completion, production and intervention (wireline, coiled tubing, workover, etc.) programmes, together with key data on the reservoir and completion techniques and equipment. Particular aspects of drilling and other operations may need to be considered separately, including detailed studies of equipment and procedures in certain cases.

The main criticism of the way blowout hazard studies are often conducted is that often the blowout frequency is purely based on historical blowout statistics. This may be seen as 'numbers magic' with no other benefit than that it produces a number. It has no intention to be specific in relation to equipment and procedures being used for the well under consideration.

The challenge in modelling a blowout hazard is to conduct an analysis which reflects the actual equipment and procedures that are used. The models that are often utilised are generic and are unable to distinguish between different platforms, systems and operators. Such aspects are overcome by the BlowFAM® methodology [5], but only partially. A more detailed approach, 'Kick-risk' is available for more in-depth studies [6, 7] . Kick-risk has the significant limitation that only technical failures are considered, whereas human errors and organisational aspects are not covered. The severe limitation of this aspect is illustrated by the Macondo blowout (see Chap. 5). Kick-risk has not been further developed for more than ten years. The following should be included in an analysis of blowout hazards:

- Well blowouts (non-environmental effects)

  - consequences related to ignition and subsequent effects are calculated as for leakages of flammable substances.

- Well blowouts (with respect to environmental loads)

  - calculation of releases
  - calculation of release duration
  - spill drifting calculation
  - calculation of environmental effects.

Integration of HOF into QRA is further discussed in Sects. 15.6.4–15.6.6. This is still a subject where there are various ongoing initiatives, and further research work will be forthcoming in the next few years.

14.3.2   Process Hazard Study

A process system hazard assessment considers the various process components, and the different outflow conditions that may arise in the event of accidental releases. The barriers, isolation, and area classification aspects are further considered, in order to determine different situations that can arise, including ignited and unignited cases. The performance of active and passive safety systems in the various situations is also considered. The results of these studies feed into the event trees developed for the various systems, areas, and modules. The modelling of the process hazard itself is usually fairly straightforward although it may be rather difficult to realistically model the operation of process control and isolation systems.

Figure 14.6 shows a typical release rate development as a function of the time, including the influence of isolation (ESD) and depressurisation (BD). The hazards related to process systems are discussed further in Chap. 8.

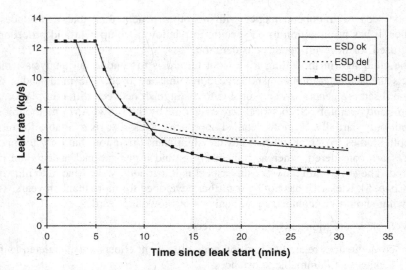

Fig. 14.6 Illustration of flow rates with and without ESD and BD

14.3.3 Riser/Pipeline Hazard Study

The different flow conditions arising from a leak or full rupture in a riser or pipeline have to be assessed as input to the fire load calculations. When the leak is subsea, the behaviour of the gas plume in the water also has to be assessed. The main modelling problems are associated with underwater leaks and releases of two phase and three phase fluids.

A critical aspect in process hazard modelling (as for riser, pipeline and blowout hazards) is the ignition probability modelling. The following should be the main stages in modelling leakage of flammable substances:

- Calculation of release frequencies (see also Chap. 6)
- Calculation of release (amounts, rates, duration, etc.)
- Calculation of spreading of leakages
- Calculation of ignition potential
- Fire load calculation
- Explosion load calculation
- Response calculation (sometimes this may be undertaken in separate studies).

The release of gas from a full rupture of a pipeline or riser may, depending on cross section and pressure, be up to several tons per second in the initial stage of an accident representing a devastating release of energy. Figure 14.7 shows a typical release rate development for a riser rupture as a function of the time. Isolation and depressurisation do not usually have any significant effect in these circumstances.

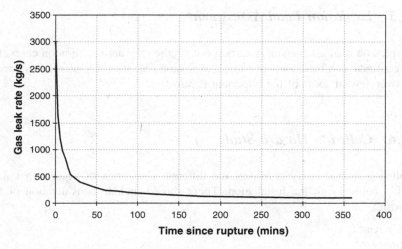

Fig. 14.7 Illustration of flow rate caused by gas riser rupture

14.3.4 Fire Load and Smoke Assessment

A fire load assessment includes consideration of all relevant fire scenarios, depending on the conditions in the modules and systems. Typically, such an assessment might include:

- Jet fires
- Pool fires
- Flash fires
- Burning blowout
- Riser fires
- Fire on sea surface
- BLEVE.

The part of this assessment pertaining to the process systems will include an assessment of the leak and outflow conditions that may arise from the different process components. This is based on the contents of the process equipment, as well as an identification of leak sources and their characteristics.

Overall smoke generation and its effect are also included in the assessment. The modelling of smoke impact on escape ways and shelter area is an aspect of the modelling where further development is needed. Section 7.4 presents a brief overview of some of the important parameters.

14.3.5   Explosion Load Assessment

An explosion load assessment is carried out for the relevant deflagration cases, for semi-confined, confined, and unconfined explosion scenarios. Chapter 8 presented a brief overview of some of the important parameters.

14.3.6   Collision Hazard Study

A collision hazard study considers the different categories of vessels that may possibly collide with the installation. These should include consideration of the following:

- Merchant
- Supply
- Standby
- Shuttle tankers
- Fishing
- Naval
- Submarines
- Mobile drilling units
- Barges and other offshore units
- Floating production units.

The collision assessment considers the likelihood that the various vessel types will collide with the installation, where the contact is likely to be, what energy levels may be involved, and what the overall consequences might be.

Collision hazard modelling is probably the area where the available models are most satisfactory although even in this case, further improvement could be achieved. The analysis of external impacts in general (collision, falling loads, helicopter crash on installation) should consist of the following stages:

- Calculation of energy distribution
- Calculation of load distribution
- Calculation of impulse distribution
- Response calculation, including both elastic and plastic response (this may also be carried out in separate studies).

Chapter 9 presents a brief overview of some of the important parameters.

14.3.7   Dropped Object Hazard Study

A dropped objects hazard study considers the locations of the cranes (including any derricks), the loads handled, and the lifting areas, in order to determine where loads may possibly be dropped. There are three different hazard mechanisms that have to be considered:

- Loads dropped on process equipment which may result in 'Loss of Containment' events.
- Loads dropped on buoyancy compartment of floating platforms which may result in critical loss of buoyancy, in the worst case leading to capsize or sinking of the platform.
- Loads dropped on subsea structure and equipment which may result in damage and possible 'loss of containment' from subsea equipment, including pipelines.

Consideration of the loads from objects falling on equipment and structure above water level is in the physical sense a relatively simple modelling exercise. When part of the fall is in the water however, the modelling is more complicated, due to the fact that the impact with the sea surface and the trajectory through the water are both influenced by several mechanisms which are significantly more complex to model.

The trajectories and speeds during the various phases of the fall have to be considered, with particular emphasis on the fall through the water where the influence of currents also needs to be considered. The strength of subsea installations (templates, pipelines, etc.) is also considered, and the results are presented as drop probabilities, hit probabilities and damage probabilities. These data form the input to the overall event tree analysis.

Available models for the assessment of dropped objects hazards are generally too simplistic in the sense that differences in operational procedures and crane protection are not usually taken into consideration.

Section 11.1 presents a brief overview of some of the important parameters.

14.3.8   Structural Failure Study

Structural response studies cover a wide variety of studies which are aimed at predicting the performance of the structural elements when subjected to accidental loads arising from accidental events such as explosion, fire, extreme weather etc. When evaluating the effect of fire on the structure the increase in temperature as a function of time is determined for different parts of the structure. Appropriate software is available to make these calculations. In some cases the structural response following the temperature rise in the structure may be judged qualitatively, based on the structural analyses already performed. In other cases, a non-linear structural analysis may be required. This may be accomplished by structural analysis software, specifically developed for this purposes, such as USFOS® [8]. Sections 7.7, 8.4.6, 9.5 and 11.4 present brief overviews of some of the important parameters.

14.4 Analysis of Critical Risks

Critical risks i.e., those risks that are shown to make a significant contribution to the total risk, need to be analysed further, by detailed consideration of the factors which provide protection from these hazards. The following subsections present a brief overview of the tasks involved in this work.

14.4.1 Barrier Study

The reliability, availability, and survivability of the barriers are analysed qualitatively and quantitatively, the latter should be more often by means of Fault Tree Analysis.

Analysis of availability and survivability of emergency systems is an important step in a risk analysis, primarily connected to hazards involving release of hydrocarbons. It may also be noted that Norwegian HES management regulations require that barriers are analysed and that the status of barriers is known during operations of installations. Dependencies also need to identified, and mitigated if needed.

14.4.1.1 Reliability and Availability Study

Reliability and availability studies are frequently carried out by means of Failure Mode and Effect Analysis, or by Fault Tree Analysis. Special attention will have to be given to potential sources of so-called 'common mode' or 'common cause' failures. Common mode failures arise when the backup or redundant systems are affected by the same mode of failure to the extent that all systems fail simultaneously. Common cause failures are failures in which otherwise well separated systems have one vital element in common, the failure of which causes all the systems to fail. Cables that have common routing, common lube-oil systems are typical examples that may have such effects.

Reliability studies are in the present context mainly related to the performance of barriers. Such studies are also carried out in order to assess the production or transport availability or regularity of a production or transport system. Monte Carlo simulation [4], is a frequently used approach for such studies.

Reliability studies of barriers on offshore installations are usually quite limited in scope, and not suitable for extensive analysis of possible dependencies. From nuclear risk assessments, there is a software tool, RiskSpectrum® (see Sect. 15.5), which could also be used for offshore installations in order to perform a comprehensive analysis of dependencies between barriers and barrier elements.

14.4.1.2 Survivability Study

Survivability analysis is often rather difficult to carry out. In its simplest form, the survivability may be considered according to an assessment of which systems and equipment are exposed to accidental loads. This would be relatively easy to determine. A more sophisticated approach would also take into account how the systems will respond to impulse or thermal loading. The degradation (if any) of its performance as a function of time should be considered.

Such analysis is difficult and requires a very detailed study of loads and responses. It also has to consider a very large number of different scenarios, possibly with relatively small differences between them.

A simplified assessment needs to err on the safe side, in other words, be conservative with regard to the performance of these systems and their accidental loading. From a design point of view, such an approach may be acceptable, because it gives a design margin. This approach introduces an over-prediction of risk into the analysis. This may affect the analysis of possible risk reducing measures.

14.4.2 Assessment of Safety Critical Systems

Analysis or evaluation of safety critical systems is an important part of the escalation analysis, and may also be carried out as part of an assurance activity for these systems.

An escalation analysis should, as a minimum, include a classification of the safety critical systems based on their vulnerability to accidental events. A comprehensive analysis should include identification and analysis of the mechanisms of failure of these systems and their dependencies, in relation to relevant accidental events. Emphasis should be given to analysis of the total system and dependent failures should be integrated into the analysis of the safety critical systems.

14.4.3 Detailed Probability Study

A more in-depth study of the probabilities associated with accident sequence development may be required. This may involve detailed modelling of:

- Ignition probability
- Frequency of leaks
- Frequency of dropped objects
- Probability of escalation.

The modelling of the relevant systems is presented mainly later in this chapter. The event tree study will need updating once the probability study has been updated.

14.4.4   HOF Integration

Quantified risk assessment has traditionally been focused on technical systems and capabilities. Much less attention has been focused on the other elements of a system, the human operators, and the organisational factors, HOF. An illustration of the relationship between the different factors is shown in Fig. 14.3. Some changes have developed over the last few years, both in the UK and in Norway. The trend towards more extensive use of floating production systems suggests that operational aspects of safety will be more important in the future, in order to mitigate hazards and control risks.

There has been an ongoing effort in this field for several years, it was 15 years ago mainly focused on 'post-accident' research (see for instance [9, 10]). Predictive work has been done, for instance, by Kirwan [3], Vinnem and Hauge [11], Vinnem et al. [12] and Gran et al. [13]. The BORA project (see Sect. 15.3.1) has developed an approach for integration of operational causes into quantitative risk assessment, in order to enable improved risk control in operational scenarios.

14.4.5   Detailed Consequence Study

For the critical risk factors, more refined consequence (load and response) analysis may be required. This would, for instance, include tasks such as:

- Calculations of fire and/or explosion loads using advanced CFD (Computational Fluid Dynamics) techniques.
- Physical model testing of certain aspects such as explosion loads.
- Non-linear structural response analysis of accidental loads.

14.4.6   Revised Event Tree Study

The results of the detailed barrier study, probability and consequence studies need to be reflected into the event trees, which also may need updating. RiskSpectrum® (see Sect. 15.5) would be a strong candidate for such an analysis.

14.5   Analysis of Different Risk Dimensions

14.5.1   Impairment Analysis

The input generated from the hazard analysis and the analysis of the critical risks is subsequently used in several studies related to the different dimensions of risk. The most fundamental of these is the impairment analysis (see further discussion in Sect. 15.10.6), in order to assess the frequency of impairment of the safety functions. Usually, the following studies are performed in this respect:

- Escape ways impairment study
- Shelter area/TR impairment study
- Evacuation impairment study.

14.5.2   Fatality Risk Analysis

Another type of analysis where the input generated from the hazard analysis and the analysis of the critical risks is used, is the fatality risk analysis. Usually, the following studies are performed in this respect:

- Immediate fatality risk study
- Escape ways risk study
- Shelter area risk study
- Evacuation means availability study
- Evacuation risk study
- Pick-up and rescue risk study
- Overall fatality risk summation.

Analysis of fatality risk is discussed in Chap. 12.

14.5.3 Analysis of Environmental Spill Risk

The analysis of environmental spills is normally focused on oil spills, although release of gas (burned or unburned) to the atmosphere is sometimes considered. The methodology is largely based on an assessment of flow rates (as a function of time) and the duration of possible outflow scenarios. A brief overview is given here, supported by a further discussion in Chap. 18, presenting the derivation of probability distributions for the extent of environmental impact.

The analysis of environmental risk from oil spills in North European waters is a relatively new application of risk assessment techniques. Application of risk assessment techniques for environmental impact is more mature in other parts of the world, but not necessarily in association with oil spill impact.

The analytical approach to analysis of marine environmental risk caused by oil spills is still somewhat immature and undergoing further development. The current steps in the analysis of environmental risk are the following:

- Analysis of causes of blowout and calculation of blowout frequency
- Analysis of alternative spill scenarios
- Spreading analysis (oil drift simulation)
- Coast line impact analysis
- Analysis of damage to Valued Ecological Components (VECs).

14.5.4 Analysis of Asset Risk

The analysis of asset risk may involve one or both of the following:

- Damage risk, related to equipment and structures
- Risk of delayed or deferred production.

These calculations are often simple, subjective prediction of overall damage cost or delay in operation. It may however be possible to carry out detailed studies of elements of damage and delay, and to quantify the overall effect based upon detailed assessments. Such an approach would be quite similar to the approach used for the detailed analysis of immediate fatality consequence (see Chap. 15).

The summation of risk is often based upon the classification of damages or delays into a few categories, and then summing up the occurrence frequency of events that fall into these categories.

14.6   Sensitivity Analysis

Sensitivity studies could be claimed to be the main purpose of most QRA studies if the overall objective of the work is the comparison of alternative solutions or identification of risk reducing measures. One of the main efforts in planning and executing a QRA should be directed towards achieving efficient sensitivity studies, in such circumstances. It has been found that sensitivity studies are more effective and refined if attention is given to defining in detail the scope of the studies. When studies are planned with sensitivity modelling in mind, the modelling becomes more focused.

There is an increased focus on sensitivity studies, especially from authorities. NORSOK standard Z-013 [1], has explicit requirements for sensitivity studies. There are two kinds of sensitivity studies that shall be carried out:

- Sensitivity studies to illustrate the effect of variations in various parameters in the risk model
- Sensitivity studies to illustrate the effect of introducing risk reduction measures.

Chapter 16 discusses presentation of risk results from QRA studies, including presentation of sensitivity studies.

14.7   Limitations of Risk Analysis

QRA has certain limitations that need to be appreciated when planning such studies, as part of the establishment of the context (see Sect. 3.4). The limitations of a risk analysis should usually be stated explicitly in the study. Limitations on the use of risk analysis will also result from the way the general requirements are adhered to. The following general aspects tend to limit the general applicability of the results:

- There has to be a sufficiently broad basis of relevant data for the quantification of accident frequency or accident causes.
- The data used usually refers to distinct phases and operations, and therefore the results should not be used for other phases and operations.
- The depth of the analysis in the consequence and escalation modelling determines the applicability of the results. Such considerations are typically made for the systems and functions that are involved in the escalation analysis. If, for instance, the modelling of passive fire protection is very coarse, then the study should not be used to determine what the optimum choice of passive fire protection should be. This may seem as an obvious fact, but failures to observe such limitations are not rare.

The level of precision in the results should not be higher than justified by the calculations, data and models used in the analysis. This implies that risk cannot be expressed on a continuous scale when the calculation of either probability or consequence (or both) is based on categories.

Another severe limitation of QRA is the ability to analyse installation specific aspects. More extensive use of models to analyse the causes of accidents will help to have studies that are more specific to the installation in question. It is sometimes said that a QRA of an offshore installation is representative of an average installation operated in an average way, but with an overall shape and module layout of the installation in question. This aspect underlines the importance of the ability to represent specific details in the analytical models, to an extent that differences can be reflected.

14.8 Use of Software

There are two types of software relevant to risk assessment studies; specialised software packages which perform detailed physical/statistical analysis of a phenomenon, and software suites that can undertake an entire study.

The specialised software packages are available for many of the physical modelling steps, such as gas dispersion, fire modelling, explosion modelling, collision modelling, etc. These packages are limited in scope, but may be very important in order to predict realistic loads, effects and responses.

The all-including software suites are different. These have replaced the spreadsheet based models used previously, which could be very complex and correspondingly difficult to review. The main purposes of having a software suite can be as follows:

- Use of fixed model code (as opposed to spreadsheets) prevents risk of it becoming corrupted.
- Standardised data handling reduces risk of error.
- Safe transfer of data, and ease of documentation.
- No human interface between model output and analysis.
- Documented QA trail.
- Transparent linkage of models and ease of documentation.
- Effective update of studies can be made.
- Modification of study structure is independent of the code.

A number of software suites have been introduced during the last 15–20 years, but there are none that have established themselves as industry standards, or as the industry accepted way of conducting QRA studies. This is somewhat surprising, because such standardisation has occurred in many other disciplines, such as Computer Aided Design. Actually, the tendency towards such standardisation was stronger some years ago. Some of the main software tools are presented briefly in Appendix A.

For QRA studies for nuclear power plants, there is a de facto industry standard software, which has almost half of the marked, the RiskSpectrum® software package (see Appendix A). This has been used to some extent for analysis also of offshore installations, but is certainly not established as industry practice so far.

14.9   Data Sources

This section presents data sources that are typically used in offshore QRA studies. Only the main sources are shown, largely limited to those in the public domain. A number of papers, reports, etc., may also be used from time to time, for specific subjects. Some comments are also provided on the use of data sources, with respect to availability etc.

Modelling and data sources are often closely coupled, in the sense that more refined models require more extensive data sources. The OGP (now IOGP) Data Directory [14], takes care of this to a certain extent, in the sense that modelling is briefly outlined.

14.9.1   Types of Data Sources

This overview focuses on generic data sources, while dealing only briefly with internal company sources. The following are the types of data sources that are used in QRA studies:

- Generic data sources
- Accident statistics
- Failure databases (loss of containment, etc.)
- Equipment failure databases
- Physical properties of various substances
- Company internal accident and incident databases.

It would often be expected that accident or failure frequencies are higher than average in certain operational phases or under certain conditions. It should be noted that the data sources (and models) currently available usually do not allow differentiation between such differences.

14.9.2   Blowout Frequency

The data that are relevant in relation to blowouts are mainly the following:

- Blowout occurrences depending on activity performed and type of well
- Flow path and failure category
- Flow rate and duration

Table 14.2 Sources of occurrence data for offshore blowouts

Data source	Coverage	Availability	Comments
WOAD®	All accidents offshore, worldwide	Available on disk: annual licence fee required	Biannual statistical analysis report may be purchased
SINTEF Blowout data base	All blowouts worldwide	Available on disk: annual licence fee required	Annual statistical analysis report issued to members
IFP Blowout data base	All blowouts worldwide	Unknown	
MMS Events File	All significant accidents on US shelf, also oil spills	Biannual report	Available on disk through WOAD® [15]
Offshore Blowouts Causes and Control	All blowouts worldwide	Available as a book from Gulf Publishing	Holand, 1997

- Time to ignition for blowouts
- Type of ignition (fire vs. explosion).

Relevant data sources for blowout frequencies are shown in Table 14.2. The availability of the factors detailed above, is relatively good from these data sources. There are nevertheless some problems with use of blowout data:

- Information about the causes of blowouts is too sparse, to the extent that information which can be used to prevent accidents is not extensively available.
- Data is used for areas wider than just the North Sea, in order to have a sufficiently large data basis. This may sometimes cause uncertainty about what equipment and procedures were in place as barriers at the time of the blowout. Some over-prediction of the blowout risk may have occurred due to this.

The SINTEF Blowout database [16], in 1998 attempted to distinguish between cases where two barriers were used and those where either only one was used, or when one of the barriers was not available for some reasons.

14.9.3   Process System Leak Frequency

The frequency of leaks from process components is an aspect where operators could have a considerable amount of data internally, mainly for small leaks. Some operators have these data available, others do not have a systematic registration of such data. All operators of platforms on the Norwegian Continental Shelf have to report gas leaks to PSA. Also a voluntary reporting scheme to the Norwegian Oil and Gas was in operation for some years. Neither of these two collection systems, however, are sufficiently detailed to enable this data to be used for risk analysis purposes.

PSA (previously NPD) has collected data systematically for hydrocarbon leaks since 1 January 1996. The data is published through the RNNP project in the annual report [17]. The data is most detailed for the period after 1 January 2001.

The UK Health and Safety Executive has, since 1992, operated an extensive data collection scheme based on mandatory reporting by all UK operators [18]. This system which was set up as a result of the recommendations in the Piper Alpha Enquiry is the best data source for process system leaks. The database covers all offshore installations operating on UK Continental Shelf. Both leak and ignition data, as well as equipment data (for the volume of exposure) are contained in the database. An annual statistical report is available on the web [18]. Use of installation specific data is briefly discussed in Sect. 14.10.

Chapter 6 has discussed the statistics of HC leaks in Norway and the UK mainly, and the causes and circumstances of leaks.

14.9.4   Riser/Pipeline Leak Frequency

The recognised source in this field is the AME Loss of Containment Report ('PARLOC' study [19]), which was updated in 2001. There is quite a lot of data available, but it is unfortunate that the report has not been updated since 2001, especially for flexible risers/flowlines. The flexible pipes that are the data basis for the 2001 version of the report are flexible risers fabricated in the 1980s and 1990s. This was when the flexible pipes were a new and partly unproven technology. These pipes experienced some failure modes that since then have been eliminated, also through selection of other materials. The author saw some internal reports around 2010 which demonstrated that the failure frequencies of flexible pipes fabricated after the year 200 were substantially lower than the failure frequencies for flexible pipes fabricated in the 1980s and 1990s. In fact at that time, no failures had occurred to flexible pipes fabricated after year 2000.

Incidents in the Norwegian section are published annually in the RNNP report [17].

14.9.5   Vessel Collision

The most extensive source on vessel collisions in the North Sea is the report by J. P. Kenny. This study was based on the UK Department of Energy's records, now held by HSE. Unfortunately, this report is getting completely outdated.

HSE has published regularly studies on collision incidents. The latest edition was published in 2003, and covered data until 2001 [20]. HSE has also published incident data for tanker collisions with FPSO vessels, which is a special form of vessel collision as the tankers are in attendant mode.

There have been several collisions between passing merchant vessels/ field-related traffic (disregarding shuttle tankers) and offshore installations since 2000. In particular, there has been a large number of collisions with field-related traffic. Some examples are:

- Fish factory ship Marbella (2880 GRT) collided with Rough accommodation platform (UK sector) on 8 May 2002, at about 09:35 in the morning. The contact was limited to a glancing blow, and the damage to the structure was superficial. The vessel, which had an ice strengthened hull, sustained significant damage but not life threatening in the bow above the water line, and was able to return to port by its own power. The collision occurred in bad visibility due to fog, at full or near full speed, and no pre-warning was given on the installation before the hit, but no injuries occurred.
- The 4,900 dwt supply vessel Far Symphony hit the mobile drilling unit West Venture drilling on the Troll field in the Norwegian North Sea sector, on 7 March 2004, at 02:48 h, with no wind and good visibility. The impact occurred at half speed (7.3 knots) and no pre-warning had been given on the installation. Both structures sustained damage, but not critical, and both could manoeuvre to shore by their own power for repairs. The supply vessel was brand new and the installation was a modern semi-submersible unit. An old installation might not have taken the high energy impact (about 40 MJ) with insignificant damage. The vessel was operating with auto-pilot engaged and course directly for the installation. Two crew members on the bridge failed to realise the threat even when entering the safety zone around the installation. Although the vessel was attending the installation, the collision scenario had certain parallel aspects with a passing vessel collision.
- The 3,120 dwt supply vessel Bourbon Surf hit the Grane fixed production platform in the Norwegian North Sea on 9. July 2007 at 0735. Both the captain and the first mate were present on the bridge when the vessel approached the installation, but were distracted with preparations for off-loading. When the captain realised that they were so close to the platform, it was too late to avoid the collision. The impact occurred with 1–2 knots. The damage was minimal to both the Grane installation and the vessel.
- In the early morning of 8th June 2009, the vessel Big Orange XVIII was on its way to the 2/4-X facility on the Ekofisk field to carry out well stimulation. The autopilot was not deactivated before the vessel entered the safety zone, and since the autopilot was active during the approach, the planned course changes were not executed in the manner that the duty officer on the bridge expected. The vessel avoided a collision with Ekofisk 2/4-X and Ekofisk 2/4-C, passing under the bridge between these facilities. Also a collision with the jack-up living quarters facility COSL Rigmar was avoided. However, the vessel ultimately collided with the unmanned water injection facility Ekofisk 2/4-W. At the time of the collision, the Big Orange XVIII had a speed of 9.3 knots. No personnel

suffered physical injury; however, the incident did cause significant material damage, to both the facility and the vessel.

There have been several less serious collisions in addition to these most serious events. An old, somewhat more spectacular accident, occurred in 1988, when a German submarine collided with a jacket structure (Norwegian sector) well below the surface, but again little damage was caused to either the structure or the submarine. Nevertheless, the West German state paid 80 million NOK (1988 value) for repairs to the fixed platform structure.

14.9.6 Falling Objects

The most extensive public domain source of data regarding falling objects is WOAD® [15]. This database, however, only contains the most severe cases of falling loads, where significant damage to the installation resulted. There are also other 'one-off' limited studies that have been performed over the years, but there is no single source with distinct authority.

HSE has also published overview of accidents and incidents to production installations [21], and mobile drilling units [22], during the period 1980–2003.

14.9.7 Marine Accidents

'Marine accidents' are accidents affecting the stability and floatability of vessels and non-fixed installations (semi-submersibles, barges, etc.). The main sources of data regarding marine accidents are the following:

- Lloyds' List for general shipping, including offshore installations and vessels. Lloyds of London have several publications that give such information, for example the Weekly Casualty Report.
- WOAD® [15], has accident data on offshore related vessels and other floating structures. WOAD® data base is available in two forms:

  - Statistical analysis report, issued biannually.
  - Electronic version, database updated at least once per year.

The report by HSE on accidents and incidents to production installations and mobile drilling units during the period 1980–2003 also covers marine accidents [21, 22].

14.9.8   Utility Area Accidents

There is no single authoritative source of data relating to accidents in utility areas. Such accidents are mainly fires (and other miscellaneous types of accidents) in engine rooms, workshops, living quarters, etc. The available sources of data relating to such occurrences are:

- Risk level project annual reports [17]
- HSE reports for mobile drilling units and production installations (see Sect. 14.9.7)
- WOAD® [15], has accident data on the severe accidents in utility areas.

14.9.9   Helicopter Accidents

National authorities are the best sources of statistics on helicopter accidents. The UK and Norwegian Civil Aviation Authorities publish annual statistics including details of accidents, incidents, and flight hours. Also OGP (now IOGP) has published such statistics. The risk level project [17], publishes data on incidents and traffic volumes. An assessment of the available risk levels associated with helicopter transportation of personnel was presented in Chap. 13.

14.9.10   Occupational and Diving Accidents

There is no single authoritative source of data relating to occupational and diving accidents. National authorities are the best sources for statistics on these accidents. The available sources for such data are:

- Risk level project annual reports [17].
- HSE annual reports.

A summary of the available data was presented in Chap. 12 and Appendix B.

14.9.11   Ignition Probability

The most recent ignition probability model is referred to as MISOF, which is short for Modelling of Ignition Sources on Offshore oil and gas Facilities [23]. The MISOF model was developed by a joint industry project in 2016 and is being updated in 2018 [24].

As emphasized in the MISOF report, the model is to a large extent based on analysis of the statistics of releases and ignited leaks at installations in the North Sea from 1992 to 2017. In addition, knowledge about the physical properties of the ignition phenomena has been applied where such knowledge has been available. For instance, the models for ignition due to exposure to gas turbine air intakes and diesel air intakes are based on an assessment of the behaviour of the machinery when exposed to flammable fluid.

One challenge with the model is the availability of relevant data. There have been few ignited hydrocarbon leaks recently, and therefore the statistical information about ignited leaks is limited. When it comes to hydrocarbon leaks above the 0.1 kg/s leak rate, the last ignited leak in the Norwegian sector occurred on 19th November 1992. In the UK sector, three ignited leaks have been found relevant, while two additional ignited leaks have been found partially relevant. In addition, information about 23 ignited leaks was used to understand the equipment categories and the ignition mechanisms.

In the PLOFAM report [25], it is emphasized that the PLOFAM leak frequency model and the MISOF ignition probability model should be used together and not combined with other models. Using the models together ensures consistent interpretation of scenarios that have an impact on both models.

Before MISOF, DNV GL has from time to time done some work around ignition sources [26, 27]. A comprehensive study aimed at establishing a model for ignition probability was conducted in 1996–98. This study included considerable data, including data from the UK HSE and their annual reports on leaks and ignitions in the UK offshore operations. The study was later updated in 2005 and 2006 [28], but has now been replaced with the MISOF model.

14.9.12 Safety System Reliability

There is no single authoritative source of data to be used in assessing the reliability of safety systems. OREDA® (2009) has considerable data, especially on fire water pumps, isolation valves, etc. Significantly less data are available for detectors, control panels, etc.

The Norwegian oil and gas Association has published Guidelines for application of IEC 61508 [29] and IEC 61511 [30]. This guideline [31], also has some typical availability values for application by the offshore industry (see Sect. 15.5.3).

SINTEF has also published the PDS method handbook [32], and the PDS data handbook [33]. The latter book contains reliability data for safety instrumented systems.

Lastly, the Risk Level Project [17], has collected availability data for selected barrier elements since 2002 (see further details in Sect. 14.10.2).

14.9.13   Data Sources for Reliability Analysis

Reliability analysis of safety systems is often used to replace actual experience data when this is not available. Such data may be found in the following sources:

- OREDA®; Offshore Reliability Data Handbook [34]
- PDS method handbook [32] and data handbook [33]
- NPRD: Nonelectronic Parts Reliability Data [35]
- IEEE Standard 500 [36].

OREDA® and the PDS handbooks are limited to offshore applications, while the other two sources are general, with no specific offshore application. OREDA® on the other hand has traditionally been focused on reliability and availability of production systems and equipment. Also, the sources mentioned in Sect. 14.9.12 have data that may be relevant.

14.9.14   Data for Fatality Modelling

Data for fatality modelling includes data to be used in assessing immediate fatalities, as well as fatalities during escape, evacuation, and rescue. Fatalities due to occupational accidents should also be included. Such data is discussed in Chap. 12.

14.10   Use of Installation Specific Data

14.10.1   Generic Versus Installation Specific Data

The installation specific data in question is mainly frequency of occurrence data relating to failures, leaks, etc. Data on the failure of safety systems are usually also available. The discussion below is mainly related to the frequency of leak data, but may easily be transformed to safety system failure data. It is important to use installation specific data whenever possible in risk analysis studies for installations in the operations phase. This is due to the following:

- Generic data often have rather high failure rates (such as for loss of containment). If the installation can show lower values, with a reasonable degree as assurance, then these may be used (alone or in combination with the generic data), to predict a more representative (and actually in many circumstances, lower) risk level.
- Use of installation specific data will usually give greater credibility to the analytical work performed.

It will be important to assure that the installation specific data collected is as complete as possible, in order that the data is credible and representative. Thus, one should review the completeness and attempt to ensure that there is no under-representation or systematic errors. Installation specific data may be used in several ways, the following being the most important:

- As the only source of data for frequency calculation.
- In combination with generic data for frequency calculation.

If a significant amount of installation specific data is available (say more than a dozen occurrences applicable to the actual conditions), then this data may be used alone, being sufficient to give a prediction of the overall failure frequency. Generic data may still, however, have to be used, in order to assess the effect of special conditions, differences between types, etc. An illustration of such use is presented in Sect. 14.10.4.

14.10.2 Installation Specific Data from RNNP

The Risk Level Project for Norwegian offshore and onshore petroleum operations has requested the operators of Norwegian installations to collect data systematically for a period of up to ten years (offshore installations). The following data are available:

- Number of hydrocarbon leaks as a function of time, classified according to leak rate (kg/s)
- Performance data (fraction of failure) of selected barrier elements during test as a function of time (see Fig. 14.8)
- Performance measurements (response times) for muster drills as a function of time.

Leak data (failure of containment barrier) is available from 1 January 1996; the records from 2001 are, however, more detailed. Consequence barrier performance data are available from 1 January 2002 for most of the companies, for the following barrier elements:

- Gas detectors
- Fire detectors
- Riser/pipeline ESD valves
- DHSV
- Christmas tree valves
- Pressure safety valve (PSV)
- BDV
- Deluge valves.

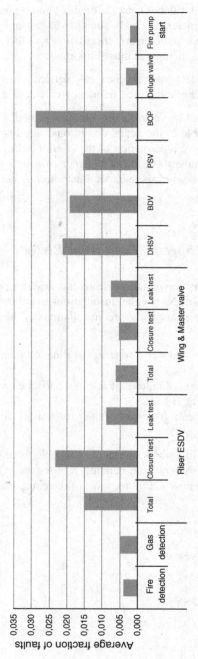

Fig. 14.8 Fraction of failures for selected barrier elements, 2005

There is a clear trend towards using the available data more extensively for installations in operation, when performing or updating risk assessment studies.

The differences between installations are quite extensive. Some installations have so few failures that the data from own operations are insufficient. There may then be a need for combined specific and generic data.

With respect to use of the data from RNNP, the following should be noted: the data presented are values of x/n, where x is the number of failures, during n tests. The 'probability of failure on demand' (PFD) may then under certain preconditions be expressed as $x/2n$. The most important preconditions are a constant failure rate and that the component is 'as good as new' after replacement/repair.

14.10.3   Combination of Specific and Generic Data

Generic data may have to be combined with the installation specific data if the volume of specific data is insufficient in relation to the objectives. One generally accepted way to perform such combination is by Bayes' analysis. This approach allows an average to be calculated by weighting the generic and the installation specific data according to the available volume of data. The following example illustrates this approach. Let us assume we have the following data:

Generic database

'a_g'   number of occurrences
'b_g'   number of exposure hours.

Installation specific database

'α'   number of occurrences
'β'   number of exposure hours.

For this illustration, we assume that a random failure model can be applied i.e., that the failures are assumed to be Poisson distributed with a constant failure rate, λ. We have to use the available data to give a prediction of 8. Then we may produce the following values for the occurrence frequency:

$$\text{Generic frequency: } \lambda_{generic} = \frac{a_g}{b_g} \tag{14.1}$$

$$\text{Installation specific frequency: } \lambda_{install} = \frac{\alpha}{\beta} \tag{14.2}$$

$$\text{Weighted (combined) frequency: } \lambda_{weighted} = \frac{\alpha + a_g}{\beta + b_g} \tag{14.3}$$

This is based on the following: A priori and a posteriori values are expected values in Poisson distributions, where a prediction interval may be expressed using a Chi-square distribution. This implies that the width of the interval is proportional to the inverse of the number of degrees of freedom i.e., when the number of observations increases, then the degree of freedom increases, and the width of the confidence interval decreases. Thus the weighted frequency will have the smallest confidence interval, due to the size of the data basis.

There are several difficulties involved in using plant specific data, the most severe often being that insufficient details relating to the incident are available. Usually, however, the observations from one's own installation are few in number, and thus it is often possible to get more details by speaking to the people involved. In the case of leaks, one may have to search out data about leak rates which was not recorded explicitly at the time of the incident.

The leak rate itself is in any case usually not directly available, and thus indirect observations are used to back calculate the leak rate. In the case of an unignited gas leak for instance, the leak rate may be calculated from the pressure of the leak source and the size of the gas cloud, if this may be found from gas detection records.

14.10.4   Example, Combination of Data

The use of installation specific data may be illustrated by a simple example. The following generic gas leaks data for a certain type of valve is assumed:

Valve years reported	3724
Number of leaks reported	
0.1–1 kg/s	25
1–10 kg/s	3
>10 kg/s	1

In the generic datab ases, it is common that leaks smaller than a certain limit are not reported at all, or are reported in an incomplete manner. This results in the observations not being very useful. The internal, platform specific, database is usually much more limited. Typically, only small leaks are observed. The following platform specific observations may be assumed:

Valve years reported	455
Number of leaks reported	
<0.1 kg/s	36
0.1–1 kg/s	2

Leak frequencies for both the generic and the platform specific data sets are calculated first. These are shown in Table 14.3.

This situation is typical, in the sense that in the only category where comparison is possible, the lowest leak rates, the platform specific frequency is somewhat lower than the generic data (here 35%). At the same time, it is a considerable difference in the number of observations that form the basis of the calculations. This is clearly demonstrated if confidence intervals are considered, as shown in Fig. 14.9.

The ratio between upper and lower limits in a prediction interval is 1.66:1, for the generic data, whereas the corresponding value for the platform specific data is 4.8:1. (Corresponding values for 90% prediction interval limits are 1.92 and 7.7.)

Figure 14.9 shows both a mean predicted frequency and an interval for the generic leak frequencies, the platform specific frequencies and the combined (Bayes') frequencies. The intervals are 80% prediction intervals, implying that it is 80% probability that the future observations will fall in the interval, if there are no changes in the operational conditions.

The next step is to consider a combined data set according to the Bayesian approach, which according to Eq. (5.3) may be predicted as follows:

Table 14.3 Traditional calculation of generic and platform specific leak rates for valves

Leak rate category (kg/s)	Failure rate values (per valve year)	
	Generic data	Platform specific data
0.1–1	6.7×10^{-3}	4.4×10^{-3}
1–10	8.1×10^{-4}	0
>10	2.7×10^{-4}	0

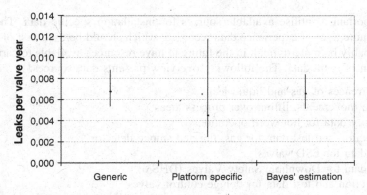

Fig. 14.9 Prediction intervals (80%) for alternative leak rate calculations

Table 14.4 Traditional prediction of generic and platform specific leak rates for valves

Leak rate category kg/s	Failure rates (per valve year)		
	Bayes prediction	Generic data	Platform specific data
0.1–1	6.5×10^{-3}	6.7×10^{-3}	4.4×10^{-3}
1–10	7.2×10^{-4}	8.1×10^{-4}	0
>10	2.4×10^{-4}	2.7×10^{-4}	0

Valve years reported	3724 + 455 = 4179
Number of leaks reported	
0.1–1 kg/s	25 + 2 = 27
1–10 kg/s	3 + 0 = 3
>10 kg/s	1 + 0 = 1

The data basis is then available in order to calculate the combined frequencies according to Eq. (5.3). The results are shown in Table 14.4, which also repeats the values for the generic and platform specific calculations, for easy comparison.

The Bayesian approach also produces adjusted generic frequencies for the leak categories which have not occurred during the time recorded on the platform. The reductions to the generic frequencies in this example are quite limited, reflecting the small number of leaks that had been observed on the platform, thus having a limited statistical significance.

14.10.5   Data Sources for Installation Specific Data

14.10.5.1   Overview

It is important to utilise available sources for installation specific data. There is usually quite a lot of experience data recorded, which should be used first of all. It will probably be more difficult in the future to have resources available to carry out collection of extra data. The following overview presents data sources:

- Occurrences of gas and liquid leaks
- Crane movements, lifting over process areas
- Activity data for hot work operations
- Inspection and test data for gas, fire and smoke detectors
- Test data for ESD valves
- Test data for Downhole Safety Valves (DHSV)
- Inspection and test data for deluge control valves
- Test data for fire water pumps
- Downtime records for fire water pumps

- Injury statistics for personnel onboard
- Passing vessel traffic pattern.

Sometimes it is required that existing data recording schemes are modified, in order to ensure that the data collected is useful for the creation of input data for risk assessments. This had to be done by several of the operators in order to enable reporting of failure data for barrier elements to the Norwegian Risk Level Project.

It is sometimes argued that the data sources indicated above have limited applicability, since data cannot be observed for all main components of a system. If we consider emergency shutdown for example, the entire system consists of detector units, cabling, signal processing units, software, output signal transfer and processing, instrument power supply and ESD valves. Only the sensors (detectors) and the ESD valves of all these units are tested regularly, thus providing relevant data. This should not however prevent the use of these data sources. They do not provide all data which could be used, but they do provide some of the most crucial data. It could also be argued that these sources provide input for those elements which have the highest contribution to system unavailability.

14.10.5.2   Example

Table 14.5 shows a typical example of the test data that might be available for deluge control valves on an installation. Tests may be carried out on each valve every 6 or 12 months, which could give the pattern shown in the table. A total of 118 tests are carried out during the year, implying that each valve is tested once. Four tests are assumed to fail, implying that the average availability during the year is 96.6%. Different periods may be used for averaging of the data, for instance 3 months or 12 months. Figure 14.10 shows the tests and the failures, as well as the accumulated 3 months' average and the annual average value.

Table 14.5 Overview of test data for deluge control valves

Week no.	No tested	No failed
8	22	1
17	15	0
24	35	2
34	27	1
45	19	0
Total	118	4
Average availability		98.3%

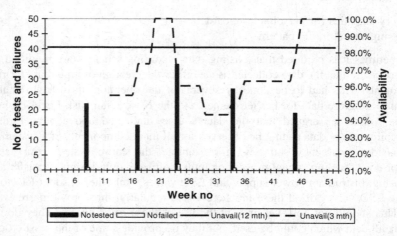

Fig. 14.10 Deluge tests and failures and average availability values (3 months and 12 months averaging)

14.11   Use of Risk Analysis Studies in Life Cycle Phases

This section presents an overview of the use of or risk analysis studies (qualitative and quantitative) in the various life cycle phases. The recommendations are consistent with the prevailing requirements of NORSOK Z-013 [1].

14.11.1   Analyses During Concept Development

Table 14.6 presents an overview of the main risk analyses which should be conducted during concept development phases, including their timing and main purposes.

Table 14.7 presents an overview of the additional analyses which should be considered during development phases. When performed their timing and main purpose should be as given in Table 14.7.

14.11.2   Analyses in Operations

Table 14.8 presents an overview of the main analyses which shall be conducted during the operations phase, including their timing, main objectives and focal points.

Table 14.6 Summary of main risk analyses during field development phase

Analysis	Timing	Main purpose
ERA	Selection of main concept, pre-conceptual phase	• Comparison and ranking of field development concepts • Optimisation of chosen concepts • Identify all major hazards
Concept TRA	When layout drawings and PFD's have been made Before issue of invitation to tender. Before submission of PDO	• Assessment of compliance with acceptance and design criteria • Identify functional requirements as input to design specification • Establish dimensioning accidental loads
Design TRA	When layout drawings, P&ID's for process and essential safety systems have been made	• Verification of design • Check of compliance with overall RAC (assumptions for safe operation) • Establish performance standards and requirements from assumptions and premises in analysis

Table 14.7 Summary of optional additional risk analyses during field development or operational phase

Analysis	Timing	Main purpose
Detailed risk analysis and analyses in connection with design change proposals, handling of deviations and project phases (TRA extension)	After concept risk analysis	• Evaluate special risk aspects on the basis of performed risk analysis in order to give design input • Evaluate how changes etc. affect risk • Evaluate effects of deviations from statutory requirements
Quantitative/qualitative studies (FMEA, HAZOP, etc.)	When appropriate or requested during engineering phases, in order to evaluate systems designs	• Identification of required improvements in system design
Integrated risk and EPA of fabrication and installation	Prior to decision on concepts for fabrication and installation	• Provide input to concept and methods for fabrication and installation • Identify operational and environmental limitation to be observed during fabrication and installation
As-built TRA	When final design of systems	• Update TRA with as-built information and reflect any changes to risk levels • Provide input to operations

Table 14.8 Summary of main risk analyses during operations phase

Analysis	Timing	Main purpose
TRA updates	During operations phase	• Update of risk levels due to experience, modifications, model improvements, changes in criteria, operational mode, manning level, maintenance philosophy, etc.
Risk analysis of critical operations, including SJA	Planning of the operations	• Identification of hazards • Identification of risk reducing measures, in order to achieve safe job performance
TRA adaptation for operations phase	Regularly during operations phase	• Ensure that the risk level is kept under control • Ensure that operational personnel are familiar with the most important risk factors and their importance • Identify and follow up of assumptions made in earlier phases • Update DSHA
Integrated TREPA of modifications	Prior to deciding on details of modifications	• Evaluate the operational phase with the modifications implemented • Identification of hazards • Identification of possible risk reducing measures
Integrated TREPA of modification work	Prior to deciding on the way in which the modification shall be implemented	• Evaluate the operational phase during modification work • Identification of hazards • Identification of possible risk reducing measures

14.12 Execution of Quantified Risk Analysis

14.12.1 Quality Aspects

Quality requirements for QRA studies are presented in the NORSOK Guidelines [1]. One of the most important requirements in the Norwegian legislation is that risk tolerance criteria shall be established prior to commencement of the risk analysis. Some of the weaknesses that often are experienced in QRA studies are presented in Sect. 14.13 (see also Sect. 3.4) about establishing the context.

14.12.1.1 Use of QRA in HES Management Context

An important aspect to consider is how QRA is used as part of HES management i.e., what is the QRA used for, and when is it used. Some important aspects in this context are the following.

Risk analyses should be planned from consideration of the relevant activities, in order to ensure that the QRA can be actively and efficiently used in the design and implementation of the activities. This determines how early in a project QRA work may be started, bearing in mind the need for sufficient input data. This question needs to be addressed in the following situations:

- New projects i.e., development of new fields and installations, including concept selection, layout development, design of technical systems and planning of operational procedures.
- Modification of existing installations and facilities, including change of operational mode or field configuration.
- Extensions and tie-ins into existing installations.

There are other situations where risk assessment should be reviewed or considered, but where QRA may not always be the most suitable tool. These situations include:

- Organisational changes
- Decommissioning and disposal of installations.

Changes in manning level may introduce factors such as the number of people exposed to accidents, which may make the use of a QRA appropriate. There are other aspects, however, which in the past were not suitably addressed by means of a QRA. These include changes in procedures for normal operation or emergency conditions, working environment conditions, etc. This may change in the future with the development of the BORA methodology (see Sect. 15.3.1).

14.12.1.2 Prerequisites Before QRA Is Started

Requirements relating to the execution and use of risk assessment should be stated in a way which ensures that the results form correct and adequate input to the decision-making process. This implies that a number of aspects need to be clarified before a risk analysis is started:

- The purpose of the risk analysis has to be clearly defined and be in accordance with the needs of HES management.
- The target groups for the results of the analysis have to be identified and acknowledged. The target groups will usually include HES management personnel, operational and/or engineering management. The offshore work-force also has a legitimate, regulatory based, right to have an understanding of risk aspects and to receive results of the studies.

- Plans should also be made for the inclusion of operational personnel in the analysis in order to ensure realism in the models and assumptions.
- The relevant decision criteria for the activity have to be defined.
- The choice of an appropriate analysis method is made partly on the basis of the scope of the study and its limitations, which all need to be clearly defined.
- The sensitivities that may be foreseen should be defined as input to the modelling, such that the model can be tailored to the sensitivities.

One comment which could be made, based upon repeated observations of failure, is that the planned use of the results from the QRA should be considered in detail. Sometimes a study does not give the results that were expected, and in these situations it is very advantageous if consideration has already been given to relevant courses of action. It may be very difficult to relate rationally to unexpected results, if no prior attention has been paid to such possible circumstances.

14.12.2   Documentation of Assumptions and Premises

Assumptions and premises for an analysis are important because these often have a considerable influence on the results, and because they are the best way to make a QRA study a 'living document'. Typical examples of assumptions and premises include:

- Activity levels on the installation, such as the number of crane movements, extent of hot work activities, etc.
- Manning level, distribution, work schedules
- Operation of safety systems
- Performance of emergency response systems and actions.

Sometimes it may be argued that assumptions are more important in a QRA than the results, at least in the long run. Results are important for the immediate follow-up, but the necessary actions have to be taken in a relatively short time span. The assumptions on the other hand constitute, perhaps most typically, the 'living' element of a study. The following are important aspects in relation to assumptions:

- Assumptions need to be identified at an early stage and discussed with the organisation in charge of the installation, for verification and acceptance.
- Assumptions need to be collected and presented systematically, and to appear in the documentation where they are relevant. It is also advantageous to indicate what the likely effects of changes to assumptions will be. Extensive cross-referencing is important.
- Assumptions need to be followed up and reviewed regularly when the study is completed, in order to make the study 'living'.

It may be a good solution to split the assumptions into three categories, because the persons most directly involved in the review and follow-up will be different:

- Design basis
- Operational assumptions
- Modelling assumptions.

14.12.3 Typical Study Definitions

Approximate manhour budgets and time schedules for some typical studies are shown in Table 14.9.

Table 14.9 Typical risk assessment studies during field development projects

Study type	Typical study definitions			Comments
	Study scope (installations)	Budget (manhours)	Duration (months)	
Concept Safety Evaluation	Entire installation at design stage	1,500	3	Very dependent on level of details covered
TRA (detailed QRA)	Entire installation during design or operation	2,500	5	Very dependent on level of details, size, complexity
Collision risk study	Platform structure at design stage or operating installation	500	2	Including probability analysis, consequences and sensitivity of risk reducing measures
Dropped objects risk assessment	Platform areas Structure/hull Subsea installations	400	2	Scope including probability analysis and consequence analysis
HAZOP	Process systems Utility systems		2	Manhours depend on team size as well as size and complexity of systems analysed
Blowout risk study	Well drilling, completion, production and intervention programme, equipment and procedures	400	2	
Escape and evacuation risk study	Escape ways evacuation and rescue means and procedures, including external resources	700	2	For all relevant accidental and environmental conditions. Qualitative and quantitative study

(continued)

Table 14.9   (continued)

Study type	Typical study definitions			Comments
	Study scope (installations)	Budget (manhours)	Duration (months)	
Fire risk assessment	Active and passive fire protection systems, including supply and power systems	400	2	For all accidental conditions, including explosions followed by fire. Qualitative and quantitative studies
Explosion risk study	Overpressure loads and venting areas	500	3	For all leak and ignition sources. Includes use of FLACS software
Installation phase risk assessment	Entire structure during installation at the field	500	2	Depending on hazards and installation complexity. Qualitative and quantitative studies
Risk assessment of modifications	Modifications to existing installations			Depending on hazards, complexity, modifications

14.13   Challenges Experienced with QRA Studies

This subsection discusses briefly some of the challenges that we often have experienced when reviewing QRA studies. Where relevant, references are made to other chapters in this book.

14.13.1   Ethical Failures

Ethical failures can ruin completely the management of major process based on risk assessment. Reference is made to the discussion of these aspects in Sect. 3.12, illustrated through a case study where risk-based management of major hazard risk failed completely.

14.13.2 Hazard Identification

Hazard identification is often performed in an unsystematic manner or with insufficient attention to operational aspects and/or combination of failures or errors. The consequence of this is a lack of assurance that hazard identification is complete.

The criteria that are used in order to screen out non-critical should be documented. This is seldom done. The hazards that are not considered further should be well documented including the evaluations that have been made for these hazards. This is often not done at all.

14.13.3 Analysis of Risk

The requirements for the analysis methodology presented in NORSOK Z-013 [1], are the most extensive requirements that are publicly known. There are requirements relating to use of models and tools, as well as documentation of premises and assumptions.

One of the requirements that are rarely satisfied is the required documentation of assumptions and premises. Often the documentation of design related assumptions is quite extensive, whereas assumptions relating to operational aspects are less well covered. What usually is the weakest is the documentation of analytical assumptions. There are usually many analytical assumptions which need to be documented, in order for the analysis to be transparent.

Although design assumptions often are well documented, sometimes unnecessary assumptions are made. It is not uncommon that design assumptions are made even for an installation that has been in operation for several years, possibly because the analysis team is unfamiliar with system capacities or has not made the effort to clarify such aspects. Usually, such aspects could be eliminated as sources of uncertainty. Keeping such unnecessary assumptions also gives a bad impression of the work that has been performed.

When software packages are used to carry out the analysis, the documentation is sometimes insufficient. It is easy to argue that the documentation of the analysis is embedded in the software files, such that full paper copy documentation is not warranted. But this makes it impossible for any third parties to review the analysis, and is as such insufficient documentation of the work.

Another aspect is the selection of cases for which detailed consequence calculations are made. These cases need to be representative of the risk picture, but there is no obvious way in which to select these. Very different selections can be found in different studies, illustrating that different approaches are used.

The documentation of the applicable principles is often non-existent or at least weak. The challenge is to determine what is representative. It may not be representative to choose randomly one low, one medium and one high release rate. At least, the leak rates to be used should be adjusted according to design capacities, for instance reflecting the leak rate at which escalation will occur.

Several reviews of studies were commissioned by PSA some years ago, in order to establish to what extent the studies were in accordance with all applicable requirements. One of the findings from these reviews is that the studies sometimes have unclear definitions of essential parameters, for instance the definition of 'Group Individual Risk'. Sometimes this leads to errors in the calculation of risk contributions.

One aspect where there has been significant improvement during the last few years is the use of installation specific data for installations that have been in operation for some time.

The analysis of installations that have been in operation for some time should be different from that for new installations. One aspect is illustrated by the so-called 'BORA' methodology (see Sect. 15.3.1). This approach was not made available until the end of 2006, and may be taken into more extensive use in the future. Another aspect is related to the modelling of accident sequences following a leak. This modelling should focus on the parameters that may be influenced in the operations phase, which are quite different from those that may be influenced in the design phase.

Blowout risk has come into focus as a consequence of the Macondo blowout. The analysis of possible causes of failure of well control is very often superficial or not existing at all. Current models are not capable of reflecting human and organisational failures, which were demonstrated to be the most critical by the Macondo blowout.

The analysis of marine hazards is normally very superficial and not at all suitable in order to identify possible risk reducing measures. PSA has on several occasions made comments about this, but the industry appears to be slow to alter their practice. This was discussed in some detail in Chap. 10.

The event sequence analysis of consequences of hydrocarbon leaks often is too coarse, and not sufficiently focused on the performance of safety systems (barrier elements). The modelling should also reflect the type of risk reducing measures or alternatives that are being considered.

For the most important safety systems (barrier elements) the analysis should include mechanisms of failure. This should be done in a way which allows analysis of dependencies and common mode and common cause failures. This is virtually never done sufficiently extensively.

14.13.4 Presentation of Analysis Results

Presentation of results from the studies is often an aspect that is given insufficient attention. This is discussed in more depth in Chap. 16, which also provides suggestions for how an extensive result presentation should be, including documentation of sensitivities and evaluation of uncertainty.

14.13.5 Identification of Risk Reduction Measures

If the hazard identification is incomplete, then this also limits the potential for identification of risk reduction measures.

Further, as also indicated in Chap. 16 and in [37], the analysis has in many instances too much focus on demonstration of adherence to risk tolerance criteria. The emphasis by the analysis team is in these circumstances often very limited when it comes to use of the study for extensive identification of risk reduction measures.

One aspect that is important for the identification of possible risk reducing measures is that the risk modelling represents explicitly the operational and design aspects. This aspect has been discussed in [38]. There may be an impression that the QRA studies made for UK installations has a higher tendency to have such models that are not easily associated with operational and design aspects.

14.13.6 Use of Study Results in Risk Management

The use of QRA study results in risk management is the main topic in the book by Professor Aven and the present first author [37].

14.14 The RISP Method

14.14.1 Introduction

The RISP method has been developed to allow for consistent use of industry experience rather than more analyses to support robust design of offshore facilities as an alternative to traditional QRAs. In the initial phases, the ideas behind the RISP method was developed in a project 'Formålstjenlige risikoanalyser' ('Expedient Risk Analyses') run by Norwegian Oil and Gas, NOROG until the spring 2017 [39]. Later, the work has been continued in 2018 and 2019 by a joint industry project [40].

The text in the following sections is to large extent based on the RISP work-group 1 report issued in the spring 2019 [40], where one of the authors of the book participated. The authors are grateful to the RISP project[1] for allowing us to include information about the project in its development phase. When the book was fina-lised, the RISP method was still being developed, and thus, there may have been

---

[1]The sponsors of the RISP project are ConocoPhillips, Vår Energy, Lundin Norway, Total, AkerBP, Equinor and Wintershall. The contractors involved are Gexcon, DNV GL, Lloyds Register, Lilleaker Consulting, Aker Solutions, Proactima and Safetec.

changes. For details about the RISP method, it is referred to the RISP report [40], as well as subsequent reports developed by the RISP project.

The overall objective of the RISP project has been to further develop the principles and ideas provided by the NOROG project into methods and guidelines, and establish a new common industrial practice. This practice should describe how various decisions in a development project are to be based on general and specific knowledge about the incidents that the installation may be exposed to (such as leaks, fires and explosions).

The methods and solutions included in the RISP method is adapted to the knowledge and information typically available at the time when the specific decisions of interest are normally made. The decision support provided shall be sufficiently robust, meaning that the recommendations given should not be subjected to scrutiny, reconsiderations or reassessment later in the project, provided that the basis for the decision support has not been changed throughout the project. This will minimise the need for late design changes, when e.g. more detailed information is available. An as-built total risks analysis/quantitative risk analysis (TRA/QRA) will thus not be required within the new industrial practice. Verification shall ensure compliance with the validity envelope of the new approach, and that any changes in assumptions made during the development project are taken into account. Some challenges with the RISK method are discussed in Sect. 14.14.5.

14.14.2   Background

Risk analyses have played, and still play, a key role in the safety work of the petroleum industry and have given the industry detailed and broad knowledge about risk factors and design principles. However, the models and tools need input data on a very detailed level and, in many cases, there is a mismatch between (a) the need for input and the time it takes to set up and use the tools, and (b) the information and time available at the time of making key decisions. Consequently, the decision support often arrives too late. In the cases where a well-proven design is used, the benefit from performing detailed analyses may not be worth the wait as the risk sources are already known through experience from similar previous developments.

Experience and insight gained throughout the years from making analyses have barely impacted the way analyses are made. In general, 'everything' is looked at anew each time, the knowledge acquired from incidents that may occur and how plants can be optimally designed is not sufficiently utilised or reflected in the way the analyses are performed.

A main recommendation from the NOROG project [39], was that during a development project, traditional quantitative risk analyses should for well-proven designs, as a main rule, be replaced by simplified assessments. This should be done to provide the best possible support for decisions being taken on an on-going basis. Thus, the emphasis on detailed calculations of total risk, and measurement against risk acceptance criteria such as FAR and 1×10^{-4}, should be changed. Rather than continuing to seek very detailed risk descriptions, the aim in the future should be to provide better decision support at the right time when the developed concept is well known. This is also in line with the definition of risk given in Norwegian regulations.

The NOROG project drafted several principals and ideas for how to better deal with the above-mentioned factors. These ideas and principles has been further matured and specified in the RISP project.

It is expected that the RISP method will comply with the intentions of the existing regulatory framework defined by the Norwegian PSA. However, if it is perceived that the formulations in the existing regulatory framework are preventing an expedient solution, the RISP method could disregard the current regulatory framework. No obvious discrepancies with the existing regulatory framework have been identified during the course of developing the RISP method, but the practical assessment of compliance has to be adjusted when using the new RISP method. E. g., the Facility regulation §11 requirement stating that '...*dimensioning accidental loads/actions or dimensioning environmental loads/actions with an annual likelihood greater than or equal to 1×10^{-4}, shall not result in loss of a main safety function*' will not be quantitatively addressed, but will instead be indirectly complied with since the robustness of the new RISP designs will be based on experience from similar existing designs that all comply with the requirement.

The RISP method is developed for use in the development phase. This means that it is intended to give risk informed decision support from the 'concept definition and optimisation' phase through the 'construction' phase in development projects. The main focus is to provide robust safety input during the early design phases.

14.14.3 The RISP Method in Brief

In RISP, the quantitative risk analysis is replaced by using prescriptive requirements in standards and proven in use experience to demonstrate an acceptable risk level. The benefit of RISP is that the safety level can be achieved more efficiently and provide risk informed decision support in a timely manner for designs within the validity envelope of the RISP method. The challenge of an experience based and prescriptive approach is however that demonstration of acceptable safety level must be achieved differently from traditional developments made by means of quantitative probabilistic criteria (such as main safety function impairment). A prescriptive and experience-based method will not evaluate risk directly towards quantitative criteria. Experience based RISP models will therefore have to produce design support ensuring at least the same level of safety as the quantitative models. Hence, the RISP models must be acknowledged by all stakeholders as equivalent and equally robust as a traditional QRA applying the methods in NORSOK Z-013 [1], and criteria set by the facilities regulations [41].

The formal RISP activities start with the extended HAZID, referred to as a HAZAN. A HAZAN report will, in addition to producing a comprehensive hazard register, also document the risks and barriers compared to an acknowledged safety standard (including uncertainties) and conclude whether or not a RISP analysis is appropriate. If the HAZAN concludes that the development (assessed for each RISP model) is within the validity envelope and hence appropriate for RISP, all risk informed decision input can be expediently produced by undertaking the simplified RISP methods. These methods will be further developed in the next phases of the RISP project.

The HAZAN consists of five steps:

- Step 1: Describe characteristics of the suggested development
- Step 2: Identify and analyse initiating events: Hazards/uncertainty factors
- Step 3: Evaluate and demonstrate strength of knowledge
- Step 4: Check predefined RISP criteria
- Step 5: Decision: Use RISP criteria or proceed with special studies.

For details about each of the steps, and the subsequent activities after the HAZAN, reference is given to the RISP report [40].

In practice, designing an offshore facility is an iterative process. This means that even though the HAZAN process is presented as sequential steps 1–5, it may not always be performed sequentially. What is important in the end is that all steps are sufficiently documented to ensure the decision to use RISP method or to proceed with special studies is well-founded. It is essential that the HAZAN is performed by personnel with relevant and sufficient competence.

For each area/hazard combination, if the RISP criteria are met, the suggested design of the area is RISP with regards to the considered hazard. If some RISP criteria are not met for particular hazards/areas, special studies will have to be performed for these hazards/areas.

Step 5 includes a decision. In case it is decided to proceed with the RISP criteria, this means that the predefined list of RISP criteria must be implemented as constrains in the design of the area. Since the RISP constrains may affect many disciplines and future decisions, the resulting constrains should be anchored at a sufficient level in the project organization.

In case special or unique uncertainties have been identified in Step 1, this may influence on the recommendation on how to proceed. In some cases additional studies can be carried out to reduce the uncertainties. In extreme cases, it can be recommended not to use the RISP method at all. Management of change must also include a continuous verification that all maturing/refinement during the development project remains within the defined RISP constraints of design. As an example, if one of the RISP criteria is that the aspect ratio should be within certain limits, the decision to use the RISP method implies that the aspect ratio limit becomes a limitation that any future changes will have to adhere to. Such limitations must be followed up in later phases.

If the following criteria are met, the project can be defined as RISP:

- Documentation of the HAZAN process has been performed with the quality as expected.
- All RISP criteria are met. If not, relevant special studies have been performed and provided sufficient knowledge for the RISP criteria not being met.
- Selected concept is in accordance with the above and reflected in the safety strategy, or sufficient special studies have been performed to close uncertainty gaps.
- It has been qualitatively demonstrated that all barrier functions are met by use of barrier elements/ performance measures.

14.14.4 Benefits of the RISP Method

The main benefits that can be achieved by use of the RISP method are as follows.

14.14.4.1 Better Use of Existing Knowledge

The RISP method has focused on aggregating relevant knowledge from previous analyses of similar designs, allowing the designer to base the robustness of their new design on already existing experience from similar proven designs rather than basing the robustness on repeating an analysis of the new design.

14.14.4.2 Speed-Up of Development Process

The new RISP method is intended to allow the designers to reach robust design decisions earlier in the development project. This can potentially also reduce the execution time of the development process without compromising on safety.

14.14.4.3 Reduced Uncertainty Related to Late Changes

Basing the design on experience from previously proven developments will potentially allow for early establishment of robust design accidental loads against major accident hazards, and at the same time remove the need for late design phase scrutiny and rigorous verification of the design and its robustness. This will restrict uncertainty in the development projects and reduce the likelihood for significant design changes towards the end of the development projects.

14.14.4.4   Improved Consistency

The RISP method is founded on open and available robustness principles based on industry experience, and the same experience models are consistently applied for all similar developments. This is expected to increase the consistency in design and robustness between the different designers and risk analysts.

14.14.4.5   Early Development Stage Information of Risk Drivers

The simplified RISP models will represent valuable information about risk drivers also in the feasibility and concept selection phases of the developments. This can improve the ability to understand possible safety challenges with concepts discussed in these early development stages.

14.14.4.6   Clearer Responsibility for Changes Affecting Risk Level

The RISP models intend to clarify the effect of important decisions, and hence provide a framework for management of change. This may also contribute to robustness since the responsibility for change of risk may be more transparent. In addition, the link between decisions and their risk impact provided by the RISP models will be independent of the development phase of the project and this will provide a clearer framework for the safety process.

14.14.5   Challenges of the RISP Method

One of the main challenges applying a RISP method is ensuring commitment from all stakeholders. This is relevant both for the justification of the method (documented and prequalified) but most of all the by-in for specific projects. This includes consensus that demonstration of safety level is achieved based on defined RISP criteria.

An important part of the RISP method is to enable the decision makers to understand the consequences of the decisions they make. Such consequences may be perceived as more visible using RISP, hence more demanding for stakeholders who may then be reluctant to use a RISP approach. On the other hand, the clear link to consequences is also a strength of the RISP method.

In RISP, the traditional quantitative risk analysis to document and demonstrate an acceptable safety level is replaced by a simplified experience-based risk assessment. The result may to some extent be a conservation of known designs.

It is expected that in many projects the design premises, detailing of the premises and way of implementation of the premises will be challenged from a cost optimisation point of view. This may imply that the validity envelope of RISP is challenged and that traditional risk assessment methods are applied. Constrains of staying within RISP may be challenging during the whole development process. Not all stakeholders may be equally motivated to stay within RISP.

Another challenge is that decisions made in the early phases with respect to major accident hazards will have to be robust to consider later changes. The extent and quality of the work in an early phase is hence crucial to obtain valid design premises that are not overly conservative.

Another argument against the RISP method is that it is suitable only for conventional platform concepts. The main approach in the future is to use unconventional concepts (see also Chap. 22). If so, the RISP method will have a limited applicability, but this does not affect the method as such.

A more fundamental objection to the RISP method is the obligation according to the regulations for the installation owners to seek continuously for improvement of health, environment and safety (see also Sect. 3.16). This implies that what was 'good enough' for the last development project is not necessarily satisfactory for a new project, even in the case that the concepts are quite similar. The reservoir characteristics may be different, the possibilities for future tie-ins may be different, as well as other contextual differences.

The ALARP principle (see Sect. 1.5.1) is also relevant in this context, requiring that the owners and the operators seek for further risk reduction actions beyond the satisfaction of the risk tolerance criteria. A risk analysis is considered to be the main basis for an ALARP evaluation. ALARP evaluations are supposed to be conducted regularly throughout the project, implying that a risk analysis would be needed at different stages of the project. It is not clear how the RISP method can be combined with such requirements.

References

1. Standard (2010) Risk and emergency preparedness analysis, Z-013. Standard Norway, Oslo
2. Kirwan B, Ainsworth LK (1992) A guide to task analysis. Taylor and Francis, London
3. Kirwan B (1994) A guide to practical HRA. Taylor and Francis, London
4. Rausand M, Haugen S (2020) Risk assessment: theory, methods and applications, 2nd edn. Wiley
5. Rosenberg T, Nielsen T (1995) Blowout risk modelling. In: Proceedings of the 14th international conference on offshore mechanics and arctic engineering, Copenhagen, Denmark, 18–22 June 1995. ASME Press, New York
6. Andersen LB (1998) Stochastic modelling for the analysis of blowout risk in exploration drilling. Reliab Eng Syst Saf 61:53–63
7. Nilsen T, Sandøy M, Rommetveit R, Guarneri A (2001) Risk-based well control planning: the integration of random and known quantities in a computerized risk management tool. SPE68447
8. Amdahl J, Eberg E, Holmås T, Landrø H et al (1995) Ultimate collapse of offshore structures exposed to fire. In: Proceedings of the 14th international conference on offshore mechanics and arctic engineering, Copenhagen, Denmark, 18–22 June 1995. ASME Press, New York

9. Bea R, Moore W (1994) Reliability based evaluations of human and organisation errors in reassessment and requalification of platforms. In: Proceedings of the 13th international conference on offshore mechanics and arctic engineering, Houston, USA, February 27–March 3, 1994

10. Bea R (1995) Quality, reliability, human and organisation factors in design of marine structures. In: Proceedings of the 14th international conference on offshore mechanics and arctic engineering, Copenhagen, Denmark, 18–22 June 1995. ASME Press, New York

11. Vinnem JE, Hauge S (1999) Operational safety of FPSOs, MP3: riser failure due to inadequate response to rapid wind change. NTNU, Trondheim

12. Vinnem JE, Bye R, Gran BA, Kongsvik T, Nyheim OM, Okstad EH, Seljelid J, Vatn J (2012) Risk modelling of maintenance work on major process equipment on offshore petroleum installations. Loss Prev Process Ind 25(2):274–292

13. Gran BA, Bye R, Nyheim OM, Okstad EH, Seljelid J, Sklet S, Vatn J, Vinnem JE (2012) Evaluation of the risk model of maintenance work on major process equipment on offshore petroleum installations. Loss Prev Process Ind 25(3):582–593

14. OGP (2010) Quantitative risk assessment, datasheet directory. Report no.: 434-20. OGP, London

15. DNV (1998) Worldwide offshore accident database. DNV, Høvik

16. SINTEF (1998) Blowout database. Safety and Reliability Department, SINTEF, Trondheim, Norway

17. PSA (2018) Trends in risk level on the Norwegian Continental Shelf. Main report, Petroleum Safety Authority, Stavanger (in Norwegian only, English summary report)

18. HSE (2018) Offshore hydrocarbon releases statistics. http://www.hse.gov.uk/offshore/statistics.htm

19. AME (2003) Pipeline and riser loss of containment data for offshore pipelines (PARLOC) 2001. Advanced Mechanics & Engineering Ltd., Guildford

20. HSE (2003) Ship/platform incident database (2001). HMSO, London

21. HSE (2005) Accident statistics for fixed offshore units on the UK Continental Shelf (1980–2003), RR349. HMSO, London

22. HSE (2005) Accident statistics for floating offshore units on the UK Continental Shelf (1980–2003), RR353. HMSO, London

23. Lloyd's Register (2016) Modelling of ignition sources on offshore oil and gas facilities—MISOF. Report for Norwegian Oil and Gas Association (Statoil, ConocoPhillips and Total E&P Norge AS). Report no.: 106364/R1, rev. Final, 25 November 2016

24. Lloyd's Register (2016) Modelling of ignition sources on offshore oil and gas facilities—MISOF(2). Report for Norwegian Oil and Gas Association (Equinor, ConocoPhillips and Total E&P Norge AS). Report no.: 107566/R2, rev. Final, 20 November 2018

25. Lloyd's Register Consulting (2018) Process leak for offshore installations frequency assessment model—PLOFAM(2). Report no.: 107566/R1, Rev: Final, Date: 06.12.2018

26. DNV (1981) Causes and consequences of fires and explosions on offshore platforms. Statistical survey of Gulf of Mexico data. Report no.: 81-0057. DNV, Høvik

27. DNV (1993) Causes and consequences of fires and explosions on offshore platforms. Statistical survey of Gulf of Mexico data. Report no.: 93-3401. DNV, Høvik

28. Scandpower Risk Management (2006) Ignition modelling in risk analysis. Report no.: 27.390.033/R1. Scandpower, Kjeller, Norway

29. IEC (2010) IEC 61508—functional safety of electrical/electronic/programmable electronic safety-related systems. IEC, Geneva

30. IEC (2018) IEC 61511—functional safety—safety instrumented systems for the process industry sector. IEC, Geneva

31. Norwegian Oil and Gas (2018) Norwegian oil and gas 070—application of IEC 61508 and IEC 61511 in the Norwegian petroleum industry (recommended SIL requirements). Stavanger

32. Hauge S, Kråknes T, Håbrekke S, Jin H (2013) PDS method handbook, 2013 edn. Sintef

33. Håbrekke S, Hauge S, Onshus T (2013) PDS data handbook, 2013 edn. Sintef

34. OREDA (2015) Offshore reliability data handbook, vols 1 and 2, 6th edn. SINTEF, Trondheim
35. Reliability Analysis Centre (1991) NPRD, nonelectronic parts reliability data. System Reliability Centre, New York
36. IEEE (1984) Guide to the collection and presentation of electrical, electronic and sensing component reliability data for nuclear power generating stations. Wiley, Hoboken
37. Aven T, Vinnem JE (2007) Risk management, with applications from the offshore petroleum industry. Springer Verlag, London
38. Vinnem JE, Pedersen JI, Rosenthal P (1996) Efficient risk management: use of computerized model for safety improvements to an existing installation. In: 3rd international conference on health, safety and environment in oil and gas exploration and production, New Orleans, USA. SPE paper 35775
39. Norwegian Oil and Gas (NOROG) (2017) Project 'Formålstjenlige risikoanalyser ('expedient risk analyses'). Results and suggestions for further work, version 6 (in Norwegian only)
40. Dalheim J, Dammen T, Sagvolden T, Sæternes S, Røed W (2019) JIP: risk informed decision support in development projects (RISP). Main report, workgroup 1—risk management. Report no.: 107522/R1, revision Final A, 1 February 2019
41. PSA (2017) Regulations relating to design and outfitting of facilities, etc. in the petroleum activities (the facilities regulations), last amended 18 December 2017

Chapter 15
Analysis Techniques

15.1 Hazard Identification

Identification of hazards was shown in Chap. 5 to be the first step in a QRA. This is often called **Hazard Identification** or HAZID. The purpose of hazard identification is to:

- Identify all the hazards associated with the planned operations or activities
- Create an overview of the risk picture, for planning the further analysis work
- Provide an overview of the different types of accidents that may occur, in order to document the range of events, which give rise to risk
- Provide assurance, as far as possible, that no significant hazard is overlooked.

It is therefore important that hazard identification provides a good overview, which can be reviewed by a number of people with different experience. The following methods may be used to identify hazards:

• Check lists	Lists developed by specialists to assist in the review of the planned operations
• Previous studies	Lists of hazards from similar studies are often used as a starting point for a new study
• Accident and failure statistics	Lists and case stories which resemble check lists, but which are based on actual failures and/or accidents. An example is shown overleaf
• Hazard and Operability Study	A technique to identify in detail sequences of failures and conditions that may cause accidents
• SAFOP	A technique to review procedures in order to identify sequences of failures and conditions that may cause accidents
• Preliminary Hazard Analysis	A technique often used as an initial screening study, but may also be used alone. PHA is described in the subsequent subsection
• Comparison with detailed studies	Detailed studies used in similar situations may be used to identify which sequences that may give rise to hazardous situations

© Springer-Verlag London Ltd., part of Springer Nature 2020
J.-E. Vinnem and W. Røed, *Offshore Risk Assessment Vol. 2*, Springer Series in Reliability Engineering, https://doi.org/10.1007/978-1-4471-7448-6_15

Section 2.3 has presented an overview of accidents on the Norwegian Continental Shelf, which may also be used as input to hazard identification.

The level of detail to be considered in hazard identification is sometimes uncertain and the approach to be adopted has to be determined prior to commencing the work.

Hazards should be identified for equipment as well as operations. For the hazards associated with equipment, there are three levels of detail:

- Equipment level: All individual equipment items, valves, instruments, vessels, etc. are identified separately as possible hazards.
- Subsystem level: System level All subsystems, such as separation stages, compression stages, etc. are identified separately as possible hazards.
- System level: All systems, such as separation, compression, metering, etc. are identified separately as possible hazards.

It is obvious that the number of hazards identified on each level will decrease at lower levels of detail. At equipment level in the order of 500–1000 hazards may be identified for a large installation. At system level perhaps only 20 hazards would be identified while 50–100 hazards may be identified at subsystem level.

The main problem working at equipment level is the high number of hazards created, most of which will be similar apart from the equipment identified. Thus the overview is very easily lost. System level on the other hand may be too coarse, and distinctions and important differences may easily be lost. The subsystem level is therefore normally the most suitable.

For hazards associated with operations, there is only one level, as each operation has to be considered in detail.

The most difficult aspect of the hazard identification is to ensure that significant hazards are not overlooked. This is a challenge to achieve. Structured analytical techniques that could assist in achieving this objective have been searched for, but so far without success. It may be argued that performing very detailed HAZOPs may be able to achieve the objective. However, the resources required to complete such a programme would be prohibitive.

Consider the following example. More than 25 years ago, a semi-submersible platform experienced uncontrolled ballasting operations, to the extent that severe listing developed and the crew members were considering whether to evacuate or not. They achieved control before it became necessary to perform evacuation. It was later found that the root cause of the problem was a minor fire in one leg of the platform, which had resulted in heating up of the hydraulic fluid used in the control system for the ballast valves. The return lines were too narrow to relieve the additional pressure generated by the heat sufficiently rapidly, thus causing uncontrolled valve operation. When the fire was extinguished, the problem disappeared. The critical question to consider is whether hazard identification could have identified such a hazard.

15.1.1 HAZOP

HAZOP is an analytical technique used to identify hazards and operability problems. The technique is being applied generally to any situation involving the interface between hardware, software and operators, although initially developed for evaluation of process plants. Fault Tree Analysis The approach may also be used in order to identify hazards.

In HAZOP analysis, an interdisciplinary team uses a systematic approach to identify hazards and operability problems occurring as a result of deviations from the intended range of process conditions. An experienced team leader systematically guides the team through the plant design using a fixed set of 'guide words' which are applied to specific 'process parameters' at discrete locations or 'study nodes' in the process system. For example the guide word 'High' combined with the process parameter 'level' results in questions concerning possible 'high-level' deviations from the design intent. Sometimes, a leader will use check lists or process experience to help the team develop the necessary list of deviations that the team will consider in the HAZOP meetings. The team analyses the effects of any deviations at the point in question and determines possible causes for the deviation (e.g. operator error, blockage in outflow etc.), the consequences of the deviations (e.g. spillage of liquid, pollution etc.), and the safeguards in place to prevent the deviation (e.g. level control, piped overflow, etc.). If the causes and consequences are significant and the safeguards are inadequate, the details are recorded so that follow-up action can be taken.

Access to detailed information concerning the design and operation of a process is necessary before a detailed HAZOP analysis can be carried out and thus it is most often used at the detailed design stage after preparation of the P&IDs or during modification and operation of existing facilities. A HAZOP analysis also requires considerable knowledge of the process, instrumentation, and operation either planned or actual, this information is usually provided by team members who are experts in these areas. A HAZOP team typically consists of five to seven people with different background and experience in such aspects as engineering, operations, maintenance, health safety and environment and so forth. It is normal for the team member who leads the analysis to be assisted by another, often referred to as the secretary, who records the results of the team's deliberations as the work proceeds.

The HAZOP relates to the following process parameters: Flow, temperature, pressure, level, react, mix, isolate, drain, inspect, maintain, start-up, shutdown. The HAZOP guide words focus the attention upon a particular aspect of the design intent or a process parameter or condition:

- No (no flow)
- Less (less pressure, flow, etc.)
- More (more temperature, flow, etc.)
- Reverse (reverse flow)
- Also (additional flow)

- Other (flow)
- Fluctuation (flow)
- Early (commencement).

Reporting is particularly important from a HAZOP, in particular with respect to documentation of actions that have been agreed. An efficient secretary is therefore essential. There are also several software packages available in order to assist in the administration and management of the HAZOP. More extensive documentation of the HAZOP may be found in Crawley et al. [1] and Lees [2].

15.1.2 PHA

Preliminary Hazard Analysis is an analytical technique used to identify hazards which, if not sufficiently prevented from occurring, will give rise to a hazardous event. Typical hazardous energy sources considered include high-pressure oil and gas, other high-temperature fluids, objects at height (lifted items), objects at velocity (helicopters, ships), explosives, radioactive materials, noise, flammable materials, toxic materials etc.

Preliminary Hazard Analysis is often used to evaluate hazards early in a project being undertaken at the conceptual and front end engineering stage. It does not require detailed design to be complete but allows the identification of possible hazards at an early stage and thus assists in selection of the most advantageous arrangement of facilities and equipment. The general process adopted involves the following steps:

- definition of the subsystems and operational modes
- identification of the hazards associated with the particular subsystem or operation
- definition of the particular hazardous event resulting from realisation of the hazard
- estimation of the probability of the event occurring and the possible consequence of each of the hazardous situations, and then using a particular rule set to categorise the probabilities and consequences
- identify and evaluate actions to be taken to reduce the probability of the hazardous event occurring or to limit the consequence
- evaluate the interaction effect of different hazardous events and also consider the effects of common mode and common cause failures.

Preliminary Hazard Analysis is undertaken in a structured manner usually using some form of table. Each hazardous event that has been identified for the particular subsystem or operation is studied in turn and recorded in one line of the table arriving at a 'risk rating' either for that particular hazardous event or the subsystem or operation.

15.1.3 SAFOP

Safe Operations (SAFOP) study is an adaptation of the HAZOP technique for analysing work processes and procedures in order to identify and evaluate risk factors. SAFOP is a powerful tool for risk assessment of new (planned) or changed operations and is applicable for all activities where a procedure will be used, such as process interventions, material handling, crane operations, maintenance, marine activities. The SAFOP checklist as described by Lloyd's Register Consulting—Energy AS (formerly Scandpower Risk Management [3]) has the following guidewords (for marine operations):

• Preop. checks	Necessary equipment, tugs not available on schedule
	Necessary equipment checking/testing not performed
• Weather	Unclear weather restrictions or unexpected deterioration of weather (abortion of operation). Weather forecasting, low temperatures
• Current	Problems related to strong, unexpected currents
• Position	Object, grillage, tugs or vessel not in correct position
• Power	No power or insufficient power (tugs, electrical, hydraulic, air)
• Equipment	Malfunction or lack of equipment
• Instruments	Malfunction or lack of instruments
• Responsibility	Undefined/unclear responsibilities (tugs, vessel, port)
• Communication	Malfunction or lack of communication equipment. Communication lines, noise, shift changes
• Execution	A work task is executed in a wrong way, timing, speed
• Procedures	Missing or unclear procedures
• Visibility	Can the operator(s) see sufficiently?
• Movement	Objects, tugs or vessels move in an uncontrolled way
• Stability	Unstable conditions
• Tolerances	Tolerances for positioning, etc.
• Interfaces	Wrong, contamination, corrosion, marine growth, etc.
• Stuck	Movement cannot be performed
• Rupture	Rupture of critical equipment, overloading
• Access	Insufficient access/space on tugs, vessel, port
• Escape routes	Sufficient, checked against requirements, protected
• Contingency	Back-up procedures/equipment not available
• Other	Other items not covered by the above guidewords
• Impact	Impact between objects, squeezing (personnel)
• Drop	Drop of objects from a higher level
• Fall	Fall of personnel to lower level
• Energy release	Electric, pressure, heat, cold, radioactive
• Toxic release	Release of hazardous substances

15.1.4   Bow-Tie

The Bow-tie methodology is a process which can be used to effectively demonstrate how a facility's Safety Management System can be implemented. It assists companies/operators in the analysis and management of the hazards and risks to which their business is exposed, and through the use of graphics, display and illustrates the relationship between hazards, controls, risk reduction measures and a business's HSE activities.

Bow-ties, Fig. 15.1, depict the relationship between hazards, threats, barriers, escalation factors, controls, consequences, recovery preparedness measures and critical tasks. Bow-ties have become a preferred tool in many circumstances, in order to illustrate the relationship between various factors. The most well-known tool for this purpose is THESIS, originally conceived by Shell International and now jointly owned and developed by ABS Consulting Ltd and Shell International. (see also Appendix A). THESIS is fully aligned with the new CCPS/Energy Institute concept book "Bow Ties in Risk Assessment" [4].

The relationship between all the involved aspects as mentioned above has been an area of fault or weakness in many organisations—using the bow-tie method can help to display all the interactions and links that are often found to be loosely related over a number of various documents.

A bow-tie is essentially a combination of the traditionally used fault and event trees, whereby the fault tree constitutes the left-hand side of a bow-tie and the event tree the right-hand side.

What a bow-tie presents in addition however, are the 'barriers' in place that prevent 'threats' from releasing a hazard and 'recovery preparedness measures' that reduce the severity of the hazard consequences.

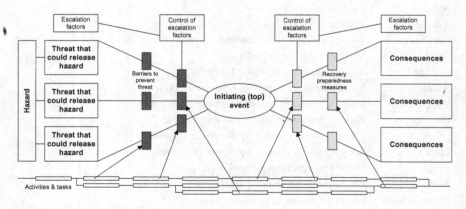

Fig. 15.1 A typical bow-tie display

15.1.5 Cyber Risks in HAZID/HAZOP/PHA/Bow-Tie

Hazard identification and review techniques have traditionally been focused on risk to personnel, the environment and the assets. This should be expanded in order to include a cyber-check [5] as an additional step. Each scenario is reviewed in a cyber-check to determine if a path that may be utilized in a cyber-attack exists, which implies that it is "hackable".

The next step is to review all of the safeguards to determine which are hackable and which are not. Any computer controlled instrumented function is generally hackable, but many safeguards are not hackable. If at least one safeguard for each deviation is not hackable, then the deviation cannot be generated through a cyber-attack. If all safeguards are hackable, then the deviation can be generated through a cyber-attack, and the consequences need to be considered. If the consequences are non-critical, then no additional safeguards are required. If the consequences are critical, then additional non-hackable safeguards are required.

Cyber-checks need to focus on the possibility that one carefully planned attack can affect several barrier functions and/or elements, to the extent that the 'defence-in-depth' concept may be jeopardized. Sometimes it may be sufficient to take out two or three barrier elements to cause a major accident. This could be difficult to consider extensively when considering one barrier element or one scenario at the time.

15.2 Cause, Probability and Frequency Analysis

Cause, probability, and frequency analysis techniques are used in QRA in order to determine many different parameters, such as:

- Potential causes that may lead to accidents.
- Frequency of initiating events.
- Conditional probability of failure of safety systems, in the case of an accident.
- Probability that operating and/or environmental conditions are especially adverse.
- Probability that a particular severe accidental consequence occurs.
- Probability that personnel are present in an area when the accident occurs.

For quantitative purposes there are many tools that may be used in order to calculate probability or frequency, including simulation methods, theoretical modelling, and formal methods such as Fault Tree Analysis and Event Tree Analysis. Frequencies are often based on statistical analysis of failure and accident data. Failure Mode and Effect Analysis may also be employed for qualitative analysis. A brief overview of the most important methods is given below.

15.2.1   Fault Tree Analysis

There are several good textbooks available which provide instruction on Fault Tree Analysis (FTA). In-depth introduction may be found in these sources, a brief introduction is provided below:

- Rausand and Haugen [6]
- Vesely et al. [7]
- Aven [8].

Fault tree analysis is a logical, structured process that can help identify potential causes of system failure, such as causes of initiating events or failure of barrier systems.

The technique was developed to identify causes of equipment failure and was used primarily as a tool in reliability and availability assessment. The fault tree is a graphical model displaying the various combinations of equipment failures and human errors that can result in the occurrence of the hazardous event, usually referred to as the top event. The strength of the fault tree technique is its ability to include both hardware failures and human errors, and thereby allow a realistic representation of the steps leading to a hazardous event. This allows an holistic approach to the identification of preventive and mitigative measures, and will result in attention being focused on the basic causes of the hazardous event, whether due to hardware or software.

FTA is particularly well suited to the analysis of complex and highly redundant systems. For systems where single failures can result in hazardous events, single-failure-oriented techniques such as FMEA and HAZOP analysis are more appropriate. For this reason fault tree analysis is often used in situations where another hazard evaluation technique, such as HAZOP analysis, has pin-pointed the possible occurrence of a hazardous event which requires further investigation.

The output of a fault tree analysis is a failure-logic diagram based upon Boolean logic gates (i.e. AND, OR) that describes how different combinations of events lead to the hazardous situation. A large number of fault trees may be necessary to adequately consider all the identified top events for a large process plant, and the analyst needs to exercise judgement when selecting the top events to be considered.

The fault tree illustrated in Fig. 15.2, shows some indicative causes of why a LAN server in an office may be stolen. This simple example focuses on either random theft or planned theft, in the latter case both the order and knowledge have to be available. The following are characteristics of a fault tree:

- Top event: Event D0
- Gates: G1; G2
- Undeveloped event: D1
- Basic events: D3; D4

Fig. 15.2 Fault tree
illustration

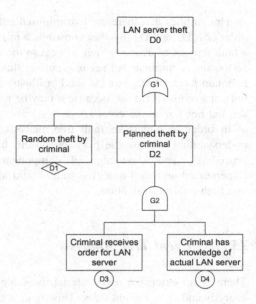

The two gates are different, as shown by the graphics in the diagram, and may be characterised as follows:

• Gate G1:   OR gate, Boolean OR, output occurs if any of the input events occur.
• Gate G2:   AND gate, Boolean AND, output occurs if all the input events occur.

The events are also different, as shown by the graphics in the diagram. All events are shown as rectangles, with different coding below. The differences may be characterised as follows:

• Undeveloped event D1:   diamond, causes not developed further.
• Basic events D3; D4:   circle, lowest level of fault tree, where reliability data is
                             applied.

The top event D0 occurs if any of D1 or D2 (or both) occurs. Event D2 occurs if both of D3 and D4 occur.

By reviewing the fault trees, it is possible to identify the different combination of failures or errors which give rise to the hazardous event. The different failure combinations may be qualitatively ranked depending upon the type and number of failures necessary to cause the top event. Inspection of these lists of failure combinations can reveal system design or operational weaknesses for which possible safety improvements can be considered by the introduction of additional barriers.

It is easy to observe from Fig. 15.2 that the top event will occur in case of the following event combinations:

• D1
• D3 and D4.

This implies that there are two minimal cut sets in Fig. 15.2; D1 is a cut set of order one, D3 and D4 together constitute a minimal cut set of order two. A cut set is a fault tree set of events which will cause the top event to occur if all events in the set occur. A minimal cut set is a cut set that cannot be reduced further and still maintain its capability as a cut set. For illustration; the set {D1; D3, D4} is a cut set, but not a minimal cut set, because it may be reduced further. {D1; D3} is also a cut set, but not a minimal cut set.

In order to undertake fault tree analysis, it is necessary to have a detailed understanding of how the plant or system functions, detailed process drawings, procedures, and knowledge of component failure modes and their effects. Experienced and well-qualified staff should always be used to ensure an efficient and high-quality evaluation.

15.2.2 Event Tree Analysis

There is no extensive text material available for instruction in the construction, analysis and use of event trees. This topic is therefore discussed at some length in Sect. 15.4 below.

15.2.3 Failure Mode and Effect Analysis

Failure Mode and Effect Analysis is a simple technique that does not require extensive theoretical description, but should rather be based on practice in conducting such studies. Useful descriptions and overview may be found in the following:

- Rausand [4]
- Stamatis [9].

15.2.4 Statistical Simulation Analysis

The best known simulation technique is the so-called Monte Carlo method, which is described in several textbooks. This topic is therefore not repeated here, interested readers may be pointed to:

- Rausand [4]
- Ripley [10].

15.2.5 Analytical Methods

A typical example of an analytical approach is the modelling of collision frequency, which is discussed in Chap. 9.

15.3 Operational Risk Analysis

The offshore petroleum industry has for a long time invested considerable resources in engineering defences, or barriers, against fire and explosion hazards on the installations. The performance of barriers is to some extent followed up through performance standards and Key Performance Indicators, though often not extensively. Safety systems are usually addressed on a one-by-one basis, not allowing dependencies and common mode/cause failures to be identified.

Half of the leaks from hydrocarbon containing equipment occur in connection with manual activities in hazardous areas, during which engineered defences often are partially inhibited or passivated, in order not to cause disruption of stable production. The occurrence of these leaks is a clear indication that system and human defences relating to containment of leaks are not functioning sufficiently well during these operations. There is a strong need to understand better the performance of barriers, particularly non-technical, during execution of manual activities.

15.3.1 BORA Methodology

In a paper presented at ESREL 2003 (Vinnem et al. [11]), operational risk assessments were discussed. It was concluded that there is a clear need for improvement of the analysis of barriers. These aspects form the outset for an extensive research activity called the BORA (Barrier and Operational Risk Analysis) project [12]. A PSAM7 paper [13] gave some preliminary observations and introduced a proposed approach.

Two case studies with modelling and analysis of physical and non-physical barriers on offshore production installations have been carried out. Barriers intended to prevent the incident occurring along with those intended to eliminate/reduce consequences are included, and particular emphasis is placed on barriers during execution of operational activities. The results from the studies should enable both industry and authorities to improve safety through:

- Knowledge about performance of barriers and improvement potentials
- Identification of the need to reinforce the total set of barriers, especially during operational activities

- Identification of efficient risk reduction measures for barriers, together with effective modifications and configuration changes.

The analysis has been quantitative as far as possible. Barriers are in general characterised by reliability/av Availability, functionality and robustness. All of these performance measures are addressed. The Norwegian regulations require that dependencies between barriers shall be known. The analysis is therefore performed such that, where relevant, common cause or mode failures and dependencies between barrier elements are accounted for.

The BORA project has proposed a methodology in order to analyse failure of operational barriers, as outlined in Ref. [11], and presented in detail by Aven et al. [14], which presents the BORA methodology as well as the sources for scoring of RIFs. The methodology has three main processes:

- Qualitative analysis of scenarios, basic causes and RIFs
- Quantification of average frequencies/probabilities
- Quantification of installation specific frequencies/probabilities.

This is shown in Fig. 15.3. Also the sources for the installation specific quantification of frequencies and probabilities are presented in Fig. 15.3. The following sources are available:

- TTS/TST verifications
- MTO (Man, Technology and Organisation) investigations
- RNNP (Risk Level Project) questionnaire surveys
- RNNP barrier performance data
- Detailed assessments (Expert input)
- General background studies.

The TTS/TST verifications [15] are focused on technical and documentation aspects of barriers. These verifications were developed by Statoil (now Equinor), and the approach has been adopted by several Norwegian offshore operating companies in Norway. MTO investigations [16] are investigations with special emphasis on human and organizational aspects that have been conducted for many accidents and incidents in the past few years, mainly by or on behalf of the Petroleum Safety Authority (PSA) in Norway. RNNP (see Sect. 22.3) is a project conducted annually by PSA for the entire Norwegian Continental Shelf [17], which for the purpose of the BORA methodology has two applicable activities:

- Biannual questionnaire survey
- Annual collection of barrier performance data.

The questionnaire survey has extensive questions relating to working environment factors as well as a number of aspects relating to perceived risk and safety culture. The barrier performance data [15] is concerned with a selection of barrier elements, most of which are technical barriers.

Traditionally, the event modelling in QRA starts with loss of containment as the initiating event, and the barriers to limit the potential consequences of the leak are

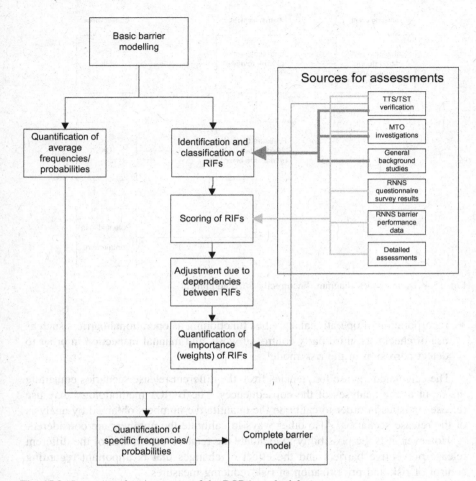

Fig. 15.3 Summary of main aspects of the BORA methodology

modelled. In the BORA project we want to visualise the barrier elements in place to prevent the leak itself. For this purpose 'barrier block diagrams' have been developed for different conditions which may cause loss of containment. For the case 'loss of containment due to incorrectly fitted equipment' (see Fig. 15.4).

The basic risk model in the BORA project may be seen as an extended QRA-model, with several extensions compared to typical offshore QRA studies:

- Event trees and fault trees are linked in one common risk model.
- Detailed modelling of the loss of containment barrier, including initiating events reflecting different causes of HC release and safety barriers aimed to prevent release of HC.

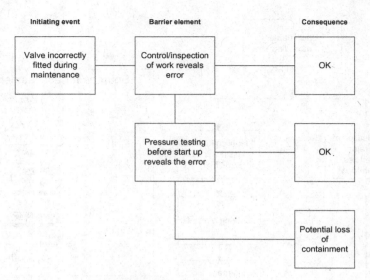

Fig. 15.4 Barrier block diagram, 'incorrectly fitted equipment'

- Incorporation of operational activities functioning as operational barriers such as use of checklists, third party control of work, and manual inspection in order to detect corrosion in the risk model.

The calculated release frequencies from the different release scenarios constitute the input to the analyses of the consequences. The BORA methodology may use release statistics in order to calibrate the quantitative numbers obtained by analysis of the release scenarios. Also other ways to calibrate the numbers are considered.

However, it is the possibility to evaluate the relative importance of the different release preventive barriers and the effect of changes that is important regarding control of risk and prioritization of risk reducing measures.

Recently, there has been some work in order to improve the BORA methodology (see for instance [18]).

15.3.2 Bayesian Belief Network

The use of Bayesian belief networks (BBN) is gaining popularity among risk analysts as they are flexible and well suited to taking the performance of human and organisational factors (HOFs) into consideration, and they provide a more precise quantitative link between the performance of risk influencing factors. Jensen [19] and Pearl [20] present this approach.

Recently a methodology called Hybrid Causal Logic (HCL) has been developed, allowing Bayesian belief networks to provide input information to fault trees and

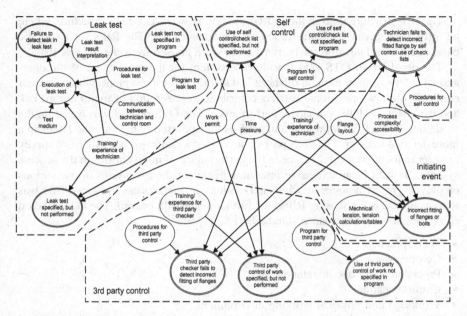

Fig. 15.5 Bayesian belief network for example (Reprinted from Reliability Engineering and System Safety, Vol 94/2, Røed et al., On the use of the hybrid causal logic method in offshore risk analysis, 11p, Copyright (2009), with permission from Elsevier)

event trees. The basic approach is presented by Mosleh et al. [21], and some suggestions for application to the offshore industry are presented by Røed et al. [22]. Figure 15.5 shows a simple illustration of the Bayesian belief network.

The example case is the following accidental event, 'release due to incorrect fitting of flanges or bolts during flowline inspection'. The assembling of the flowlines occurs after inspection, but prior to start-up. The event sequences caused by the initiating event are presented as a barrier block diagram in Fig. 15.4. There are three barrier functions to prevent the initiating event to occur. The technician carries out self control after assembling the flowlines, followed by independent (third party) control. Finally a leak test is carried out prior to start-up.

15.3.3 Risk_OMT Project

The Risk_OMT program builds on previous work in the BORA and OTS projects in addition to the analysis of questionnaire survey data and leaks [23]. One of the main aspects of the BORA project was to address the barrier situation in detail when operational activities are carried out. A list of 10 activities and conditions associated with hydrocarbon leak risk was established during the work with activity indicators [24].

The RIF structure used in the BORA modelling was a one-level structure, where all RIFs were given the same structural importance [12]. The RIFs for a specific basic event in a fault tree for barrier failure were chosen from the general list.

The objective of OTS [25] is to have a system for the assessment of the operational safety condition on an offshore installation or onshore plant, with particular emphasis on how operational barriers contribute to prevention of major hazard risk and the effect of HOFs on barrier performance. The OTS project was based on the modelling in the BORA project, and gave us the opportunity to study the HOFs in a more detailed manner. The OTS project developed a dedicated questionnaire survey in order to focus on work practice in the performance of interventions in the process systems, as well as an extensive interview guide for the interviews of the various employee groups associated with interventions in the process systems. This was used as the basis for scoring RIFs (see further discussion below). The OTS method comprises seven performance standards (PSs [23]):

- Work practice
- Competence
- Procedures and documentation
- Communication
- Workload and physical working environment
- Management
- Management of change.

A RIF can be defined as an aspect of a system or an activity that affects the risk level of the object [26]. It is in principle a theoretical variable; however, how to measure this variable may or may not be specified. One hypothesis is that risk control can be achieved through the control of changes in RIFs. The conditions for this hypothesis to be true are that:

- All relevant RIFs are identified
- RIFs are "measurable"
- The relationship between RIFs and risk is known.

RIFs are introduced to represent the "soft" relations between HOFs and the parameters in the fault and event trees. RIFs are considered in a given context as the "true" underlying properties of the system being analysed. RIFs are however not known in the analysis, and thus we treat RIFs as random quantities. This is assumed to be influenced by RIFs which are modelled by Bayesian Belief Networks.

Formally, we use the term *score* to denote the summarised information regarding RIFs from interviews, surveys and so on. A score is thus treated as a realisation (observation) of the true underlying RIF. This corresponds to an arrow from the RIF to the corresponding score in the BBN. The scoring system is based on characters A–F, where A corresponds to best industry practice and F corresponds to an unacceptable state with respect to the actual RIF. RIFs are structured on two levels, where the level "closest" to the basic event level represents those factors that directly influence the failure probabilities. Examples of such RIFs are time pressure, competence, work motivation and design. RIFs at the second level are of a more managerial nature, such as competence management, management of change and strategic task management.

The basic model of operational barriers is illustrated in the upper part of Fig. 15.6. The leak scenarios caused by the initiating events B1–B4 are modelled together, as they are all result from the work operation "Work on isolated depressurised equipment". The initiating events B1–B4 associated with the maloperation of flanges, bolts and valves, are defined as follows [27]:

B1: Incorrect blinding/isolation
B2: Incorrect fitting of flanges or bolts during maintenance
B3: Valve(s) in incorrect position after maintenance
B4: Erroneous choice of installations of sealing device.

A detailed analysis of a substantial data set of actual leaks that occurred in the Norwegian sector during 2006–2010 was carried out in order to test the realism of the model representation in the Risk_OMT project. Figure 6.23 has presented an overview of the main work process phases during which leaks occur. Nine of the 34 leaks occurred during the actual execution of maintenance (or modification) work, five during preparations, 12 during reinstatement and eight after the process plant had been started up.

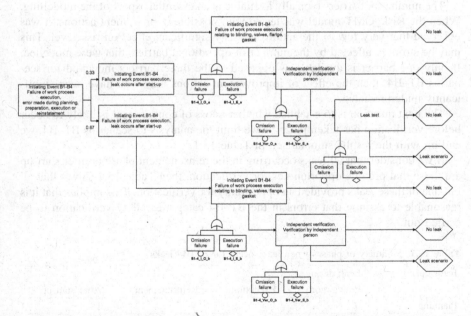

Fig. 15.6 Improved modelling for leak scenarios B1–B4 [29] (Reprinted from Reliability Engineering and System Safety, Vol 113, Vinnem, On the development of failure models for hydrocarbon leaks during maintenance work in process plants on offshore petroleum installations, 10p, Copyright (2013), with permission from Elsevier)

The overall message from Fig. 6.23 is that most leaks occur before leak tests are carried out during the reinstatement phase. It is therefore interesting to consider these leaks in more detail. We will thus consider leaks separately for faults that occur during planning; preparation, execution and reinstatement. Each of the bars in Fig. 6.23 is presented in more depth in separate diagrams.

The scenarios B1–B4 fall into the 'latent leak condition' category, implying that typical errors (such as finishing the work with a valve in an incorrect position) may not be revealed until pressure is reapplied either during or after start-up, or during the actual work process, possibly because of some unforeseen external influence on the system.

Figure 6.23 shows that only eight of the 34 leaks resulted in leaks after the start-up of the process equipment in question. In the case of scenarios B1–B4, the corresponding ratio is seven out of 22 leaks. This implies that in about two thirds of leaks a loss of hydrocarbon containment will occur before leak testing, which corresponds to an additional independent verification in the model.

The important implication of this is that the two 'additional' barriers (not considering the correct execution of the work process by the relevant discipline which obviously is the primary barrier) are not relevant in the majority of the circumstances. A system that should be reliable with three barriers is in the majority of the cases reduced to a maximum of two barriers, and maybe only one in many cases, i.e. only the execution of the work process itself.

The number of barriers typically available is an essential aspect of the modelling. When the Risk_OMT model was tested with possible improvement actions, it was observed that very few of the actions had any significant effect on risk level. This may be strongly affected by the number of operational barriers that were modelled. If only one barrier is available as opposed to the three barriers modelled for scenarios B1–B4, then the effect of improvement actions may well have been significantly underpredicted.

The next question is to establish whether a loss of containment is likely to occur before verification has taken place. If we limit the analysis to scenarios B1–B4, we end up with the results summarised in Table 15.1.

If we consider that all leaks occurring in the reinstatement phase or after start-up are those that provide opportunities for verification, then Table 15.1 shows that all but one of these leaks provided an opportunity for verification. This implies that it is reasonable to assume that errors in the B1–B4 categories allow verification to be carried out.

Table 15.1 Summary of phases when leaks occur for B1–B4 leaks

Error during	Leak during			
	Preparation	Execution	Reinstatement	After start-up
Planning	0	1	4	1
Preparation	0	0	0	1
Execution	0	0	5	3
Reinstatement	0	0	5	2

There is hence a need to revise the work process modelling in the Risk_OMT project based on the discussions above. This is presented in Fig. 15.6 in the form of a barrier diagram.

The upper part of Fig. 15.6 is identical to the model which assumes that two additional barrier elements are always available, corresponding to leaks occurring after start-up. This corresponds to one third of leaks, with two additional barriers available in addition to the correct performance of the work tasks. For two thirds of leaks, the leak test barrier is not available, and only one additional barrier is available in addition to the correct performance of work tasks.

The details on Risk_OMT modelling are discussed in Refs. [25] and [28]. The improvements are discussed in Ref. [29].

15.4 Event Tree Analysis

15.4.1 Basics of Event Tree

An event tree is a visual model describing possible event chains which may develop from a hazardous situation. Initiating events (sometimes called top events) are defined and their frequency or probability of occurrence calculated. Possible outcomes from the initiating event are determined by using a list of questions where each question is answered 'yes' or 'no'. The questions will often correspond to safety barriers in a system such as 'isolation failed?' The method therefore reflects the designer's way of thinking.

The probability of alternative outcomes is calculated for each question which forms a branching point in a logic diagram. These branching points are often called the 'nodes' of the event tree. The probability or frequency of alternative end events (also often called terminal events) is calculated based on the probability or frequency of the initiating event and the conditional probability associated with each branch. End events may be gathered in groups having similar consequences to give on overall risk picture.

The event tree is quite similar to a cause consequence diagram although the latter uses more text and a few more graphical symbols. The cause consequence diagram is somewhat easier to read, but significantly less information can be compressed into one sheet. This may be part of the reason why event trees appear to be preferred. From event trees the following are often performed:

- Frequency calculation for consequence classes
- Sensitivity analyses (effect of variations of some parameters)
- Identification of major contributions to each consequence class.

In addition to frequency/probability prediction, an event tree may also be used for direct calculations of consequences. A simple way to carry out a fatality risk assessment, is to assign a number of fatalities to the branching points (in case of branching one way), and these are summed to find the number of fatalities for the end events. The most typical way to calculate consequences is to carry out separate calculations associated with the different branches and/or terminal events.

The theory on which the event tree methodology is based is very simple and requires only limited explanation. The following sections outline both the theory and the practical application of event tree analysis.

15.4.1.1 Accident Sequence Modelling Accident Sequence Modelling

One of the most crucial tasks of QRA (and also probably the most difficult) is the modelling of the potential accident sequences. This is demonstrated by the incident involving the maloperation of ballast valves due to build-up of pressure in hydraulic system return lines, as a result of a fire (see page 3). In most situations it is a challenge to identify the possible hazard, and to accurately represent the possible accident sequences. The following are the main difficulties in such modelling:

- The process is normally highly time dependent.
- Escalation involves complex interactions between different processes and different equipment.
- Human intervention may sometimes have extensive effects on the development.
- Small differences in circumstances may often lead to vastly different final scenarios.

Dynamic situations are probably the main challenge. Tools and approaches need to be able to reflect dynamics in the most accurate way, in order to achieve realistic modelling. It is recognised that an event tree model is usually too static a tool to be really suitable for detailed analysis of accident sequences and the dynamics of such a process. Very little effort however has so far been devoted to the development of alternative tools and approaches. One such alternative, PLATO®, is briefly described in Sect. 15.6.1.

15.4.1.2 Event Tree Illustration Event Tree Illustration

The event tree used for initial illustration (Fig. 15.7) is an event tree for evaluation of evacuation from a platform. The initiating event in the event tree is assumed to be an event which requires evacuation from the platform, e.g. a blowout, a large fire etc. From this initiating event, different scenarios may develop, depending upon the circumstances. The different circumstances are described to the right of the event tree, in the form of a number of questions relating to the nodes.

The first question considered is whether precautionary evacuation from the platform has been performed. If this is the case, then we move to the left along the first branch of the event tree, otherwise we move to the right.

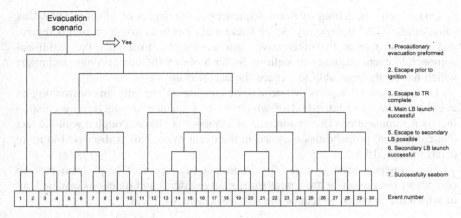

Fig. 15.7 Event tree for escape and evacuation

The second question is whether escape has been performed prior to ignition. Obviously, if precautionary escape has been performed, this question is superfluous. For the left branch from the first question, this second question is therefore not considered. However, for the right branch it is relevant.

In this way, we can continue through the event tree, splitting the scenarios into more and more detailed scenarios, depending on alternative outcomes to intermediate situations.

15.4.1.3 Sequence of Events Sequence of Events

The analysis of accidental scenarios includes the following elements in relation to hydrocarbon leaks:

- Modelling of leaking media
- Event sequence analysis, including ignition and barrier modelling
- Escalation modelling
- Impairment modelling
- Consequence modelling.

The analysis of these processes is extremely complex on offshore platforms. In fact offshore platforms are the most difficult objects to analyse for accidental event development. Onshore petrochemical and chemical plants are more complex in relation to the process design, but could be considered simpler, due to spacing between units, and the way plants are laid on the ground level. Large offshore platforms often have 3–4 levels of equipment with different kinds of interaction. Also nuclear plants (and even space vehicles) are simpler than the largest offshore installations with respect to escalation of accident consequences, although they are more complicated with respect to the work processes.

Due to this, modelling of event sequences is the aspect of offshore QRA that causes most of the uncertainty. Some R&D work has been going on in this regard, and there are one or two alternative options available to replace the traditional approach to event sequence modelling. So far however there is no single technique which has really been able to replace the use of event trees.

There are several aspects that need to be considered carefully, in constructing an event tree, the most important of which is the sequence in which the escalation factors are considered. The importance of sequence is closely coupled with the fact that conditional probabilities are used in the event trees. This is discussed in more detail in Sect. 15.6.2.

The sequence issue is especially important in respect of leaks from process equipment because there are a number of safety systems and functions installed, all of which are intended to reduce the risk associated with leaks.

15.4.1.4   Node Branching Rule Node Branching Rule

Another aspect which may be mentioned is that the rule of branching in two mutually exclusive sequences from each node (binary output from nodes) is sometimes broken, in order to save space. Consider the following alternatives with respect to ignition of small gas leak:

- Immediate ignition (fire is implicit)
- Delayed ignition causing explosion
- Delayed ignition causing fire
- No ignition.

If the standard rule of dual branches from each node is followed, this leads to three nodes being required:

- No ignition, versus
- Ignition, which splits into

  - Immediate ignition, versus
  - Delayed ignition, which splits into

  Explosion
  Fire.

These three nodes will occupy a lot of space in a graphical representation of the event tree and a more condensed presentation is possible if one node is 'allowed' to have all four sequences as outputs.

Figure 15.8 shows a simple example with binary division in each node, whereas Fig. 15.9 shows the same example redrawn such that one node has three outputs. These two diagrams further show that event trees may be drawn horizontally as well as vertically.

Fig. 15.8 Event tree example
with binary division

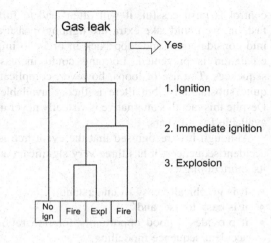

Fig. 15.9 Event tree example
with one combined node

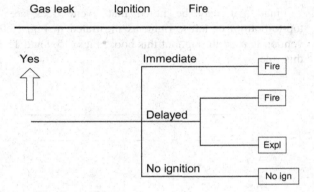

The requirement that all outputs are mutually exclusive is valid in all cases. For instance, when considering failure or success of evacuation, the outcomes are classified in binary states as either 'failed' or 'success', as shown in the event tree, Fig. 15.7.

When the standard rule of only two output branches from each node is applied strictly, then there will always be one more end event than there are nodes in the tree. When more branches are allowed from each node, then the number of end events may be smaller than the number of nodes.

15.4.1.5 Loops in the Tree Loops in the Tree

Since sequence is an important aspect, one might assume that loops in the event tree could be quite useful. In the case of fire for instance, a typical node question is whether automatic systems are capable of controlling the fire. If the automatic control is unsuccessful, it will often lead to further escalation. But in a looped fashion, we could take extra fire fighting measures (activation of manual control) into consideration, and loop back in order to improve on the chance that further escalation is prevented. Looping could increase the realism in the modelled sequences. The use of loops, however, complicates the calculation of frequencies quite substantially, but there is theory available also to cover this aspect [30]. Despite this fact this alternative is virtually never used, although the theory has been available for 30 years.

Although it is recognised that the event tree is far from ideal for modelling of accident sequences, it has three very significant advantages, which compensate for its shortcomings:

- It is graphically easy to understand,
- it is easy to use, and
- it provides a good opportunity for integration of reliability analysis into the accident sequence modelling.

Finally, it may be noted that event trees are commonly drawn either top-to-bottom or left-to-right, as illustrated in Fig. 15.9. The top-to-bottom convention is used throughout this book, Figs. 15.9 and 15.10 being the exceptions to this rule.

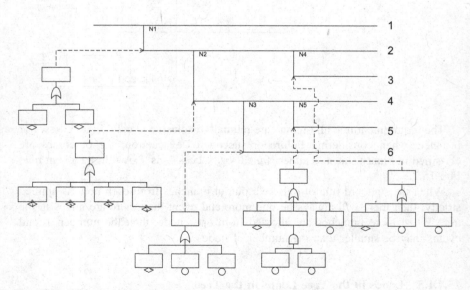

Fig. 15.10 Illustration of integration between event tree and fault trees

15.4.1.6 Probability and Frequency Calculation Probability and Frequency Estimation

The event tree can also be used for quantification of the likelihood of different scenarios. Probability values can be assigned to each branch and in this way we build up a tree of conditional probabilities. If we return to the evacuation example again (Fig. 15.7), we may assume that the probability of precautionary evacuation being performed is 0.6. This means that the probability that precautionary evacuation is **not** performed will be 0.4. Secondly, given that precautionary evacuation has not been performed, we may assume that the probability of escape before ignition is 0.8, as the conditional probability. The total probability of escape before ignition, given no precautionary escape, then becomes $0.4 \times 0.8 = 0.32$.

By continuing this logic through the tree, we can arrive at probabilities for the terminal events in the event tree. If in addition we multiply with the frequency of the initiating event, we arrive at the frequency for each terminal event.

15.4.1.7 Combination of Event Trees and Fault Trees Probability and Frequency Estimation

The RiskSpectrum® software is outlined in Sect. 15.5, noting that it allows an integrated analysis of event trees and fault trees. A sketch showing the principles of such integration is outlined in Fig. 15.10.

15.4.2 Major Hazard Scenarios

Major Hazard Scenarios The main use of event trees in offshore QRA is for modelling accident sequences from hydrocarbon leaks and other major hazards. The following are the main types of hazards for which event trees are used:

- Blowouts
- Hydrocarbon leak events from process equipment
- Hydrocarbon leak events from riser
- Fires in utility systems, mud process and quarters
- Structural and marine accidents.

Separate event trees could be developed for each relevant leak category and for each piece of equipment. The number of event trees would therefore be very substantial for a large platform and it is therefore necessary to eliminate trees and parts thereof that are not really required, in order to avoid losing the overview.

The discussion in this section is focused on hydrocarbon leaks, including blowouts.

15.4.3 Initiating Event Frequency

Initiating Event Frequency The frequency of initiating events is shown in the event tree. Event trees are often presented for the following categories of leaks:

Process Leaks:

- Small leak
- Medium leak
- Large leak.

Riser and Pipeline Leaks:

- Small leak
- Medium leak
- Large leak
- Full Bore.

Blowouts:

- Full flow
- Reduced flow
- Different flow paths/location of release.

The number of categories may obviously change, depending on the circumstances of the analysis. The leak categories may be based on:

- Mass flow, often in kg/s.
- Dimensions of the leak area, (often using an equivalent diameter circular hole).

There is a unique relationship between the gas composition, the pressure, the mass flow and the area of opening. A leak classification frequently used is:

- Small leaks, 0.1–1 kg/s (sometimes from 0.05 kg/s)
- Medium leaks, 1–10 kg/s
- Large leaks, >10 kg/s.

In order to illustrate typical occurrence frequencies, the following values could be observed for gas leaks from one installation during 10 years of operation:

- Large leaks; none
- Medium leaks; 1
- Small leaks; 19
- Over 250 registered seepages and other leaks below 0.1 kg/s.

Another way to illustrate frequencies is from the Risk Level project, which reports the following average frequencies during the 10 year period 1996–2005:

• Large leaks (>10 kg/s):   0.0069 leaks per installation year
• Medium leaks (1–10 kg/s): 0.151 leaks per installation year
• Small leaks (0.1–1 kg/s):   0.45 leaks per installation year.

15.4.3.1 How to Divide into Categories? How to divide in Categories?

One potential problem associated with use of either of the two systems of categorising leaks is that it may not truly reflect actual situations. This may be highlighted by considering how escalation may be modelled (this phenomenon is
sometimes called 'artefact').

When leaks are grouped in categories, common characteristics are calculated for
each of the categories. Thus for small leaks, the flame length of jet fire may be 3 m,
and for a medium leak, 17 m. If the distance to the next section of process
equipment is 7 m, then the flame from small category leaks will not impinge on the
next section of equipment, whereas flames from medium sized leaks will always
impinge.

This, however, is an artificial situation brought about by grouping leaks and
giving them a single representative size. In actuality the larger leaks in the small
leak category may have a jet flame length of over 7 m and thus would give rise to
escalation. A logical system for categorising leaks would define the smallest leaks
as those below a size which causes jet fire impingement and subsequent escalation.
The next category of leak would be those that cause escalation to the next section
due to jet fire impingement. The principles are illustrated in Fig. 15.11.

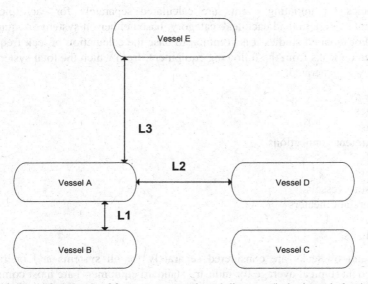

Fig. 15.11 Simplified sketch of five process vessels and distances (in horizontal plane)

It is assumed that all these five vessels are installed in the same area. This illustration is simplified in order to demonstrate the principles, in the sense that only the distances between vessels are illustrated. Instruments and piping may result in the real separation distances between vessels being shorter. With respect to process segments, the following is assumed:

- Vessels A and B belong to the same ESD segment.
- Vessels C and D belong to the same ESD segment, which is different from the segment which Vessels A and B belong to.
- Vessel E is a separate ESD segment from all the other vessels.

The leak categories should now be determined on the basis of jet fire flame lengths in relation to the distances between the vessels. The leak categories may be defined as follows:

- Since Vessels A and B belong to the same segment, the distance L1 is not applicable as basis for these definitions.
- The distance L2 is used as the lower limit for significant leaks, smallest category.
- The second category is based on the distance L3, which gives the lower limit for the category.
- The third category should be based on the distance to a fire wall (not shown).

15.4.3.2   Leak Frequencies for Selected Categories Leak Frequencies for selected Categories

Frequencies for initiating events are calculated separately for each piece of equipment or system, and each leak category, based either on system or equipment values. For detailed studies it is common to base the calculation of leak frequency in an area on leaks from the following equipment, from which the total system leak frequency is generated:

- valves
- flanges
- bends
- instrument connections
- welds
- piping
- pressure vessels
- coolers and heaters
- risers
- pipelines.

Gas and oil leaks are considered separately for all systems and operations. Generic data (typical average for industry standard equipment) are most commonly used. Installation specific data should be used, whenever available, as discussed in Sect. 14.10.

The approach indicated here is the traditional approach where leak frequencies are calculated based on an equipment count i.e., without taking operations into consideration. The BORA project has developed a general approach in order to take activities and operations into account. This was outlined in Sect. 14.10.

For blowouts, the following operations are considered separately, although the models for each activity is very simple:

- shallow gas zone drilling
- exploration drilling
- well testing
- development drilling
- completion of production wells
- completion of injection wells
- regular production
- wireline operations
- coiled tubing operations
- snubbing operations
- workover operations.

The distinction is also often made between wells with regular deviation and so-called horizontal wells (with sometimes very long horizontal sections), High Pressure/High Temperature wells and wells with completion in multiple reservoir zones ('multibore' wells).

15.4.4 Nodes in Event Trees

Event tree probabilities are provided at each branching point (node) in the event trees. Typically the following aspects are considered:

- Detection of leaks
- Ignition
- Emergency shut down, blowdown, flaring
- Fire fighting system
- Explosion and fire
- Extent of escalation of accidental effects.

This list only shows the main categories that are considered and further categorisation may be required, in a detailed event tree. In a detailed event tree the following active and passive safety systems and functions world be covered by the logic nodes:

Safety Systems Reliability:

- ESD system, including valves
- Blowdown valves
- Gas detection
- High Integrity Pressure Protective System
- Fire detection
- Smoke detection
- Fire fighting, automatic and manual.

Passive Fire Protection:

- Escalation (mainly depending on passive fire protection)
- Ignition time and location.

There is some discussion as to whether all safety systems should be reflected in the event trees as separate nodes or not. Some analysts would claim that not all safety systems need to be reflected separately in the event trees. They will claim that it is most efficient in many circumstances, to combine several systems into one node, to avoid the event tree being too unmanageable.

The opposite view is that more focus is put on those safety systems that are reflected explicitly as nodes in the event tree, and that this will help in meeting the regulatory requirement to document the effect of barrier system failures. It will often be most efficient to find a compromise between these two extreme positions.

Let us illustrate a case where there is a node stated as 'Closure of ESD valves', which then would include implicitly the following barrier elements; ESD valves; ESD logic as well as auto gas detection and manual gas detection sub-functions. The probability of failure to shut the ESD valves can be calculated for this node in the following manner (if the elements and sub-functions are independent):

$$P^f_{TOT} = P^f_{ESDV} + P^f_{ESDL} + P^f_{GASDET} \cdot P^f_{MANDET} \tag{15.1}$$

where

P^f_{TOT} probability of failure to shut the ESD valves

P^f_{ESDV} probability of failure of the actual ESD valve itself

P^f_{ESDL} probability of failure of the ESD logic

P^f_{GASDET} probability of failure of gas detection

P^f_{MANDET} probability of failure of manual gas detection.

Equation 6.1, may be valid for many similar cases. It should be noted that this equation assumes independence between [automatic] gas detection and manual detection. The individual elements of Eq. 6.1, may be calculated by Fault Tree Analysis or based on operational experience (or a combination).

The importance of the correct sequence by which the nodes are considered has already been pointed out. It could be mentioned that one typical error in this context is that ignition of a gas leak is considered as the first node in the tree, prior to consideration of leak detection. But the probability of ignition is highly dependent on whether the leak has been detected or not. The first node should therefore in most cases be concerned with the detection.

15.4.5 End Event Frequency

The calculation of end event frequencies is mathematically straightforward, just involving multiplication of the initiating event frequency by the appropriate conditional probabilities. The amount of calculations may, however, make the use of computerisation necessary. The following relationship between frequencies and probabilities may be observed:

- Initiating event: Usually given by its frequency.
- Nodes: Probabilities are always used, principally these are conditional probabilities.
- End events: Have the same dimension as the initiating event, therefore usually frequency.

The end event frequency may be expressed as:

$$\lambda_j = \lambda_i \cdot \prod_K P_k \qquad (15.2)$$

where

λ_j frequency of end event j
λ_i frequency of initiating event in the tree
p_k conditional probability of branch k
K set of branches that defines the path from initiating event to end event j.

The initiating event frequency is usually considered to be constant, assuming for instance a Poisson distribution of the occurrence of events. With this assumption, a simple relationship between probability and frequency exists, as shown below.

If the annual frequency of small gas leaks is λ_i, then the probability of at least 1 gas leak in a one year period, may be expressed as:

$$P(at\ least\ 1\ leak) = 1 - e^{-\lambda_i t} \approx \lambda_i t \qquad (15.3)$$

The approximation is valid only if the probability is lower than 1% (the error at 10% is 0.05), the first expression is always valid. The probability of no gas leaks in a year, is (with the same condition for the approximation):

$$P(0\,leaks) \approx 1 - \lambda_i t \tag{15.4}$$

Equations 6.3 and 6.4, may be used for the end events as well as for the initiating event.

The end, or terminal events in the tree, are sometimes called the 'accidental events'. The frequency of the end events are often multiplied by the impairment [conditional] probability (in range 0.0–1.0) in order to determine the impairment frequency i.e., the frequency of events which the safety functions are not designed to sustain.

$$\lambda_{imp,j,l} = \lambda_j P_{imp,j,l} \tag{15.5}$$

where

$\lambda_{imp,j,l}$ impairment frequency for end event j

$p_{imp,j,l}$ conditional probability of impairment for safety function l for end event j.

15.4.6   Gas Leak in Process Area

Hydrocarbon leaks are analysed to consider different fire and explosion scenarios. Event trees are often constructed quite simplistically, but may also be more sophisticated.

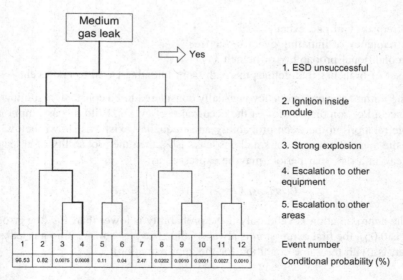

Fig. 15.12 Event tree for medium gas leak, with Piper Alpha sequence highlighted

Figure 15.12 presents a simple event tree for process system medium sized leaks in the range 1–10 kg/s. The sequence of events in the Piper Alpha accident (see Sect. 4.15) has been marked with a thicker line in the event tree. This event tree considers only one safety system, the ESD system. The nodes following the safety system node involve the consideration of ignition inside the module as well as different mechanisms of escalation including strong explosion.

The nodes (branching points) in the diagram are focused on the following safety systems and important safety aspects:

- ESD system availability
- Ignition
- Explosion
- Escalation to nearby equipment
- Escalation to other areas.

The conditional probabilities of the terminal events are also shown. These reflect typical conditions on a relatively modern production platform on the NCS.

It could be observed that the sequence of events in the Piper Alpha accident is not particularly probable on a modern platform in the North Sea, due to the probability distribution used. It would be expected that the probability of this particular sequence would be higher on an old installation like Piper Alpha.

It may be observed that the Piper Alpha sequence is quite well reflected in the simple event tree shown above. In event tree terms, Piper Alpha may be characterised as follows:

- Medium gas leak.
- Operator in the area initiated ESD.
- Ignition occurred in spite of this (ESD probably not initiated until after the explosion).
- The resulting explosion was not strong (it has been back calculated to 0.2–0.4 bar).
- Escalation (probably due to fragments) was first to other equipment, setting off an oil fire.
- Escalation then subsequently resulted in riser rupture.

Although the Piper Alpha events can be quite simply modelled it will often be important to expand the hydrocarbon leak event tree into more details because only in this way is it possible to model explicitly the influence of different protective and/ or detailed systems and functions. The following example shows a detailed event tree for a medium gas leak (see Figs. 15.13 and 15.14).

This event tree has a considerably higher number of nodes than the simple event tree in Fig. 15.12, also including operator intervention. In fact it is shown that this event tree involves a small extent of 'looping' in the event tree, in the sense that 'operator intervention' is shown on a high level in Fig. 15.13 and also on a lower level, in Fig. 15.14.

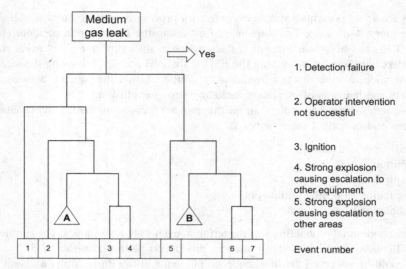

Fig. 15.13 Detailed event tree for small and medium gas leaks

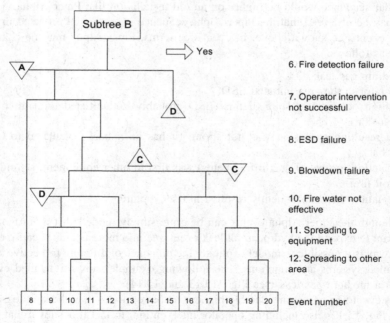

Fig. 15.14 Subtrees for detailed event tree for small and medium gas leaks

There are two subtrees shown in Fig. 15.13, A and B. Figure 15.14 is principally devoted to Subtree B, but contains in fact also Subtree A, as a subset of Subtree B.

There are two additional subtrees inside Subtree B, which are used to simplify the drawing of the subtrees. Transfer logic normally used in fault trees is used for the subtree transfers. This implies for instance that the Subtree A to be inserted into Fig. 15.13 is the part of Subtree B in Fig. 15.14, which could be denoted 'Fire detection successful' (actually the 'No' outcome of 'Fire detection failure').

The use of transfer symbols is not common in Event Tree Analysis. If the trees in Figs. 15.13 and 15.14 were used for calculations, then the transfers cannot be allowed, because the nodes may have different probabilities, according to where they are in the event tree.

The effectiveness of fire water activation (Level 10) is strongly dependent on the circumstances that prevail in the scenario, reflecting what has been mentioned earlier, that all probabilities in the event tree are conditional probabilities.

This detailed event tree is a real case, in the sense that it has been used in an actual detailed QRA, and a point has been made to present it in the way it was used. There is one aspect of this tree which is somewhat unfortunate, in the sense that so-called 'double negation' is used. This implies that when the question 'Fire detection failure' is posed, the 'No' branch actually implies a positive outcome, 'Fire detection successful'. There is also a similar double negation for 'detection failure'. It is recommended to structure event trees such that 'double negation' is avoided, and the wording of the event trees in Figs. 15.13 and 15.14 is therefore not a recommended solution.

The total number of nodes in the expanded (actually full) version of this event tree is 48, implying that there is a total of 49 terminal events in this event tree.

15.4.7 Blowout Event Tree

The discussion of blowouts in this section deals only with the effect on personnel and facilities. The modelling of aspects that determine the environmental consequences fall outside the scope for this book and are not discussed in detail.

A standard event tree is often used for the description of the relevant accident scenarios. The same tree is often used for all blowout scenarios, irrespective of the cause. The event tree is shown in Fig. 15.15, and the nodes discussed in the text below.

15.4.7.1 Node: Immediate Ignition

Ignition is regarded as 'immediate' if the leak is ignited within the first seconds (may be up to just a few minutes) after the leak occurs. In these cases ESD isolation will often have limited effect, due to the rapid development. An explosion may be less likely in these circumstances, as an explosive gas cloud may not have had the

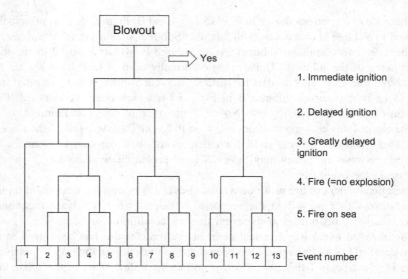

Fig. 15.15 Blowout event tree

time to form. This is not always the case, an explosive cloud may be rapidly forming in some cases. The Piper Alpha accident may illustrate this aspect, as it is likely that the explosion in this accident occurred only some 20 s after the leak started.

15.4.7.2   Node: Delayed Ignition

Ignition is regarded as delayed if it normally takes some few minutes (perhaps up to 30 min) for a leak to ignite. The possibility of strong explosion is much higher in this case, as a cloud of considerable size may have been formed before being ignited.

15.4.7.3   Node: Greatly Delayed Ignition

Greatly delayed ignition is of interest in the case of blowouts and riser/pipeline leaks, where huge clouds may be generated, and travel some distance before finding an ignition source. (Consider for example one actual case when a blowout was ignited 2–3 days after it started, by a work vessel which came into tow the wrecked platform away.)

If none of the ignition cases occur, then it is implied that the blowout is unignited. This implies that the consequences mainly are spilled oil and/or gas releases to the atmosphere. The size of the spill or dispersed cloud is completely dependent on the duration of the blowout, and may range from a few tons up to tens of thousands of tons of oil, or up to billions of m^3 for gas.

15.4.7.4 Node: Fire

On offshore platforms gas fires are often more significant than oil pool fires; the latter are dealt with in Sect. 15.4.7.6 below. Authority requirements and offshore design practices have often concentrated attention on protection against pool fires, presumably under the assumption that protection against gas fires is impossible or unrealistic.

Gas leaks may lead to jet fires, if rapidly ignited. Such fires are very heat intensive, and have a significant effect on objects in the flame. This calls for a dedicated assessment. There are no official definitions or standard regarding jet fires that are appropriate, and thus realistic fire scenarios have to be judged. The measures necessary to give adequate protection from jet fires also need to be determined.

15.4.7.5 Explosion

Explosions ('No' branch for the 'Fire' node) following a massive gas leak from a blowout may involve a substantial amount of gas. Recent R&D programmes [31] have demonstrated that under the worst case conditions very strong explosions may theoretically occur in such circumstances. The important aspects related to occurrence of explosion is whether escalation occurs or not, whether it is escalation to another segment, or to another area or deck. Usually this is not directly expressed in the blowout event trees, probably because, due to the long duration of the fire, escalation is virtually certain once a blowout is ignited. If escalation occurs instantly because of the explosion, such early escalation may be more critical, especially if it occurs prior to evacuation having been completed. The scenario could in such cases be similar to the Piper Alpha accident. Only one such scenario with corresponding severity (37 fatalities) is known from the accident records, namely an explosion and fire caused by a blowout in the US Gulf of Mexico area in 1970.

15.4.7.6 Node: Fire on Sea

In the case of an offshore platform blowout, there is always a chance that some amounts of oil may be spilled onto the sea surface without being completely burned in the air. This oil may then burn on the sea surface. If the volume of oil burning on the sea surface reaches a significant amount, then the radiation loads on the underside of the deck may be quite high. The smoke production may also prevent escape and evacuation from being completed.

Pool fires in the open are controlled by the evaporation rate from the fuel surface. The liquid absorbs energy from the flame and evaporates. The vapour will mix with the entrained air as it rises due to buoyancy effects. It is further heated to ignition and reacts generating heat. Burned gases then radiate energy until they reach some low temperature at which point they merely exchange heat with the surroundings. The main characteristics of a pool fire which are important with respect to safety are:

- duration of the pool fire
- extent of the pool fire i.e., height and diameter of the flame
- radiation heat load on objects located outside the flame
- heat load on objects enveloped by the flame.

These characteristics are strongly dependent on the geometrical conditions at the location where the oil spill occurs.

When a pool fire occurs inside an enclosure where the air supply is limited, the actual extent of air supply will determine the intensity of the pool fire.

Fire on sea may in theory also be caused by a subsea blowout from a wellhead on the seabed. A burning subsea blowout will only occur if the flow is ignited, usually by equipment on the installation. Only gas has the possibility to be ignited inside the installation, and the gas fraction will therefore influence the probability of ignition.

15.4.8   Gas Leak from Riser/Pipeline Gas Leak from Riser/ pipeline

15.4.8.1   Leak and Outflow Conditions

A sudden rupture of a high-capacity gas/oil pipeline in air (i.e. above sea level) will result in a massive release of highly combustible material. The amount of energy stored in such a line may be enormous, and an accidental release of hydrocarbons may give rise to substantial mechanical damage and/or fire. To assess the hazard it is necessary to know the time-dependent rate of outflow and the characteristics of the outflow when ignited. An example is illustrated in Fig. 15.16.

The event tree for riser leaks is usually quite simple, because there are limited possibilities for risk reduction. The best approach for the control of risk in this context is to prevent the actual occurrence of the rupture itself.

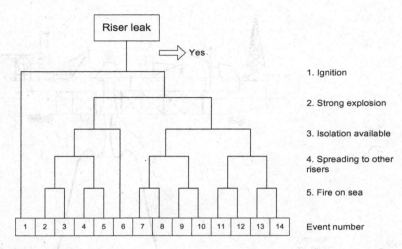

Fig. 15.16 Riser leak event tree

15.4.8.2 Ignition

The flow rate in case of a gas leak will be very high if a riser rupture occurs above the sea level; this was already indicated in Sect. 14.3.3. The size of the gas cloud will therefore be quite extensive in a very short time, in fact it could be so extensive that large parts of the cloud are above the upper explosive limit (UEL), such that ignition is unlikely.

Ignition of a leak from an oil riser is quite different from a gas leak. The crude oil is relatively incompressible and the outflow conditions will be much more affected by friction, implying that expansion will be limited to an initial 'gushing'. The possibilities for ignition are therefore much more limited.

15.4.8.3 Isolation of Flow

Subsea isolation valves were installed quite extensively on gas pipelines in the first few years after the Piper Alpha accident, and some 50 valves were installed on existing pipelines. A subsea valve will act as a barrier stopping the outflow of gas from the pipeline, even if a leak develops in the riser. A possible fire will therefore have short duration, if such a barrier is installed. After the Piper Alpha accident much attention was given to the ESD-valves located on the platform, in particular with regard to their survivability in various accidental conditions. The most extensive protection is however provided by a subsea valve location.

A subsea isolation valves is typically located 200–500 m away from the platform. The reasons for this are that:

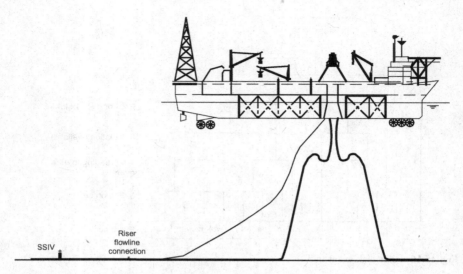

Fig. 15.17 Location of subsea isolation valves on gas pipelines

- It reduces the likelihood that the valve will be damaged by dropped objects from the platform.
- The valve will be capable of blocking not only riser leaks but also leaks in the section of the pipeline closest to the platform. This is also the part of the pipeline which is clearly most likely to develop leaks.

The disadvantage of this location is that the inventory in the pipeline/riser section between the valve and the platform will be greater and thus represents a greater risk. Figure 15.17 indicates a location of a subsea isolation valve on a gas export pipeline from an FPSO installation. With such vessels, the connection between the pipeline and the vessel is usually through flexible flowlines, which are considered to have a higher probability of leakage, compared to a steel riser. Installation of subsea isolation valves is therefore more common in these circumstances.

Possible leaks from the valve itself also have to be considered. A subsea valve implies that several potential leak points are introduced in the pipeline. This means that a gas leak may develop through the valve itself, and this leak can obviously not be stopped by the valve. In practice, it can be expected that the expected leak frequency is higher after the valve has been installed, and it is therefore important that the valve is located sufficiently far away from the platform to avoid the possibility of the development of a gas cloud around the platform in the event of a leak from the valve.

If a subsea valve is installed, then the focus in the operations phase must be on maintaining high availability of the valve, such that the probability of failure to close in an accident is minimised.

15.4.8.4 Spread to Other Risers

The consequences may be even more severe, if the accident escalates into additional risers. The fire loads may be very extensive, and if the duration of the fire is long, then the likelihood of rupture of a second riser is quite high. This was also demonstrated in the Piper Alpha accident (see Sect. 4.15).

15.4.8.5 Fire on Sea

Fire on the sea surface is important, because the support structure may be damaged in addition to the topside, as demonstrated by Piper Alpha (see Sect. 4.15).

15.5 Analysis of Barriers

15.5.1 Cause Analysis

NORSOK Z-013 [32] emphasises that the cause analysis of mechanisms that may lead to major accidents is the preferred option. It is therefore expected that fault tree analysis and other methods, such as human reliability analysis (HRA), are used extensively for cause analysis in QRA studies.

However, the use of detailed cause analysis was actually more extensive in the 1990s. Detailed cause analysis is rarely carried out even in detailed QRA studies at present. Quite often, this is replaced, for instance, by just using the unavailability that corresponds to the safety integrity level (SIL, see Sect. 15.5.3) or experience data.

The main challenge with this lack of cause analysis is that there is no basis on which to identify possible risk reducing measures, which may prevent the occurrence of initiating events.

15.5.2 Analysis of Dependencies Between Barriers

The way quantitative risk analysis in the petroleum industry has been conducted for many years makes the comprehensive analysis of dependencies between barriers impossible. Reliability analysis of barrier systems and elements is conducted to a limited extent as input to the node probabilities in event trees. These reliability studies are usually conducted separately for each node, often in a superficial manner, and without consideration of the influence from utility systems.

An exception, where comprehensive analysis of barriers is usually conducted, is when a HIPPS (High Integrity Pressure Protection System) is used, and reliability is extremely crucial. The analysis is however often limited to the pressure protection function.

In the QRA studies for nuclear power plants, it is common to perform extensive event tree and fault tree analysis, to an extent where dependencies may be analysed in detail. The most commonly used tool is RiskSpectrum® [33]. This analysis tool has event trees and fault trees in a common manner, but has the ability to transform event trees to fault trees, such that all fault trees for barriers then may be integrated into a huge common fault tree. From this overall fault tree, dependencies may be analysed in detail, using common techniques for analysis of cut sets, common mode failures and importance calculations. The RiskSpectrum® analysis tool gives the following advantages:

- Dependencies may be identified, together with common mode failures.
- Importance measures may be calculated for components, systems and failures.
- The analysis may be used to identify the requirements for barriers to be effective.
- The analysis may be used in order to identify what compensating measures are required if barrier systems are unavailable.

A pilot study was completed in order to demonstrate the advantages of application of the RiskSpectrum® tool [34]. For the installation in question, the following were found to be the systems with highest importance with respect to prevention of uncontrolled escalation of fire:

- Pneumatic power supply
- Two named electric power supply circuits.

Such results would usually never be found using traditional quantitative risk analysis.

It was further found from the pilot study that the contribution from common mode failures was lower than expected. As it was a quite limited pilot study, it is unsure whether this is an observation which has wide ranging applicability.

It is usually physical barrier elements that are analysed with the use of RiskSpectrum®. This was also the limitation used in the pilot study. The regulations, on the other hand, require that physical as well as non-physical barrier systems and elements are considered in parallel. It would be possible to extend a RiskSpectrum® analysis also to include human and organisational barrier systems and elements.

15.5.3 Analysis of SIL

According to PSA management regulations, instrumented safety systems shall be designed according to IEC 61508 [35], and this standard is used as the recommended standard for the specification, design and operation of such safety systems.

Norwegian Oil and Gas has issued its Guideline 070 [36] in order to adapt and simplify the application of the IEC 61508 and IEC 61511 [37] standards for use in the Norwegian petroleum industry.

As a basis for the given SIL requirements, typical loop diagrams for a number of safety functions are provided in Guideline 070, together with industrially verified component reliability data. The following are the functions that are covered in Guideline 070:

- Sectioning of the process
- Fire detection
- Gas detection
- Isolation of the sources of ignition
- Starting and stopping fire pumps, both manually and automatically
- Active firefighting
- Process safety
- Well safety
- Isolation of riser
- Subsea ESD isolation
- Topside and subsea HIPPS protection
- Depressurisation
- Ballasting for floating facilities
- Prevention of blowouts and of well leaks during drilling operations.

Table 15.2 Minimum SIL requirements according to Guideline 070

Safety function	Minimum SIL requirement
Process segregation	1
PSD functions (PAHH, LAHH, LALL)	2
PSD/ESD function	3
PSD function (TAHH/TALL)	2
ESD sectioning (closure of one ESD valve)	2
Depressurisation (blowdown; opening of one blow down valve); BDV	2
Isolation of topside well (shut in of one well)	3
Isolation of riser (shut in of one riser)	2
Fire detection	2
Gas detection	2
Electrical isolation	2
Release of firewater/deluge	2
Manual initiation of F&G /ESD functions from field/CCR	2
Start of ballast system for the initiation of rig reestablishment	1
Emergency stop of ballast system	2
Subsea ESD	3
Drilling BOP function	2

Guideline 070 also defines minimum SIL requirements for each safety system according to the list above (see Table 15.2), using the following four levels, including their interpretations with respect to the probability of failure on demand (PFD):

SIL4: PFD: $\geq 10^{-5}\text{--}{<}10^{-4}$
SIL3: PFD: $\geq 10^{-4}\text{--}{<}10^{-3}$
SIL2: PFD: $\geq 10^{-3}\text{--}{<}10^{-2}$
SIL1: PFD: $\geq 10^{-2}\text{--}{<}10^{-1}$

15.6   Event Sequence Analysis

15.6.1   Time Dependency

A 'one-directional' time development is often assumed when constructing an event tree. For a gas leak this typically follows the sequence:

- Leak
- Gas detection
- Isolation
- Ignition (potential)
- Fire detection
- Fire fighting
- Secondary loss of containment.

In actuality the scenario development is seldom so simple if the scenario is completely without control. Very often there will be loops, where secondary leaks, explosions and escalation of the fire occur. In practice this cannot be integrated into the event tree.

Cause–consequence analysis is another form of event tree which has the ability to show time delays between steps, and to some extent couplings or combinations. The time sequence is still assumed to be 'one-directional', however. The big advantage of the event tree method, on the other hand, is the ease in communicating the assumed accident sequence to non-analysts.

The event trees usually used in QRA are considered as 'static', in the sense that the logic of the tree, its couplings etc. are fixed by the analyst prior to conducting the actual analysis. The alternative to the static event tree is the **dynamic** event tree, which can be programmed to alter its logic and construction to reflect the modelled development of an accident. Commercially, there is only one package available for modelling of such dynamic trees, namely PLATO®, developed by Environmental Resources Management (formerly Four Elements Ltd.), London [38]. PLATO® is said to be a simulator for accident development, but may perhaps better be explained as a dynamic event tree generator.

But the dynamics has its price. What would typically be an event tree with 50 terminal events, may in the dynamic analysis have 5,000 terminals.

The dynamic event tree generator in PLATO® will develop the branches in the tree according to the results of the consequence calculations that are automatically carried out as the process is developing. In the past the high number of outcomes has apparently limited quite considerably what can be done in terms of consequence calculation for each terminal event, in order for the computing time to be realistic. It is a difficult choice to make, between representation of the dynamic tree with simplified consequence calculations, or more static (and simpler) event trees with more advanced consequence calculations. The benefits of the dynamic event trees may be lost entirely, if oversimplified consequence calculations are used. An independent review [39] found that the models that are used for combustion are suitable and sufficiently detailed for application to an offshore installation. It may on the other hand be argued that since 1997, there has been an increasing use of CFD calculations within QRA studies.

In spite of the severe restrictions on how the event tree may model the dynamics in the accident sequence, the program is still being used extensively. But it should be noted that further research and development work would be advantageous in order to improve the accident sequence modelling.

15.6.2 Node Sequence in Event Tree Modelling

The sequence of nodes in an event tree is one of the most difficult aspects, where it may be claimed that there is in fact no universal truth. It may appear that this is unimportant as node probabilities are to be multiplied anyway, according to Eq. 6.2 , But this is far from the case. The node probabilities are conditional probabilities, and the sequence will therefore be of considerable importance.

In this field no absolute rules may be stated, because it will depend on the structure of the tree, the safety systems and the functions that are involved. A suggested rule to use is the following:

- If systems and actions have a time sequence in the development, they should then be represented in the same sequence in the event tree.
- If activation of one system or function has an effect on the success of other systems, then that one system should be considered first in the event tree.

Consider the following example: Detection of a gas leak will usually result in emergency shutdown, which will isolate sections of the process plant, but also cut power to all electrical equipment which could be an ignition source. The ignition node therefore needs to follow the detection node, as the opposite would result in a gross over-prediction of the risk associated with ignited leaks.

15.6.3 Directional Modelling

Another limitation of the normal event tree is that it becomes too complicated if different flame directions are considered (applies mainly to jet fires). The event tree is often modelled using a 'typical' direction, or the most probable direction or the worst case direction. But how shall this be determined? In the case of a gas leak from a flange on a piping system, all directions along the circumference of the flange are equally likely.

PLATO®, the dynamic event tree generator mentioned above, is however also able to handle escalation due to flames in different directions.

An alternative to this approach has been chosen by some analysts who use event trees modelled in six different (Cartesian) directions, in order to provide an approximate model of reality. The advantage of this approach is that directional modelling may be accomplished with 'normal' trees using a PC, although the resulting number of event trees becomes very high. The software ASAP® performs such modelling (see Appendix A).

15.6.4 MTO

MTO analysis is primarily developed as a technique for the investigation of accidents and incidents. It may, on the other hand, also be used for analytical purposes, and a brief summary is therefore included. It may be noted that MTO investigation is the main investigation technique used by Petroleum Safety Authority Norway for investigation of accidents on the Norwegian Continental Shelf.

There are few sources available for a general description of the MTO-analysis, one of which is by Tinmannsvik et al. [16], on which the following summary is based. The method is based on HPES (Human Performance Enhancement System) from the nuclear industry, and has been developed by Jean-Pierre Bento. The MTO-analysis is based on three methods:

1. Structured analysis by use of an event- and cause-diagram.
2. Change analysis by describing how events have deviated from earlier events or common practice.
3. Barrier analysis by identifying technological and administrative barriers which have failed or are missing.

Figure 15.18 illustrates the MTO-analysis worksheet, when used in an accident investigation. The first step in an MTO-analysis is to develop the event sequence horizontally and illustrate the event sequence in a block diagram. Then, the analyst should identify possible technical and human causes of each event and insert these vertically to the events in the diagram.

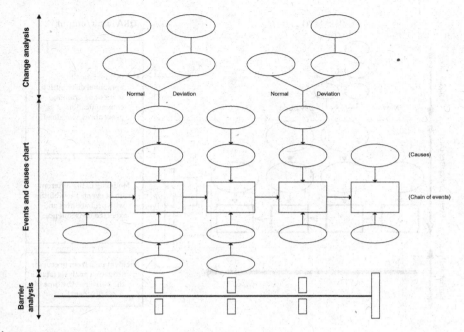

Fig. 15.18 Illustrative MTO-diagram

The development of the event sequence is often referred to as a 'timeline analysis' i.e., an analysis of the sequence of events and their timing. This is a step in the accident investigation which is common for many of the investigation techniques.

The next step is to make a change analysis i.e., to assess how events in the accident progress have deviated from normal situation, or common practice. Normal situations and deviations are also illustrated in the Fig. 15.18. Further, determine which technical, human or organisational barriers that have failed or were missing during the accident progress. All missing or failed barriers are shown below the events in the diagram. The basic questions in the analysis are:

• What may have prevented the continuation of the accident sequence?
• What may the organisation have done in the past in order to prevent the accident?

The last but important step in the MTO-analysis is to identify and present recommendations. The recommendations should be as realistic and specific as possible, and might be technical, human or organisational (Fig. 15.19).

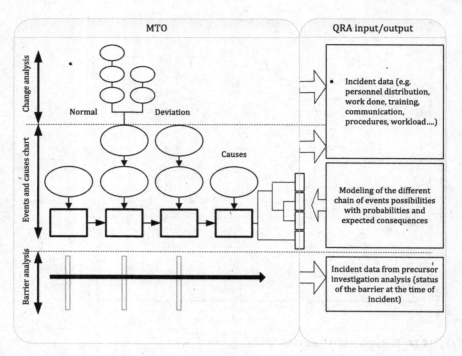

Fig. 15.19 MTO and QRA input [40] (Reprinted from Reliability Engineering and System Safety, Vol 101/May 2012, Skogdalen and Vinnem, Combining precursor incidents investigations and QRA in oil and gas industry, 11p, Copyright (2012), with permission from Elsevier)

A classification system for basic causes has also been developed, in order to enable trend analysis of accident causes. The causes are classified into the following categories:

- Working environment
- Operational organisation
- Routines for change management
- Installation management
- MMI—Man Machine Interface
- Working schedules
- Communication
- Procedures, instructions
- Supervision
- Working practices
- Competence, training.

It should be noted that the MTO-analysis is not suitable for quantitative analysis. Figures 15.20 and 15.21 show a complete MTO diagram from an actual case.

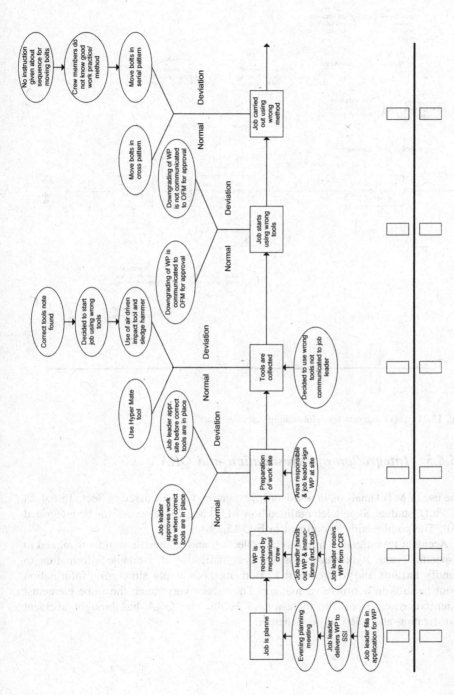

Fig. 15.20 MTO diagram for 'Hot bolting' incident, Part 1

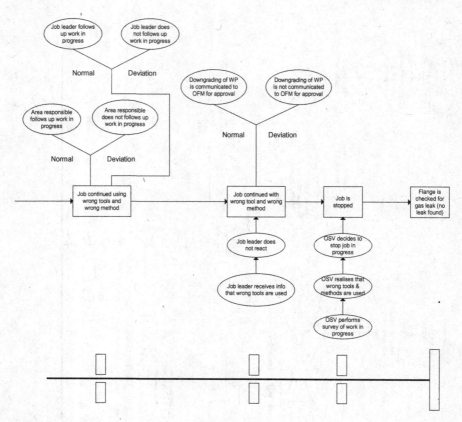

Fig. 15.21 MTO diagram for 'Hot bolting' incident, Part 2

15.6.5   Integration of Investigation and QRA

The use of MTO analysis in accident investigation was discussed in Sect. 15.6.4. In his Ph.D. studies, Skogdalen outlined how MTO and QRA studies can be integrated [40]. The combination is outlined in Fig. 15.19.

Accident investigation implies the collection and examination of facts related to a specific event. Risk analysis is the systematic use of available information to identify hazards and to predict risk. Both methods imply structuring information about hazards in a structured manner. They share very much the same elements. Extensive research on the inclusion of HOFs into QRA has brought accident investigation and QRA closer together.

15.6.6 Survey of the Extent of HOFs in QRA

Norwegian and UK legislation as well as the NORSOK standard Z–013 [30] require that QRA studies address technical aspects as well as human and organisational aspects. In his Ph.D. studies, Skogdalen assessed 14 QRA studies in order to establish to what extent these studies considered human and organisational aspects as potential causes of major accidents [41]. These studies were classified into four categories, according to the criteria related to the HOF as shown in Table 15.3.

Table 15.3 Levels of QRA HOF inclusion and corresponding requirements (Reprinted from Reliability Engineering and System Safety, Vol 96/4, Skogdalen and Vinnem, Quantitative risk analysis offshore—Human and organizational factors, 12p, Copyright (2011), with permission from Elsevier)

Levels of HOF inclusion	Requirements
Level 1	• Analysis of technical and operational factors. Operational factors are activity levels, etc. • Risk reducing measures are technical as well as operational actions, such as fewer shipping arrivals
Level 2	• Explain the importance of HOF • The influence of HOFs on different systems is partly described • Human errors are calculated separately • Interview with the crew. The results are commented on but the models and calculation are not adjusted
Level 3	• Systematic collection of data related to HOFs • QRA-models are adjusted according to findings from HOFs • Identification of the causes of errors to support the development of preventive or mitigating measures • Inclusion of a systematic procedure for generating reproducible qualitative and quantitative results (BORA, OTS...)
Level 4	• QRA is an integrated part of the risk management system • The results from QRA form the basis for daily risk management • QRA is known and accepted at all levels of the organisation • QRA is combined with indicators to show the status of the safety barriers

Table 15.4 Results of the assessment of HOF inclusion in QRA studies (Reprinted from Reliability Engineering and System Safety, Vol 96/4, Skogdalen and Vinnem, Quantitative risk analysis offshore—Human and organizational factors, 12p, Copyright (2011), with permission from Elsevier)

Level	Number of installations
1—None existing	5
2—Explained but not adjusted	6
3—Explained and the models are adjusted	2
4—Explained, models adjusted and the results are commented	0

Table 15.4 presents the results of the study in terms of how many QRA studies fall into these four categories.

Table 15.4 shows that the inclusion of HOF aspects so far is quite weak and that there is significant uncertainty about how they should be implemented in practice. Further details are presented in Ref. [38].

15.7   HC Leak Modelling

The modelling of an accidental scenario associated with gas and oil starts with the leaking medium. This may be from many sources, such as:

- pipes and associated fittings
- vessels
- pipelines/risers.

The phase of the leaking medium is the next important aspect:

- 1 phase flow i.e., gas or oil (liquid) phase
- 2 phase flow i.e., usually gas and oil (liquid) mixed
- 3 phase flow i.e., gas, oil, water.

Different models suitable for the different phase compositions and different sources (mainly reflecting the difference between outflow from a vessel or from a pipeline or pipe section) have to be used. The models are primarily aimed at determining the flow rate as a function of time.

15.7.1   Leak Statistics

The Petroleum Safety Authority [Norway] has for more than 15 years collected considerable amounts of experience data for the Norwegian Continental Shelf, in particular for hydrocarbon leaks from process equipment and operations trough the risk level project (see Sects. 6.3 and 22.3).

The data in the project has been normalised in relation to several parameters, for hydrocarbon leaks it is the number of installations and number of manhours worked on the installations. The installations have been divided into categories:

- Fixed production installations
- Floating production installations
- Complexes, bridge linked production installations
- Normally unattended installations (production installations)
- Mobile units.

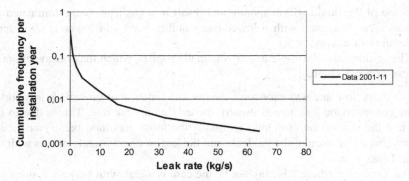

Fig. 15.22 Leak rate distribution for Norwegian leaks, 2001–2011

It should be noted that complexes are somewhat special. There may be from two to more than ten bridge linked installations, of which several may in theory handle hydrocarbons. But with respect to normalisation, complexes are counted as one installation, irrespective of how many bridge linked installations there are, and how many of them that handle hydrocarbons.

Figure 15.22 presents a cumulative leak rate distribution as average for all installation types in the entire Norwegian sector, expressed per installation years.

The leak frequencies per installation year for the categories of production installations are shown in Fig. 15.22. It may be observed that the total frequency of leaks above 0.1 kg/s is 0.37 per installation year for (average 2001–2011).

The frequency of exceeding 1 kg/s is 0.10 per installation year, whereas the frequency of exceeding 10 kg/s is 0.015 per installation year. The frequency of exceeding 50 kg/s is about 0.0025 per installation year.

15.7.2 Calculation of Leak Rates from Experience Data

One of the main parameters used in order to characterise a hydrocarbon leak, in particular gas leaks, is the mass flow rate (often called the 'leak rate'), usually expressed in kg/s. It is therefore important to be able to calculate the leak rate for hydrocarbon leaks that are observed on the installations.

A release in the liquid phase from a pressurised system will normally give a small gas cloud compared to gas release with the same mass flow rate. This implies that the probability of ignition will be lower, as will the probability of gas explosion. But there are some aspects that may imply that the liquid release may be just as dangerous as the gas leak, in some cases even worse:

- Some of the liquid will evaporate and result in a gas leak, often dominated by heavy gas fractions, with a lower flammability limit and lower ignition temperature or energy.
- The release may constitute a spray of small droplets, which may behave more or less as a gas cloud.

The mass flow rate will vary considerably as a function of time; what is reported as the characteristic leak rate is usually the maximum leak rate. This is due to the fact that the size of the gas cloud is usually the most important parameter for the hazard characterisation of the gas leak, and the largest cloud usually occurs with the highest leak rate.

One exception where this may not be the case is a leak with very short duration, instantaneous leaks. The maximum gas cloud may then occur with some time delay in relation to the release. In these cases, an 'equivalent' continuous gas leak is considered, which would give a gas cloud with the same size. The flow rate given as the characteristic leak rate is that of the equivalent continuous leak, implying that the hazard potentials should be the same.

Measurements which may be used in order to calculate the mass flow rate are typically related to the dimensions of the gas cloud, usually only the fraction above the lower flammability limit is considered. Gas detector recordings may be used to calculate the dimensions of the gas cloud, as a function of time if the detector readings allow that.

Below follow some simple illustrations of how different leak sizes will result in gas cloud of different sizes, based on some simple modelling with CFD tools, based on Ref. [42]. More detailed cases may be calculated by means of studies using CFD tools.

A leak rate of 0.1 kg/s gives a gas cloud above the lower flammability limit with a typical volume of 0.5 m^3, in the case of a free, unobstructed jet. If the jet is without impulse (diffuse leak), the volume increases to 10 m^3. The times to stable conditions are 2 s in the case of a jet, and 20 s in the case of diffuse leak. Due to the low leak rate, there is little or no difference between a leak in the open or inside a process module with limited natural ventilation. For larger leak rates, however, this distinction becomes an important parameter.

A large leak (around 10 kg/s) in open air will cause a gas cloud which is stable after less than 5 s, whereas the time would be in the order of 60 s inside a process module. The volume of the gas cloud (within flammable limits) is also much larger in the latter case, as much as 20 times larger.

As an example, the following could be considered: a large leak in a process module, starting at 10 kg/s, having decreased to 1 kg/s after 60 s. The maximum gas cloud occurs after 30 s, with a volume of 700 m^3 within flammable limits. This corresponds to a stationary leak of 4 kg/s, which would give a stable gas cloud of the same size as the equivalent stable leak scenario.

15.7.3 Modelling of Leaks

There are several factors which influence the flow modelling and influence the duration of the leak:

- Isolation of sections of the process systems into limited volumes.
- Depressurisation of one or more sections of the process system to 'limit the volume of gas or oil escaping from the leak.

The depressurisation model is the most difficult aspect. There are simple as well as complex models available for use in modelling this aspect. Multi-phase releases from pipelines and risers really require complex computational tools such as OLGA (see Appendix A).

Realistic modelling of the leak and its duration is obviously very important to determine the size and duration of any fire that may occur, and the response of the platform. Simpler models may be used for coarse evaluations.

15.8 Ignition Probability Modelling

Ignition probabilities are one of the most critical elements of risk quantification in that the risk results are normally directly dependent on the probability of ignition. There are limited accident statistics available on the subject of ignition probability, most likely because such statistics are difficult to establish following an accident involving an ignited release. It may be noted that the extent of available data for the other critical element of the risk quantification, that is leak frequency, is quite a bit better although not perfect.

Further, there is very little experimental data available, due to the difficulty or impossibility of establishing realistic values through laboratory experiments. The type of data that are available is limited to flash points, auto ignition temperatures, etc.

Ignition probability models have been published in several textbooks and papers. These models reflect leak rates and module volumes, but seldom include anything approaching design and operation details. The collection of leak and ignition data by UK HSE is the most extensive on-line data collection scheme in existence.

ESRA took the initiative in the mid 1990s to develop a common ignition model in a Joint Industry Project (JIP). The model was described in a report by DNV [43]. The data basis at the time was uncertain, because the systematic collection of leak and ignition data had been initiated only a couple of years earlier in UK, and had not been initiated in Norway.

A new JIP project was conducted ten years later. The model was based on data from the systematic collection of leak and ignition data in the UK and Norway (see Sects. 6.4.3 and 15.8.2). However, the model was at the same time substantially revised [44], unfortunately not necessarily to a better model, many experts would argue.

In the recent years another joint industry project has been carried out, financed by Norwegian Oil and Gas, in order to revise the model yet again. The revised model is presented in Sect. 15.8.5 below.

15.8.1   Experience Data

There has been no ignited hydrocarbon leak with leak rate above 0.1 kg/s in the Norwegian sector since 19th November 1992. A small gas leak, probably in the order of just above 0.1 kg/s, was ignited most likely by grinding sparks during modification work. Prior to that there was an oil leak from an export pump a few years earlier, where the leak source was also the ignition source, a failed seal on the export pump during normal operation. About 0.5 m^3 of crude oil leaked during a period of less than 2 min, implying that the leak rate was considerable, around 5 kg/s.

Comparison of the UK and Norwegian leak rates was presented in Sect. 6.4 above, where it was shown that leak frequencies per installation year are lower in the UK sector, when compared to the Norwegian sector. The ignited leaks in the Norwegian sector occurred for than 25 years ago, whereas ignitions in the UK sector have occurred also during the last ten years.

When the comparison is made for ignited leaks, the situation is opposite. For the period 1 October 1992 until 31 March 2011 the following gas and two-phase leaks in the UK sector compares to no ignited leaks in the Norwegian sector [45]:

- 604 gas/two-phase leaks >0.1 kg/s
    - Of which 2247 leaks >1 kg/s
- 9 gas/two-phase leaks >0.1 kg/s have been ignited
    - Of which 2 ignited leaks leak >10 kg/s.

15.8.2   Why Is It Difficult to Develop an Ignition Model?

There has been considerable effort in the petroleum industry to develop suitable models for the calculation of ignition probability, but it seems to be demanding to arrive at a good model that is broadly recognised as suitable and defensible.

The main reason for this uncertainty may be the limited data available for the development of suitable models. Table 15.5 presents an overview of process leaks

Table 15.5   Overview of ignited process leaks (>0.1 kg/s), North Sea, 1993–2016

Leaking system	Leak rate	Operational mode	Ignition source	Description
Fuel gas	0.85	Start-up	Burners	Excess gas ignited by burners
Fuel gas	0.15	Start-up	Turbine exhaust	Hot gases ignited in turbine exhaust stack
Gas compression	0.18	Maintenance, Hot work	Welding flame	Welding inside habitat during annual shutdown
Fuel gas	0.34	Start-up	Hot exhaust stack	Excess gas during start-up ignited by exhaust stack
LP vent tower	0.69	Normal production	Lightning strike	LP vent was ignited by a lightning strike
HP vent	2.7	Construction, Hot work	Local welding activity	Minor flame observed from vent line being welded on during shutdown and depressurised system
Separation	467	Start-up	Turbine	Leak from cooler sucked into turbine and ignited
Open drains	36.5	Normal production	Corroded heating element	Small fire in open drains tank with high water content
LP vent	0.48	Flushing	Flare	Liquid overflow from drain tank ignited by flare
Blowdown	8.0	Process shutdown	Flare	Liquid overflow from drum ignited by flare
Flowlines	0.66	Normal production	Static	Pool fire due to long lasting leak on NUI
Drains	0.34	Construction, Hot work	Welding flame	Welding unisolated closed drains pipework inside habitat during ann. shutdown whilst well was bled into it
Process	0.14	Unknown	Pump seal failure	Sparks due to mechanical disintegration

in the North Sea with an initial leak rate above 0.1 kg/s that have been ignited during a period of 24 years. Only three of the 12 ignited leaks refer to normal production, which is the phase that ignition models address. Of these three ignitions, the first is related to a vent tower and the second is related to the open drains system, neither of which is modelled in ignition models. It may thus be argued that during these 18 years, only one single ignition is relevant for ignition models, which obviously provides a poor data set for model development.

A very detailed statistical overview of ignited leaks in the North Sea (including Norwegian Sea, West of Shetland, etc.) is provided in the MISOF report [46].

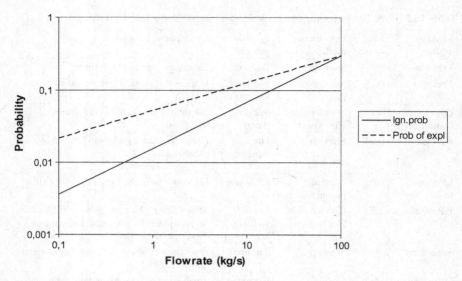

Fig. 15.23 Simplified ignition model according to Cox et al

15.8.3   Cox Model

Cox' Model Cox et al. [47] has presented a simplified model, and a framework for a more sophisticated model to be used in determining the probability of ignition. The model is based on relatively simple assumptions regarding the ignition probability for the lowest leaks, and the observed ignition probability for blowouts as the extreme (Fig. 15.23).

15.8.4   Platform Specific Modelling

Use of the Cox model results in relatively high ignition probabilities. A more fundamental problem is that no actions taken to prevent ignition are reflected in the model. It is therefore important that platform specific modelling is used, and preferably also operation specific modelling. The Cox' model however, was for a long time the only publicly available model and has therefore gained significant usage. The objectives of platform specific modelling are to reflect the following aspects:

- The probability of ignition of a HC leak, which is dependent upon the likelihood and susceptibility of the leaking medium to ignite.
- The size and concentrations of the flammable cloud i.e., the leak rate in relation to the module volume and the ventilation rate.

- Different types of equipment have different failure modes and frequencies which may be susceptible to failure that leads to ignition. The likelihood of ignition from different equipment units should therefore distinguish between equipment types, and the location of the equipment in relation to the leak.
- Ignition by manual operations (such as welding) should be considered explicitly. The same also applies to permanent ignition source s, such as the flare, burners, etc.
- The ignition probability should be expressed as a time dependent function.

A 'baseline' (or 'background') probability of ignition is considered to exist in all areas, irrespective of equipment and operations, due to miscellaneous activities and equipment that is not possible to consider explicitly.

In addition to these main technical requirements for an ignition there is also a need for a model which is not too complicated to use. Actual modelling will therefore always be a compromise.

15.8.5 Model Overview—MISOF Model

The model named Modelling of Ignition Sources on Offshore oil and gas Facilities (MISOF [46]) is aiming to provide the best available technology in industry for use in quantitative risk analysis for offshore installations located in the North Sea, and may be used in other domains if it is confirmed that the properties of the installation are equivalent with what are found generally on North Sea installations.

The ignition probability is the product of two probabilities; the probability for exposure of a live ignition source to a flammable fluid and the ignition probability in the case of such exposure. The objective of MISOF is to provide define the ignition probability given exposure to flammable fluid for the most important potential ignition sources on offshore oil and gas facilities. Thus, in order to use the MISOF model, a model for the exposure probability is required. The quality of the probabilistic exposure model is fundamental for the accuracy of the ignition probability predictions.

MISOF provides ignition source data for use with a simple exposure model, with all ignition sources are distributed evenly in space or alternatively ignition source data to be based on advanced probabilistic exposure models including the locations of the leak sources and the specific ignition sources into account.

The model parameters are to a large extent based on analysis of statistics of leaks and ignited events on installations in the North Sea starting from 1992 and including 2015. Reflection of the physical properties of the ignition phenomena is applied where such knowledge is available, but in general the parameters are based on a statistical methodology.

The model parameters have been tested by running the ignition model for two platforms located on the Norwegian Continental Shelf and comparing the output with the historical ignition probability obtained from the established North Sea

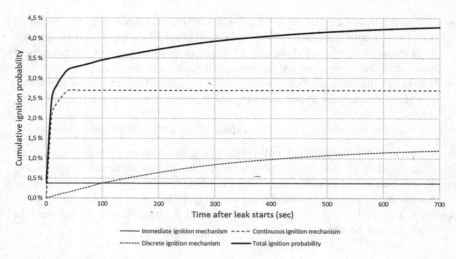

Fig. 15.24 Typical output from MISOF model

statistics. The results indicate that the MISOF model is able to produce reasonable predictions of the ignition probability relative to the observed historical ignition probability if combined with an appropriate probabilistic exposure model.

An extensive documentation of the MISOF is available [42], and the present documentation is aimed at giving an overview, but not the details. The presentation is based on the MISOF documentation. A typical presentation of MISOF results is presented in Fig. 15.24.

15.8.6 Ignition Model for Objects in Classified Areas

The model for ignition sources located inside classified areas is only valid for objects intended for use in potentially explosive atmospheres, i.e. classified areas. There are in general three Ex zones:

- Zone 0: An area in which an explosive mixture is continuously present or present for long periods
- Zone 1: An area in which an explosive mixture is likely to occur in normal operation
- Zone 2: An area in which an explosive mixture is not likely to occur in normal operation and if it occurs it will exist only for a short time.

On a high level, the ignition model consists of the two groups of models; the first for the probability for ignition before a flammable gas cloud has been formed. This is a special case where the ignition mechanism often is related to the properties of the object that the release originates from and/or the fluid that is released. The likelihood of exposure is irrelevant in this case. This group is denoted "immediate ignition".

The second group is the probability of ignition due to exposure of objects that constitute a potential source of ignition if exposed to flammable fluids. Ignition will in this case take place after start of the release. This group is denoted "delayed ignition".

These two groups are further broken down to a few categories dependent on the type of equipment and ignition mechanism. For the group "immediate ignition" there are two categories:

(i)   Immediate ignition (excluding pumps)
(ii) Pump immediate ignition.

The equipment categories under the delayed ignition group are as follows:

(i)   Rotating machinery
(ii) Electrical equipment
(iii) Other.

The starting point for modelling of the ignition probability given exposure to these different equipment categories is the generic parameter "Ignition sources in the area". Based on general classification of ignition mechanisms the following two types are used in the ignition model:

(i)   Continuous
(ii) Discrete.

The modelling of the exposure probability for these two classes are different, which is reflected in the transient exposure model. For the categories 'Rotating machinery' and 'Electrical equipment', there are three different models requiring inputs of varying resolution. The model that should be used is dependent on the available information in terms of the location and properties of the equipment. The high-level model is purely generic, i.e. there is no correlation between the value of the ignition parameter in the model and equipment properties and layout. The detailed model allows for specific modelling in terms of location of the pieces of equipment for both categories. For the 'Electrical equipment' category, the level of protection and protection method can be reflected.

15.8.7   Ignition Model for Objects in Unclassified Areas

The model for ignition in unclassified areas cover the ignition potential related to gas exposure of objects not intended for use in explosive atmospheres. The objects that are covered are:

• Gas turbine air intakes
• Combustion engines (in practice diesel engines)

- Equipment in enclosures protected by a mechanical ventilation system
- Equipment in unclassified areas
- Supply vessels
- Hot work
- Flare

It is emphasized that other sources of ignition may be relevant, and the applicability must be clarified as part of the risk analysis being performed.

15.8.8 Uncertainty Evaluation

Based on an extensive discussion in the MISOF report [42] it has been concluded by the authors of the report that the base ignition probabilities (the parameters in the MISOF model) for immediate and delayed ignition represent the **upper boundary** for the underlying ignition probability applicable for installations operating at the UKCS and NCS. This is done in order to have some robustness if the premises for the calculations at some later time should not be fulfilled. The most likely ignition probability is a factor of two less than the base ignition probability. This should be communicated in a quantitative risk analysis based on PLOFAM (see Sect. 6.13) and MISOF models to ensure well informed risk-informed decisions being made.

15.9 Escalation Modelling

Barriers are those systems and actions that prevent escalation from occurring. The importance of barriers is well illustrated by comparison of the outcome from two actual events each involving a medium sized gas leak in the compression area of a platform. The worst case, the Piper Alpha disaster on 6.7.1988 is well known, the explosion on the Brent Al pha platform on 5.7.1988 is less well known. This accident started in exactly the same way as the Piper Alpha accident. The result was a gas fire following the explosion, brought under control in some 45 min due to automatic systems, with only superficial damage to the compression module. The dramatic difference between the two events arose because on the Brent Alpha platform, the barriers functioned as intended, while on Piper Alpha they did not. Some more details about these two accidents are presented in Sect. 7.1.1.

This section discusses the modelling of barriers in the event tree, with respect to their functionality, reliability and availability as well as the survivability of the systems, sometimes called, vulnerability to accidental loads.

15.9.1 Functionality

Analysis of the functionality of the barriers involves determining whether they are capable of performing their intended function. As an example, gas detectors of the catalyst type have often been 'poisoned' by salt and other contaminations. Fire water systems may be clogged with dirt, rust and other particles, to an extent that the required fire water capacity can no longer be provided.

The analysis of functionality is a deterministic analysis of the capacity and/or capability of the system in normal operating condition, including consideration of operational premises and constraints. The results of the functionality analysis may determine the probabilistic modelling of the barrier's function.

Many aspects of functionality may be verified by testing, for instance by performing a flow test of the fire water system, involving measuring the flow rates. But it is at the same time important to distinguish between functionality under ideal test conditions and under real-life accidental conditions, where the functionality may be jeopardised by maloperation.

15.9.2 Availability and Reliability

Many of the nodes in an event tree are related to the performance of safety systems which are normally passive, or 'dormant' systems, only intended to be activated upon detection of a hazardous event or accident. Thus even though these systems are repairable, and are being maintained, they function in an accident sequence as unrepaired systems, in the sense that in a demand condition, there is usually no time for repair.

This implies that both availability and reliability are crucial aspects. Let us consider first the **availability** of a system required to operate upon detection of particular conditions. This is often called the 'on demand availability'. The state of maintenance, inspection, and/or testing will determine its availability to function as intended. Next, the **reliability** of the system i.e., the time to first failure, after the system has been activated, is also of crucial importance for some systems.

There is considerable variation between systems, as to which of these two aspects is most important or whether they have equal significance, as may be illustrated by consideration of the gas detection system. It is very crucial that the detection of a possible leak is as early as possible. The system's availability is therefore the crucial aspect when the leak starts. When detection has occurred, there is really no further use of the detection system. Its reliability is unimportant. The fire water system however, is a different matter. An immediate start is crucial when the system is activated, but it is equally important that it continues to operate as long as the fire lasts. Consequently, both the availability and the reliability are important aspects.

When both the availability and the reliability are computed, all aspects of preventive and curative maintenance, including inspection and testing, will have to be considered. Fault tree analysis is a commonly used analysis technique.

Many of the barriers (safety systems) which relate to the control of hydrocarbon systems, are automatic and cannot be negatively affected by personnel in the local control room. On the other hand some systems will require initiation by control room personnel, most typically the blowdown system, which upon actuation will depressurise the process equipment either sequentially or simultaneously. The participation of operators in the actuation process means that human and organisational factors need to be explicitly addressed in the availability and reliability studies.

In fact the importance of HOF is sometimes even more vital for barriers related to non-hydrocarbon systems. For instance, it has been shown that human errors are the main cause of failure of barriers against marine hazards to FPSO vessels [48]. It is therefore important that analysis of barrier availability and reliability is performed with due attention to the importance of HOF.

Most safety systems are periodically tested, which implies considerable experience data, if it is systematically collected and analysed. This may be used to produce installation specific availability data. Reliability data for the continued operation of the system during the course of the accident can usually not be extracted from test data.

15.9.3   Survivability

Survivability analysis may be considered to be a form of reliability analysis, except that the operating conditions are the conditions of the accident. A severe explosion will most probably damage the fire water distribution system, to such an extent that fire water cannot be supplied to an area, even though its original functional condition and state of maintenance is perfect and error free. It is however, worth considering the experience from the so-called 'large scale' explosion tests in 1996/97 [31], from which it was observed that fire water piping survived considerably higher overpressure loads than previously thought. This is briefly discussed in Sect. 9.3. Survivability is also important in relation to the integrity of process piping and equipment, as well as blowdown and flare system piping.

If a fault tree analysis is carried out, survivability considerations may be integrated into the reliability analysis. Due to the nature of the phenomena involved, testing of survivability in realistic accidental conditions is virtually impossible.

15.9.4 Node Probability

Node Probability The final value of a node conditional probability is a function of all the elements mentioned above, and may as an example, be expressed as follows for gas detection:

$$P^f_{GASDET} = P^f_{FUNCT} + P^f_{UNAVAIL} + P^f_{SURV} \qquad (15.6)$$

where

P^f_{GASDET} probability of failure of gas detection

P^f_{FUNCT} probability of gas detection not capable of functioning as intended in the specific accident circumstances

$P^f_{UNAVAIL}$ probability of gas detection unavailable due to maintenance problems

P^f_{SURV} probability of gas detection not surviving the accident conditions for the required period.

15.10 Escalation Analysis

The entire process from an initial accidental event to the final end events, determined by consideration of the performance of protective systems and the responses of equipment and structures, is sometimes called the 'escalation process'. This is the widest interpretation of 'escalation'. Under this interpretation, escalation thus involves determination of different accident sequences and the related loads and responses applicable to each sequence.

A narrower interpretation of 'escalation' is to describe it as the secondary failure of containment, due to accidental effects. This is the interpretation of 'escalation' used in this book, and the wide interpretation is replaced by the term 'accident sequence' modelling or analysis.

It may be important to carry out escalation analysis if the risk to assets is being considered. An alternative, which may be carried out independently of the escalation analysis, is the so-called impairment analysis, which involves an assessment of the frequencies of impairment of the main safety functions. Both escalation analysis and impairment analysis are focused on response to accidental loading, mainly to fire and explosion loads.

15.10.1 Modelling of Fire Escalation

Escalation of fire from one area to another is required to predict whether a fire spreads out of the original area. Secondary fire effects such as smoke or radiation stemming from the original fire are not considered as escalation. It is assumed that fires may escalate due to damage to the fire walls, by direct flow of fuel to the adjacent area or by external flames. Escalation to other areas may be due to three different escalation mechanisms:

- Heat impact from external flames
- Flames passing though penetrations and openings in the floor, walls or roof
- Failure of the segregating walls.

The 'critical duration' for external flames, is the transition point between a short duration flash fire and a stable fire. If the fire duration exceeds this critical duration, the escalation probability increases from near zero to a value dependent upon specific local conditions.

In such cases the effect of protective systems (which are focused on preventing escalation to other equipment) is limited. The failure of segregating walls, ceilings, and floors in the process areas will be strongly dependent on the loading and passive fire protection. The likelihood of structural failure due to fires may be considered in two ways:

- Coarse modelling based on simple heat transfer values
- Detailed modelling based on a comprehensive nonlinear structural analysis.

The modelling of fire escalation in a process area is a complex task, which could be a 'never ending story', unless limited in some way. Some extent of simplification has to be used. The following example, taken from a detailed QRA [49], illustrates a fairly detailed fire escalation model. In the study referenced, the fire escalation has been carried out into the following steps:

1. A non-linear structural analysis of the failure times for piping was carried out, using a range of parameters for; wall thickness, piping diameter, internal pressure, system medium and blowdown time.
2. A survey was carried out in the process areas to judge the conditional probability that fire from a certain process segment would impinge on piping from other segments. This assessment included a consideration of the size of the flame and the size of the adjacent piping.
3. An escalation probability was then calculated by considering the particular circumstances of each scenario, according to Steps 1 and 2 above. Figure 15.25 [45] presents an example of the results from the non-linear stress analysis of the piping systems under fire loads.

Figure 15.25 corresponds well with the values stated in the IOGP data directory [50], which states <5 min for failure of unprotected pipe support and connector/flange without protection, exposed to a heat flux of 250 kW/m^2. The data directory has referenced Mendonos [51].

The actual probability of escalation in a specific scenario is tremendously more complex to model, especially if the model shall be simplified, general and well documented. Some of the aspects upon which such a model will depend are:

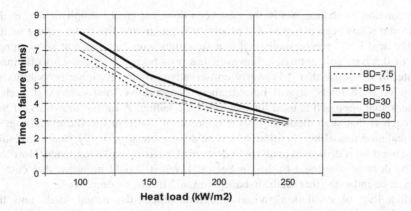

Fig. 15.25 Results from pipe failure study, times according to heat load and blowdown (BD) time

- Fire dimensions in relation to the location of other equipment
- Type of fire
- Duration of fire
- Medium inside pipes
- Effect of active and/or passive fire protection.

15.10.2 Modelling of Explosion Escalation

Explosions, as a possible source of escalation, have come very much into the focus in the recent years, mainly as a result of the so-called 'large scale' tests conducted during 1996/7 [29]. These tests found considerably higher blast loads than those that had been found in smaller scale tests, and thus brought the existing design methods into question. Explosions may lead to escalation in several different ways:

- Global structural collapse
- Rupture of explosion barriers (separating areas or modules)
- Excessive deformation of explosion barriers to the extent that they no longer form functional barriers
- Excessive deformation of decks or walls causing loss of containment in equipment units in other areas
- Excessive deformation of process equipment causing loss of containment in equipment units in other areas
- Damage to safety systems which renders them non-functional, following the explosion.

Escalation modelling has in the past been done extremely simplistically, in the sense that it has been assumed that process equipment and fire water piping would rupture at 0.3 bar overpressure, and that structures would collapse at an overpressure of 0.5 bar. This approach, however, has now been clearly shown to be inadequate, in the sense that it is overly conservative. With the higher probability of extensive blast loads, it will be extremely conservative (and costly), if such a conservative approach is used for escalation modelling. A further drawback of this approach is that such modelling is the opposite of platform specific modelling.

Escalation modelling therefore has to be done more specifically, and this results in the need for a dedicated analysis to determine realistic explosion loads. How this can be done is discussed further in Sect. 9.4, but it should be noted that current experience indicates that analysis based on CFD has to be employed.

Modelling of escalation should therefore reflect the actual loads and the capacities of the platform's structure and equipment. This may be done in either of the following ways:

- Convert the output from the explosion analysis to idealised dynamic loads which may be then used as input to response calculations. (Often a triangular pressure pulse is used.)
- Discretise the output (pressure–time curves) from the explosion calculations into linear sections which may be used as input to structural analysis software.

It is essential that the response calculations are carried out with due attention to the dynamics of the system taking account of both elastic and plastic responses and the effect of large deflections.

15.10.3   Damage Limitation

There are extensive possibilities to limit possible damage and thus limit escalation potential. It will be important that these are reflected as far as possible in the analysis, not the least because then the analysis will be capable of determining the effects of any risk reducing measures that may be considered. Limitation of damage is based upon the use of active and passive systems such as:

- Passive fire protection on structures, walls, decks, piping, and equipment
- Explosion relief systems for reducing explosion overpressure
- Active fire protection systems for cooling and/or fire suppression
- Active explosion protection systems for reduction of overpressure.

Traditionally, passive systems have been considered preferable because they are independent of activation. The main problem for active systems has been the failure to activate them in the case of an accident. There is also a trend that probability reducing measures are to be preferred over consequence reducing measures. There may sometimes be a conflict between these two principles. The focus in the following text is on how to model these systems. More thorough discussion of the possibilities for risk reduction is provided in Sect. 8.5.

15.10.3.1 Passive Fire Protection

There are several software packages to analyse the protective function of passive fire protection. These may be applied to structures as well as equipment. Given an accidental fire load and a protective shielding, the resulting temperature loading on the actual structure or equipment can be calculated with a reasonable degree of precision and assurance.

These calculations will have to be based upon somewhat idealistic conditions and often do not reflect possible mechanical failure of the fire protection material, or ageing of the material. It is considered in spite of these limitations that the accuracy of the predicted results is reasonably good.

15.10.3.2 Active Fire and Explosion Protection

The influence of active fire protection is difficult to model explicitly. It appears that rather limited research has addressed this subject, and the application of active fire protection has mainly been based on standards, regulations and industry accepted guidelines. It is possible to calculate the cooling effect of active fire protection under idealised conditions, but this is rarely done and moreover the effect of using idealised conditions has probably a large effect on the applicability of the results.

The effect of active fire protection in damage limitation is often considered rather simplistically without detailed calculations. The probable effect of this is the introduction of further conservatism in the analysis.

The same considerations also apply to the use of active explosion protection, or suppression, mainly by use of fire water deluge systems. This has recently changed as a result of the large scale test programme, and the explosion simulation CFD codes are now able to simulate the effect of water deluge systems on explosion overpressure.

15.10.3.3 Explosion Relief

Explosion relief by panels and openings in module walls, roof and floor is considered together with the actual load calculations as these two aspects are very strongly interlinked. Modern CFD codes are able to take account of explosion relief measures.

15.10.3.4 Analytical Consideration

The sections above have demonstrated that the methods to analyse accidental loads in a detailed and quantitative fashion are somewhat limited. This is further complicated by the fact that practical circumstances would play an important role in order to differentiate between what can actually happen following an accident and the extent to which damage may be caused. When an analytical capability exists, it is very often coupled with relatively idealistic considerations.

Damage due to projectiles is another aspect where detailed modelling is virtually impossible. Some coarse modelling based empirical data has been attempted, but not detailed modelling on a case-by-case basis.

These are the main reasons why sophisticated analysis of accidental damage is seldom attempted. Actually the situation is to some extent changing, in that the damage following an explosion is now becoming possible to calculate with advanced analytical tools. So far, however, these tools are not as effective as those used for fire loads. Considerable resources however need to be devoted to such studies, if they are to be effective.

15.10.4 Response of Equipment to Fire and Explosion

15.10.4.1 Fire Response Fire Response

The critical part of pipe flanges is the bolts. The critical steel temperature for flanges with ordinary bolts is approximately 450 °C, while the critical temperature for flanges with special bolts is 650 °C (Gowan [52]).

Vessels filled with flammable liquids will absorb heat during a fire. On the 'wet' part of the vessels the absorbed heat heats up and evaporates the liquid. When the fire risk is considered, it is normal to consider the effect of a hydrocarbon pool fire beneath the vessel. The pressure inside the vessel will increase as a result of evaporation of the liquid phase. If the pressure relief system for the vessel has insufficient capacity (the evaporation rate is higher than the relief rate), a BLEVE ('Boiling Liquid Expanding Vapour Explosion') may occur.

There are quite considerable difference between an empty vessel, a vessel filled with gas, and a vessel filled with liquid. In Gowan [48] this is demonstrated by reference to one specific case with 122 kW/m^2 heat load on a pipe, where the following response times (time to temperature of the steel wall reached 600 °C) resulted:

- Pipe (=14', thickness 20 mm) filled with gas: 4 min
- Pipe (=32', thickness 43 mm) filled with gas: 7 min
- Pipe (=32', thickness 43 mm) filled with liquid: 13 min.

With several test series with pool fires as basis, calculation method for the absorbed heat has been developed by API [53] based upon several series of tests with pool fires. This is expressed in the formulas:

$$q = 2.6\,FA^{0.18} \tag{15.7}$$

$$Q = 27.9 \cdot F \cdot A^{0.82} \tag{15.8}$$

where
q = average heat absorbed per m^2 surface of the wet part of vessel, kW/m^2
F = dimensionless factor

F = 1.0 for uninsulated tanks or vessels.
F < 1.0 for insulated tanks and vessels.

A = area of the wet part of the vessel, m^2, and
Q = total absorbed heat by the wet part of the vessel, kW.

This formulation is based on the assumption that the flame from a pool fire will impinge on 55% of the total surface of a spherical tank, 75% of a horizontal cylindrical vessel, and up to 9 m on the sides of a vertical cylindrical tank.

The part of the vessel that is not filled with liquid ('dry') will have a temperature rise in the steel and at high temperatures steel plates may rupture.

Table 15.6 shows the time to rupture of uninsulated steel plates as a function of the tension in the steel plates and the thickness of the plates. The values in the table are calculated based upon an absorbed heat flux of 44 kW/m^2. The steel plates are exposed on one side.

Literature often quotes 540 °C as the critical steel temperature for load-bearing elements based upon the fact that at this temperature the yield stress of steel is approximately half that at ambient temperature [54]. As a guideline 540 °C can be used as the critical steel temperature for process equipment in general. With an absorbed radiation flux of 30 kW/m2, the equilibrium temperature in the steel will after some time (depending on thickness) be 535 °C. The time to reach this equilibrium temperature varies with the thickness of the steel.

Table 15.6 Time in minutes to rupture of uninsulated steel plates exposed to a pool fire [64]

Tension in the steel plates (MPa (N/mm²))	Thickness of the steel plates (mm)	Time to rupture in minutes from start of fire (min)
70	3	5
	13	13
	25	23
140	3	2
	13	8
	25	17

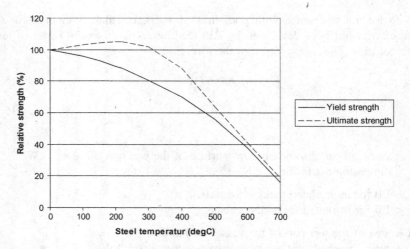

Fig. 15.26 Properties of structural steel at elevated temperatures

Another illustration of the behaviour of steel under fire loading can be found in Fig. 15.26. The diagram shows that reduction of yield strength is quite gradual. It also shows that the ultimate strength (governed by the stress–strain relationship) actually increases up to 250 °C.

Structural response of an entire system may be calculated, using non-linear finite element calculations.

15.10.4.2   Explosion Response

There is actually a considerable amount of data available regarding the response of structures, equipment and humans to explosion overpressure loads. Much of the data regarding the effect of explosions on people comes from work and experience in the military. Structural response may be calculated, using non-linear finite element calculations.

15.10.5   Tolerability Criteria for Personnel

15.10.5.1   Heat Radiation

API RP 521 [51] states a level of 6.3 kW/m^2 as permissible for exposure up to 1 min for personnel with 'appropriate clothing' [51]. For 'emergency actions lasting several minutes' 4.7 kW/m^2 is quoted as the exposure limit and 1.6 kW/m^2 for continuous exposure.

Some tests with voluntary participation of test personnel were conducted in May 2003, in order to determine if the limits based on API were too conservative or not. In general, there was insufficient data in order to conclude that the levels were too conservative; the records indicated though that somewhat longer exposure times could be accepted, without severe burns.

NORSOK Z–013 [30] refers to a document called "Human resistance against thermal effects, explosion effects, toxic effects and obscuration of vision" [55]. This document summarises all the relevant values in order to assess human response to accidental effects in relation to fire and explosion hazards.

15.10.5.2   Oxygen Content in Air

When the oxygen concentration falls from 21 to 14% by volume, respiration and pulse increase. The ability to maintain attention and think clearly is diminished and muscular coordination is somewhat disturbed [56].

15.10.5.3   Carbon Monoxide (CO)

Sax [57] quotes a lowest published 'toxic' limit of 650 ppm for 45 min exposure. Lethal concentrations are generally quoted to be higher.

15.10.5.4   Air Temperature

High air temperatures can be sustained, providing that the humidity is low. In saunas for example, temperatures in the order of 100 °C are commonly used. In desert climates temperatures can reach 50 °C or more in the summer but usually then with low humidity.

The criterion for impairment may be taken as an air temperature exceeding 50 ° C. The criterion applies mainly to TR as short term exposure of higher temperatures may be allowed during escape and evacuation.

15.10.5.5   Smoke

Smoke may hinder escape and evacuation if the visibility is reduced to such an extent that personnel are not able to orientate themselves or see whether the escape way leads to safety or not. Sometime an 'obscuration' factor is used in order to express the limitation of the visibility. The damage criterion could therefore be phrased as follows:

- The safety function is considered to be impaired when the smoke concentration is so high that the end of escape ways and corridors cannot be seen. This is sometimes translated into a minimum distance of sight, say in order of 10 m.

15.10.6 Impairment Criteria for Safety Functions

Impairment criteria are necessary in order to judge when the safety functions are unable to function adequately. The following text discusses the considerations of impairment and the main aspects to be taken into account.

It is worth noting that most of these criteria are 'soft' i.e., they are not coupled with hardware damage nor structural failure, but depend upon the effect of the incident on personnel.

15.10.6.1 Impairment of Escape Ways

The probability of the escape ways being blocked is related to the time it takes for the personnel to evacuate to the TR. It may also be useful to define what constitutes 'blocking' of the escape ways. Normally, there will be three factors which require consideration:

- Structural damage/debris
- High heat loads
- Combustion productions.

The first factor is mainly associated with severe structural impacts (collisions) or the effects of explosions.

In many scenarios the heat load will be the most important factor when evaluating the functioning of escape ways. A limiting value of 20–25 kW/m^2 is normally accepted as the greatest heat load that humans can tolerate for more than a few seconds. Lower values should be used, if exposure for longer periods is considered (see Sect. 15.10.5 above).

Impairment due to combustion products may cause impairment of larger areas. The combustion products from a fire primarily have two effects:

- Reduced visibility due to soot production
- Toxicity, primarily associated with CO and CO_2.

15.10.6.2 Impairment of Temporary Refuge (TR)

The following are the conditions constituting loss of integrity of the TR, as specified by the Health and Safety Executive [58]:

- Loss of structural support.
- Deterioration of life support conditions.
- Loss of communication and command support.
- Unusable evacuation means for those taking shelter in TR.

Impairment of the Shelter Area under Norwegian legislation (corresponds to Temporary Refuge in UK) is usually considered in the same way, except that evacuation is considered separately, not as part of the TR.

All accidental events affecting the TR are evaluated and the probability of 'impairment' of the TR for each event is calculated in the same way as for escape ways. The evaluation should include a study of possible smoke and gas ingress into the living quarters and TR.

The TR must remain habitable until the personnel inside have been safely evacuated. This means that the time the TR must remain intact is longer than the corresponding time for the escape ways leading to the TR.

15.10.6.3 Impairment of Evacuation Systems

The vulnerability of the primary evacuation system is assessed for each accidental event. There is sometimes some confusion about what constitutes the 'primary evacuation means', because companies tend to state that the helicopter is the 'primary means of evacuation'. This may often be true for precautionary evacuation, but is seldom so for emergency evacuation, especially in the event of a gas leak or fire. In these circumstances, the lifeboats must be considered the primary means of evacuation. It is vitally important that there is no confusion about what the main mode of evacuation shall be. Confusion about how to evacuate apparently contributed to the high death tolls in the Piper Alpha disaster in 1988.

The impairment assessment of the primary evacuation system is similar to escape ways. The assessment of impairment probabilities for the lifeboats takes into account factors like explosion damage, extensive heat load, fire on sea etc.

When assessing impairment of lifeboats, there are a number of factors to consider. In some scenarios, the evacuation systems themselves may tolerate the accidental loads they exposed to while the personnel who are going to use the boats are more vulnerable. Impairment of lifeboats is therefore not necessarily limited by the ability of the lifeboat to survive the accidental effects. Effects which must be considered include the following:

- Smoke effects: Toxic effects as well as reduced visibility. Smoke will obviously not affect the lifeboat itself, but personnel may be unable to use it because it is engulfed in heavy smoke, or possibly filled with smoke.
- Thermal effects: GRP lifeboats can tolerate $10\text{--}25 \text{ kW/m}^2$ without being seriously affected or losing integrity. However, if a lifeboat is exposed to high radiation levels in the range $10\text{--}25 \text{ kW/m}^2$, the temperature is likely to rise

relatively rapidly. This means that personnel inside the lifeboat may be exposed to unacceptably high air temperatures within a relatively short time.

The discussion above is primarily related to the situation where the lifeboat is still hanging in the davits on the side of the installation. After it is lowered to the sea, the inbuilt sprinkler system on the boat itself will effectively cool the lifeboat. Higher radiation levels are therefore likely to be sustainable without impairment, unless the heat loads are very high, or the exposure time is very long.

Due to the normally short time it takes to lower the lifeboats, it is considered that high heat loads, probably in excess of 50 kW/m^2, may be tolerable for this period of time. The limiting factor determining whether or not the lifeboats may be used will therefore frequently be the ability of people to enter the lifeboats. In some cases, access to the lifeboats is completely sheltered.

15.10.6.4   Impairment of Main Structure

The effects of high heat loads, explosion overpressure loads and impact loads on the main support structure (or hull structure in the case of a floating installation) have to be considered in relation to the capability of the structure to resist these loads. This topic is discussed in more details in Chap. 7.

15.10.7   Required Intactness Times for Safety Functions

The last aspect to consider in relation to impairment is the time the safety functions need to remain usable. The following aspects are part of a consideration of the required intactness times for the safety functions.

- The mustering time for the installation must be based on the number of personnel present, dimensions, etc., and be compared with the results of drills (if available). 20 min is often used as a typical mustering time (including confirmation of those missing) for emergency situations on large platforms, 10 min is sometimes used for smaller installations.
- The time necessary for search and rescue of missing/wounded persons has to be included in the required intactness times. For large platforms this time is normally in the order of 15–20 min, less for smaller platforms.
- The time required to enter and launch a conventional lifeboat is assessed to be typically around 10 min. In predicting the required intactness time, allowance is normally made for the time necessary to move to another lifeboat and to launch that, in addition to the normal 10 min launching time. Evacuation by several boats may have to be considered for larger platforms. The entire duration is usually considered to take somewhere in the range 10–30 min, depending on the circumstances.

- The time required to carry out a helicopter evacuation is usually not included as an alternative to lifeboat evacuation. Helicopters are often used for precautionary evacuation, when there is ample time available, but not for time critical emergency evacuation. The helicopter evacuation time is dependent on the mobilisation time for helicopter, their seat capacity, the time for a round trip to a suitable offloading location (often another installation), and the number of personnel to be evacuated. Several hours may be needed, if one helicopter is to take care of more than 100 persons.

It may be noted that some of the times are relatively straightforward to calculate, while others (especially the time to search for survivors) may only be subjectively predicted.

Figure 15.27 shows the recorded muster times in exercises [15] on Norwegian installations, as a function of time. Observations for each installation are shown, as well as a trend line. There is a clear relationship between the average POB and the average time needed to complete mustering, including establishing the status of personnel. When a similar correlation exercise was performed in relation to required muster time, there was no visible correlation at all. 'Required muster time' is in this context the muster time requirement that the operator has defined in the emergency management system.

In determining the time requirement for intactness of escape ways the following need to be considered. If the escape ways need to be usable for the time it takes to reach the SA (or TR), and to seek and rescue injured personnel, then the necessary time will be in the range 10–30 min, but up to 60 min for large installations.

The permissible heat loads for the escape ways may however, be based on short exposure periods, from seconds up to 1–2 min. The arguments here are that personnel will try to reach TR as rapidly as possible and thus will only be subjected to high heat loads for short durations. Such an approach however, will not allow time

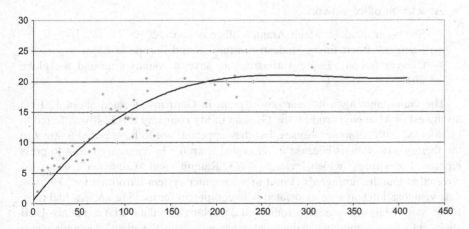

Fig. 15.27 Recorded muster times in exercises on Norwegian installations, as function of average POB

for attending to injured personnel, and survivors who may have to await assistance to reach TR (or SA).

If maximum heat loads are to be based on the presence of an escape way up to 30 min, then only very low heat loads would be permitted. This would lead to extensive protection requirements, which would be impracticable to implement.

The following required intactness times are presented as typical values for the safety functions of a small platform, based on assumptions as presented above (all including time for mustering, search and rescue as well as lifeboat evacuation):

Temporary Refuge: 40 min
Escape Ways: 20 min
Evacuation Means: 40 min
Control Room: 40 min.

15.11   Cyber Risk Modelling

15.11.1   Lessons from Incidents

Some cyber-attacks on the energy sector that are known in the public domain are the following [59]:

- Attacks on technical network

  - Stuxnet: Iran's uranium enrichment facility 2010
  - German steel mill 2014
  - Ukrainian power network 2015
  - German nuclear plant 2016.

- Attacks on office network

  - Shamoon incident: Saudi Aramco office network 2012
  - Energetic Bear: Energy industry in the US and Europe 2012 →
  - Cleaver (recon): Energy infrastructure several countries around the globe 2012.

The Gundremmingen nuclear power plant in Germany, located about 120 km northwest of Munich, is run by the German utility company RWE. It was found to be infected with computer viruses, but they appeared not to have posed a threat to the facility's operations because it is isolated from the Internet, according to press reports. The viruses, which included "W32.Ramnit" and "Conficker", were discovered at Gundremmingen's B unit in a computer system retrofitted in 2008 with data visualisation software associated with equipment for moving nuclear fuel rods, RWE said. Malware was also found on 18 removable data drives, mainly USB sticks, in office computers maintained separately from the plant's operating systems. W32.Ramnit is designed to steal files from infected computers and targets

Microsoft Windows software, according to the security firm Symantec. Conficker has infected millions of Windows computers worldwide since it first came to light in 2008. It is able to spread through networks and by copying itself onto removable data drives, Symantec said.

The 'Energetic Bear' is a Russian virus that lets hackers take control of power plants. Over 1,000 energy firms have been infected, according to media reports. The hackers obtained access to power plant control systems, and could have disrupted energy supplies in affected countries, if they had used the sabotage capabilities open to them, according to the Daily Mail.

Reference [60] refers to a case involving a new build rig heading from South Korea to Brazil in 2010. Malware was allegedly introduced from a worker's laptop and then spread throughout the rig's various networks and control systems, including the blowout preventer computer. The rig was required to shut down for 19 days until cyber security personnel had repaired the networks. Similar reports refer to a drilling rig working offshore West Africa that found itself tilting to one side after being infected with malware.

Another offshore drilling rig operating in the Gulf of Mexico lost control in 2013 of its DP systems forcing it to shut in the well and move off station. "What happened was that various operators on that [mobile offshore drilling unit] were using the very same systems to plug in their smart phones and other devices to access other materials on the Internet, which introduced malware and that resulted in a drive off," said Admiral Paul Zukunft, a US Coast Guard commandant [59].

2017 saw the first attack on industrial control systems with the TRISIS attack in Saudi-Arabia [61], where only a code error prevented the refinery from blowing up.

A.P Møller–Maersk was hit by the so-called NotPetya attack in the summer of 2017, after the company's IT system, including booking applications, was brought down by malware hidden in a document used to file tax returns in Ukraine.

15.11.2   International Standards

One of the central standard series in the cyber security field is the following suite of standards:

- IEC 62443: Security for industrial automation and control systems (IACS)

Please note the abbreviation, as it is often also used for 'International Association of Classification Societies'. The following individual documents belong to this suite of documents:

IEC 62443-1.1:   Terminology, concepts and models
IEC 62443-1.2:   Master glossary of terms and abbreviations
IEC 62443-1.3:   System security compliance metrics
IEC 62443-1.4:   IACS security lifecycle and use-case
IEC 62443-2.1:   Requirements for an IACS security management system

IEC 62443-2.2:   Implementation guidance for an IACS security management system
IEC 62443-2.3:   Path management in the IACS environment
IEC 62443-2.4:   Security program requirements for IACS service providers
IEC 62443-3.1:   Security technologies for IACS
IEC 62443-3.2:   Security levels for zones and conduits
IEC 62443-3.3:   System security requirements and security levels
IEC 62443-4.1:   Product development requirements
IEC 62443-4.2:   Technical security requirements for IACS components.

IEC 62443–2–4:2017 specifies requirements for security capabilities for IACS service providers that they can offer to the asset owner during integration and maintenance activities of an automation solution, consisting of basic process control system, safety instrumented system and complementary hardware and software. These security capabilities are policy, procedure, practice and personnel related.

Other standards, such as IEC 61508 [35] for electronic and programmable safety-related instrumented systems are also relevant to some extent.

15.11.3   Approach to Cyber-Security Risk Assessment

It should be recognized that a cyber-attack can come in many ways, and could for instance come through an administrative system (as in the Saudi Aramco attack in 2012, see Sect. 15.1.1), where there are vulnerabilities, as in the Møller–Maersk attack in 2017. One of the recent trends [61] is that hackers appear to be becoming very dedicated and can spend several months just preparing for one single attack.

If a hacker gets into the inside of the network in an oil company, he or she could get access to the control of DP systems and ballast systems on floating installations. The hacker could also take control of fire and gas detection, active fire-fighting systems, process control systems, blowdown or power generation. All of these functions may be used to initiate an event sequence with catastrophic outcomes. Especially the positioning and stability systems on floating installations are very vulnerable. But also the topside systems could be used, when we keep in mind that one single attack may influence several of the systems simultaneously.

Section 2.4 draws attention to uncertainties in risk assessment, where 'unknown unknowns' are the most critical phenomena. Cyber-attacks may be one such illustration of possible unknown scenarios, especially if several barrier elements are attacked simultaneously. This aspect is a good example of an area where limitations in human perception and imagination will influence our ability to foresee what may occur and what the challenges are.

One of the challenges of many of the safety systems is that the software solutions were developed twenty years ago or more, when there was no focus on ways to

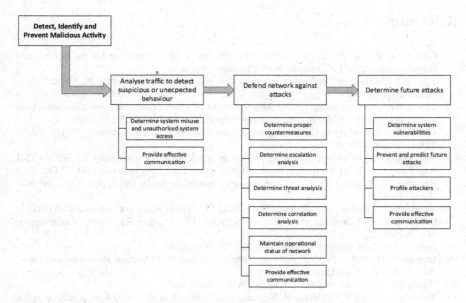

Fig. 15.28 Goal tree for a cyber operator (adapted from Ref. [63])

protect the systems. Section 15.1.5 introduces the integration of cyber-checks into for instance HAZOP studies, where all the safeguards could be checked for vulnerability to cyber-attacks. These systems did not become critical until it was allowed that that systems could be controlled from outside the facilities over the network. If it was still been allowed only to send out status information to users outside the facility, then there would not be any vulnerabilities.

Cyber-attacks may also be embedded in the safety systems as latent threats, in the sense that safety systems could be prevented from fulfilling their intended function if a process incident occurs. If hacking were to has prevent active fire fighting from being launched in the case of fire detection, then the worst case outcome of such failure to provide fire water for cooling and extinction has already been demonstrated by the Piper Alpha accident (see Sect. 4.15).

The dedicated attacks have been mentioned above. Another attack that would be different would be a so-called drive-by attack, which would be less well prepared, but could be very critical to single systems, such as the DP system.

Kott et al. [62] have focused on situational awareness in the cyber defence (see Fig. 15.28).

Section 19.8 discusses some approaches to design against cyber-attacks.

References

1. Crawley F, Preston M, Tyler B (2000) HAZOP: guide to best practice for the process and chemical industries. UK, Institution of Chemical Engineers
2. Lees FP (2004) Lees' loss prevention in the process industries, 3rd edn. Butterworth–Heinemann, Oxford
3. Lloyds Register Consulting (2004) An assessment of safety, risks and costs associated with subsea pipeline disposals, Kjeller, Norway. Report No.: 32.701.001/R1
4. CCPS (2018) Bow ties in risk management: a concept book for process safety. Wiley. ISBN: 978-1-119-49039-5
5. McGlone J (2015) Inherent safety against cyber attack for process facilities. In: SPE/IATMI Asia Pacific oil & gas conference and exhibition, Nusa Dua, Bali, Indonesia, 20–22 Oct 2015
6. Rausand M, Haugen S (2013) Risk assessment: theory, methods, and applications, 2nd edn. Wiley, New York
7. Vesely WE, Goldberg FF, Roberts NM, Haasl DF (1981) Fault tree handbook (NUREG–0492). Office of Nuclear Regulatory Research, U.S. Nuclear Regulatory Commission, Washington DC
8. Aven T (1992) Reliability and risk analysis. Elsevier, London
9. Stamatis DH (1995) Failure mode and effect analysis: FMEA from theory to execution. American Society for Quality, Milwaukee
10. Ripley BD (1987) Stochastic simulation. Wiley, New York
11. Vinnem JE, Aven T, Hundseid H, Vassmyr KA, Vollen F et al (2003) Risk assessments for offshore installations in the operational phase. In: European safety and reliability conference 2003, Maastricht, The Netherlands
12. Vinnem JE, Hauge S, Seljelid J, Aven T (2003) Operational risk analysis—total analysis of physical and non–physical barriers. Preventor, Bryne, Norway, Preventor Report 200254–03, 16 Oct 2003
13. Vinnem JE, Aven T, Hauge S, Seljelid J, Veire G (2004) Integrated barrier analysis in operational risk assessment in offshore petroleum operations. In: Proceedings of the international conference on probabilistic safety assessment and management PSAM7 and European safety and reliability conference, Berlin, Germany, 14–18 June 2004
14. Aven T, Sklet S, Vinnem JE (2006) Barrier and operational risk analysis of hydrocarbon releases (BORA–release). Part I, method description, J Hazard Mater A137:681–691
15. Thomassen O, Sørum M (2002) Mapping and monitoring the safety level: In: SPE international conference on health, safety and environment in oil and gas exploration and production, Kuala Lumpur, Malaysia. SPE paper 73923, 20–22 Mar 2002
16. Tinmannsvik RK, Sklet S, Jersin E (2005) Investigation methodology: man, technology, organisation (In Norwegian only). SINTEF. Report No.: STF38 A04422, Oct 2005. http://www.ptil.no/NR/rdonlyres/63D64078–11AA–4AC7–931F–A6A6CB790573/0/UlykkesgranskingSTF38A04422.pdf
17. PSA (2006) Trends in risk levels, main report 2017 (in Norwegian only). Petroleum Safety Authority, 26 Apr 2018
18. Zhen X, Vinnem JE, Penga C, Huang Y (2018) Quantitative risk modelling of maintenance work on major offshore process equipment. Loss Prev Process Ind 56:430–443
19. Jensen FV (2001) Bayesian networks and decision graphs. Springer, London
20. Pearl J (2001) Probabilistic reasoning in intelligent systems: networks of plausible inference. Morgan Kaufman, San Mateo
21. Mosleh A, Dias A, Eghbali G, Fazen K (2004) An integrated framework for identification, classification and assessment of aviation systems hazards. In: Proceedings of the international conference on probabilistic safety assessment and management PSAM7 and European safety and reliability conference, Berlin, Germany, 14–18 June 2004
22. Røed W, Mosleh A, Vinnem JE, Aven T (2009) On the use of hybrid causal logic method in offshore risk analysis. Reliab Eng Syst Saf 94(2):445–455

23. Kongsvik T, Johnsen SÅ, Sklet S (2011) Safety climate and hydrocarbon leaks: an empirical contribution to the leading-lagging indicator discussion. J Loss Prev Process Ind 24:405–411
24. Vinnem JE, Veire G, Heide B, Aven T (2004) A method for developing and structuring risk activity indicators for major accidents, presented at PSAM7, Berlin, 14–18 June, 2004
25. Sklet S, Ringstad AJ, Steen SA, Tronstad L, Haugen S, Seljelid J, Kongsvik T, Wærø I (2010) Monitoring of human and organizational factors influencing risk of major accidents. In: SPE international conference on health, safety and environment in oil and gas exploration and production, Rio de Janeiro, Brazil, 12–14 April 2010
26. Øien K (2001) Risk indicators as a tool for risk control. Reliab Eng Syst Saf (RESS) 74 (2):147–167
27. Vinnem JE et al (2012) Risk modelling of maintenance work on major process equipment on offshore petroleum installations. Loss Prev Process Ind 25(2):274–292
28. Gran BA et al (2012) Evaluation of the risk model of maintenance work on major process equipment on offshore petroleum installations. Loss Prev Process Ind 25(3):582–593
29. Vinnem JE (2013) On the development of failure models for hydrocarbon leaks during maintenance work in process plants on offshore petroleum installations. Reliab Eng Syst Saf 113
30. Nielsen DS (1976) The cause consequence diagram as a basis for quantitative accident analysis. RISØ National Laboratory, Denmark. Report No.: M–1374
31. SCI (1998) Blast and fire engineering for topside systems, phase 2. Ascot. SCI. Report No.: 253
32. Standard Norge (2010) Z-013 risk and emergency preparedness assessment (Rev. 3, Oct. 2010)
33. Lloyds Register Consulting (2006) Riskspectrum® software. http://www.riskspectrum.com/
34. Bäckström O (2003) Pilot project fault tree analysis for Statfjord A (in Swedish only). Stockholm; Lloyds Register Consulting. Report No.: 99161–R–005, May 2003
35. IEC (2010) Functional safety of electrical/electronic/programmable electronic safety-related systems. Part 1: general requirements. IEC 61508:2010
36. Norwegian Oil and Gas (2018) 070 guidelines for the application of IEC 61508 and IEC 61511 in the petroleum activities on the continental shelf (Recommended SIL requirements), 26 June 2018
37. IEC (2016) Functional safety—safety instrumented systems for the process industry sector. Part 1: framework, definitions, system, hardware and software requirements, IEC61511-1
38. Morris MI, Miles A, Cooper JPS (1994) Quantification of escalation effects in offshore quantitative risk assessment. J Loss Prev Process Ind 7(4):337–344
39. Jones JC, Irvine P (1997) PLATO software for offshore risk assessment: a critique of the combustion features incorporated. J Loss Prev Process Ind 10(4):259–264
40. Skogdalen JE, Vinnem JE (2012) Combining precursor incidents investigations and QRA in oil and gas industry. Reliab Eng Syst Saf 101:48–58
41. Skogdalen JE, Vinnem JE (2011) Quantitative risk analysis offshore-human and organizational factors. Reliab Eng Syst Saf 96:468–479
42. PSA (2005) Investigation of the anchor line failures on ocean vanguard, 14 Dec 2004, well 6406/1–3 (In Norwegian only) PSA, 23 May 2005. Petroleum Safety Authority, Stavanger. http://www.ptil.no/NR/rdonlyres/83A74F56–7F2D–470C–9A36–1153AADE50A7/7950/ovgrrappkomprimertny.pdf
43. DNV GL (1996). JIP ignition modelling, time dependent ignition probability model, DNV report 96–3629, Rev. 04
44. Lloyds Register Consulting (2006) Ignition modelling in risk analysis. Lloyds Register Consulting, Kjeller, Norway. Report No.: 27.390.033/R1
45. PSA (2012). Trends in risk level on the Norwegian continental shelf, main report (in Norwegian only, English summery report). Petroleum Safety Authority, Stavanger, 25 Apr 2012
46. Lloyd's Register Consulting (2018) Modelling of ignition sources on offshore oil and gas facilities—MISOF(2). Kjeller, Norway. Report No.: 107566/R2

47. Cox AW, Lees FP, Ang ML (1991) Classification of hazardous locations. Institution of Chemical Engineers, UK
48. Vinnem JE, Hauge S (1999) Operational safety of FPSOs, MP3; riser failure due to inadequate response to rapid wind change. NTNU, Trondheim
49. Vinnem JE, Pedersen JI, Rosenthal P (1996) Efficient risk management: use of computerized QRA model for safety improvements to an existing installation. In: Presented at SPE 3rd international conference on health, safety and environment, New Orleans, June 1996
50. IOGP (2010). Vulnerability of plant/structure, iogp risk assessment data directory. Report No. 434–15, IOGP, Mar 2010
51. Mendonos S (2003) Improvement of rule sets for quantitative risk assessment in various industrial sectors, safety and reliability. In: Proceedings of ESREL 2003, vol 2. Balkema Publishers
52. Gowan RG (1978) Developments in fire protection of offshore platforms. Applied Science Publisher Ltd., London
53. API (1976). Recommended practice for the design and installation of pressure–relieving systems in refineries. Part 1—Design, API recommended practice 520. American Petroleum Institute, Washington
54. American Iron and Steel Institute (1979). Fire-safe structural steel. A Design Guide, Washington
55. DNV GL/ Lloyds Register Consulting (2001). Human resistance against thermal effects, explosion effects, toxic effects and obscuration of vision. (http://www.preventor.no/tol_lim.pdf)
56. Henderson Y, Haggard HW (1943) Noxious gases, 2nd edn. Reinhold Publishing Co., New York
57. Sax NI (1984) Dangerous properties of industrial materials, 6th edn. Van Nostrand Reinhold Co., New York
58. HSE (2018) Offshore statistics & regulatory activity, report 2017, August 2018. http://www.hse.gov.uk/offshore/statistics.htm
59. Statoil (2016) Statoil's global risk management including IT security, lecture by Monica Solem at NTNU Marin Technology Dept, October 2016
60. Jacobs T (2016) Industrial-sized cyber attacks threaten the upstream sector. J Petrol Technol
61. Slowik J (2018) The rising tide threats to industrial control systems. Infowarcon. (https://infowarcon.com/wp-content/uploads/2018/11/Slowik-Rising-Tide-Threats-to-ICS.pdf)
62. Kott A, Wang C, Erbacher RF (2014) Cyber defense and situational awareness. Springer
63. Connors E, Jones RET, Endsley MR (2010) A comprehensive study of requirements and metrics for cyber defense situation awareness. SA Technologies Inc, Marietta, GA
64. API (1997) Guide for pressure relieving and depressuring systems, RP 521. American Petroleum Institute, Washington DC

Chapter 16
Presentation of Risk Results from QRA Studies

16.1 Requirements for Risk Presentation

16.1.1 Regulatory Requirements

The Norwegian legislation has the most extensive requirements for presentation of risk results, at least if worldwide legislation for offshore operations is considered. The regulations for management of health, safety and environment [1], has two sections (16 and 17) with requirements to analysis of major hazard risk and presentation of results. The text of these two sections was presented in Sect. 1.5.2. Some of the main requirements are as follows:

- Balanced and comprehensive presentation of the analysis and results
- Suitable for the 'target groups', i.e. stake holders
- Present results relating to sensitivity and uncertainty.

In addition to the above, PSA has issued a memorandum in 2018 with the title "Integrated and unified risk management in the petroleum industry" [2]. Although this publication is not legally binding, it gives some perspectives on how uncertainty should be handled in the petroleum industry. As emphasized in the document, *"Uncertainty is a key component of the risk concept. Taking account of uncertainty when choosing solutions and measures is therefore a regulatory requirement"*. Taking account of uncertainty means clarifying the strength of knowledge. Since risk assessments based upon strong knowledge should have larger impact on risk management than risk assessments based on weak knowledge, it is essential to include assessments of knowledge strength in risk assessment. From the authors' point of view, this is not done in a proper way in QRA seen in the industry today.

© Springer-Verlag London Ltd., part of Springer Nature 2020
J.-E. Vinnem and W. Røed, *Offshore Risk Assessment Vol. 2*, Springer Series in Reliability Engineering, https://doi.org/10.1007/978-1-4471-7448-6_16

16.1.2   NORSOK Requirements

NORSOK Z–013 [3], has a Chap. 5 with general requirements for a risk assessment process, with a Sect. 5.6; Establishing the risk picture, and a special Sect. 5.6.3 for quantitative risk analysis. The requirements of this subsection for presentation of risk picture and uncertainty are:

5.6.3.3 Presentation of the risk picture

For the presentation of the risk picture, the following requirements apply (if included in scope):

a) the main results and conclusions of any risk analysis shall be presented as risk for the activity in question, in accordance with the structure of the RAC and for the relevant risk elements. The risk picture shall include

    1) ranking of risk contributors,
    2) identification of potential risk reducing measures,
    3) important operational assumptions/measures in order to control risk.

b) if required, the presentation of risk picture shall include dimensioning accidental loads;
c) presentation of possible measures that may be used for reduction of risk and their risk reducing effect;
d) the analysis shall present and describe accident scenarios relevant for the assessment of the emergency preparedness (see 9.4);
e) presentation of the sensitivity in the results with respect to variations in input data and crucial premises. The basis for the chosen sensitivity analyses shall be presented;
f) the results of the QRA shall be traceable through the analysis report. It shall be possible to identify any mechanism/equipment that causes large risk contribution;
g) intermediate results shall be presented such that risk contributors can be traced through the report;
h) assumptions and premises of importance to the risk assessment results, to decisions related to future project development or with implications to operations/maintenance shall be documented;
i) assumptions, premises and results shall be presented in a way suitable as input for defining performance requirements for safety and emergency preparedness measures in later life cycle phases;
j) assumptions, premises and results for environmental risk shall be presented in a way suitable as input for the environmental preparedness and response analysis;
k) all recommendations made in the analysis shall be listed separately with references to calculations.

5.6.3.4 Sensitivity analysis

The following requirements apply:

a) sensitivity analyses shall be carried out to include

    1) identification of the most important aspects and assumptions/parameters in the analysis,
    2) evaluation of effects of changes in the assumptions/parameters, including the effect of any excessively conservative assumptions,
    3) evaluation of effects of potential risk reducing measures.

b) the input parameters to be considered for sensitivity analyses should, if relevant, include

1) total manning and personnel distribution,
2) leak frequencies,
3) probability of ignition,
4) performance (reliability, availability, functionality, etc.) of important barrier functions, systems and/or elements (technical, human and organisational) for personnel, environment and asset risk,
5) operational parameters, such as the activity levels,
6) environmental resources and their vulnerability,
7) spreading of contaminant.

16.1.3   Risk Result Presentation and Risk Tolerance Criteria

One important misunderstanding that may be clarified right away, is the following: Presentation of risk results should not be limited to the parameters for which risk tolerance criteria may be expressed.

Too often, if the risk tolerance criterion is a FAR value, a diagram like Fig. 16.1 is presented as the illustration of the results in relation to the FAR limit, without any further illustrations of the results at all.

That implies that possibly the several hundred manhours that went into modelling of consequences of fire, explosion and impacts are not utilised properly, they are only used to produce quantitative results, but all the insights into what are the likely scenarios and consequences are not used at all. This should be avoided. One way to avoid it is presented in this chapter.

16.1.4   Proposed Presentation Format

The requirements in Sect. 16.1.1 regarding the contents of the presentation may be summarised as follows:

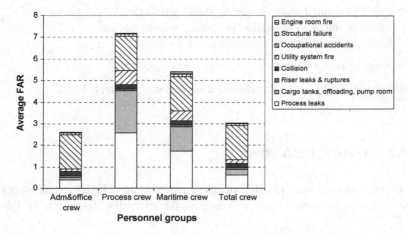

Fig. 16.1 Presentation of average FAR values for different crew groups and total crew

- Balanced and comprehensive risk picture
- Sensitivity calculations
- Evaluation of uncertainties/strength of knowledge.

These requirements are expressed in a functional manner, which leave a lot open for interpretation and evaluation by the users. It may be argued that a 'balanced and comprehensive picture' has very seldom, if at all, been presented in a QRA study report, at least in the experience of the present authors.

It should be emphasised that the objective with risk presentation should always be to provide a basis for decision-making about risk reducing measures.

Previous versions of the PSA regulations have been specifically requiring that risk due to personnel transport shall be included in the risk analysis, but this has been omitted in the latest years. In the presentation below, only fatalities on the installation are included, not the contributions from personnel transport. It may be argued this is the most logical, as the operators of offshore installations do not have any direct influence on the risk level during helicopter transport of personnel.

16.2 Presentation of Risk According to Application Area

16.2.1 Life Cycle Phases

The requirements regarding the presentation of risk results have to reflect the actual area of application of the risk analysis. This implies that the presentations used in different life cycle phases need to be tailored to the actual phase. A brief summary of phase specific requirements could be as follows:

- Design phase:
  - Special emphasis on dimensioning accidental loads (see Chap. 19)
- Operations phase:
  - Effect of limited changes during stable production
  - Effect of major modification proposals, when relevant
  - Effect of simultaneous operations, when relevant
- Decommissioning phase:
  - Special emphasis on abandonment operations (see Sect. 12.8).

16.2.2 ALARP Evaluations

One of the important uses of risk analysis results is in the ALARP evaluations i.e., in order to demonstrate effects of changes. An extensive discussion of this is however outside the scope of this book; this is discussed in some depth in Aven and Vinnem [4].

It could nevertheless be emphasised in this connection that the risk result presentation has to give an extensive input to ALARP evaluations.

16.2.3 Risk Presentation for Different User Groups

The presentation of risk results also needs to be tailored to the different user groups, with respect to the need they have for information about risk and their qualifications. The following user groups may be considered:

- Crew members in operations and maintenance crews
- Installation management, onshore staff
- Planning personnel involved in planning of well operations, maintenance and modifications in the process areas
- HES specialists.

The presentation to crew members should be such that the conditions for easy understanding are as good as possible; a suggestion is presented in Sect. 16.8. The remainder of this chapter may be considered as tailored to management needs, and to some extent also for HES specialists. The latter group usually also needs much more detailed presentations, often provided in appendices to the main report.

Chapter 21 also documents requirements for presentation of risk picture for planning purposes, relating to maintenance, inspection and modifications in the process areas.

16.2.4 Framework for Risk Presentations

The framework for the risk presentations in the remainder of this chapter is a more generic presentation of risk, where no particular life cycle phase is focused upon.

16.3 Presentation of Overall Risk

16.3.1 Main Results

The presentation of overall risk needs to focus on the overall risk level and the risk levels for particular groups and areas on the installation. The expected values, such as IR or FAR (see definitions in Sect. 2.1.4) should be stated, as well as the distribution of fatalities, such as presentation of an f–N curve (or similar). Both are shown in the following.

It should be emphasised that several risk parameters should be used in parallel, for instance PLL, IR, FAR. Different arrangements may have different effects. For instance, if the crew size is reduced, this will reduce the PLL value, but usually not the IR or FAR values.

Figure 16.1 presents the overall risk results for the FPSO concept, as an average for the total crew, as well as for the following crew groups:

- Process crew
- Maritime crew
- Admin and office crew.

Figure 16.1 also presents the contributions from the different hazard types, such as process leaks, cargo tanks, etc. In a real case, the contributions from the different hazards should be discussed thoroughly.

Figure 16.2 presents the f–N curve for the concept, including contributions from immediate fatality risk and occupational accidents, as well as escape fatalities.

The difference between the curves for 'escape' and 'all fatalities' in Fig. 16.2 corresponds to the contribution from fatalities during evacuation and rescue. In a real case, the contributions from the different hazards should be discussed, but this is omitted here, due to hypothetical data.

Figure 16.2 shows that the calculated frequency of fatal accidents is 0.01 per year, whereas the frequency of accidents with at least 10 fatalities is around 0.0001 per year. The diagram further shows that fatalities during evacuation and pick-up are the dominating contributions for accidents with three or more fatalities.

Fig. 16.2 Presentation of f–N curves for immediate and occupational accidents, escape fatalities and overall fatalities

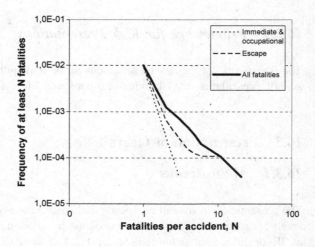

16.3.2   References for Risk Results

Reference values for risk results are useful in order to illustrate what the results imply in practice, for easier perception by non-experts. There are at least two ways this may be done; by comparison with similar installations or with other activities in the society.

One important aspect of presentation of overall risk may be to compare the risk level with other similar concepts. Figure 16.3 presents an example where one FPSO concept is compared with similar concepts.

The so-called 'current FPSO' was somewhat special, in the sense that it had several unconventional features. The comparison with other FPSOs, mainly with conventional solutions, was therefore particularly important. In a real case, the details of the comparison should be discussed, but this is omitted here.

Risk levels are sometimes compared to that of driving cars, motorcycles, mountain-climbing, etc. Preferably, these comparisons should be made relative to industries that are comparable, such as onshore petroleum facilities, chemical industry, mining, etc.

One simple way to illustrate what the FAR level implies is the following: The FAR value implies the number of fatalities in 100 million manhours. The total number of manhours on Norwegian production installations and mobile units for the period 2014–17, was approximately 106 million manhours i.e., around 100 million manhours for practical purposes. Before 2014, 100 million manhours was typically performed in three years; the period 2011–2013 had app. 96 million manhours. This implies that the experienced FAR value for all practical purposes may be taken as the number of fatalities that has occurred on the Norwegian

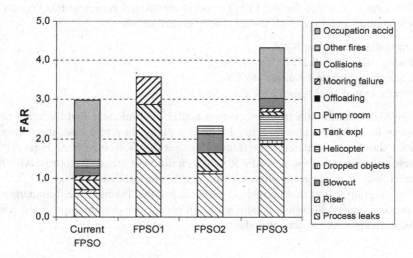

Fig. 16.3 Comparison of actual concept with alternative FPSO concepts

Continental shelf on production installations during a three/four year period. The overview in Appendix B of the number of fatalities on production installations during some four 3–year periods is as follows:

- 2001–03: 1 fatality
- 2004–06: 0 fatalities
- 2007–09: 1 fatality
- 2010–12: 0 fatalities
- 2013–15: 2 fatalities
- 2016–18: 1 fatality.

In this list, the Turøy accident in 2016 with 13 fatalities has been excluded since it was during transport. All the other fatalities are due to occupational accidents, no fatality in a major accident (see description in Sect. 2.1.3) on the installations, excluding transport, has occurred since 1985.

16.4 Presentation of Risk Contributions

16.4.1 FAR Contributions

The presentation of risk contributions is perhaps one of the areas where the most substantial improvements are required. Very often, the presentation of contributions is limited to what is shown in Fig. 16.1. There are usually many more perspectives which are valuable to present, in order to obtain a good insight into the important risk parameters, as a basis for identification and evaluation of potential risk reducing measures. Several different illustrations are presented below.

The total PLL value for the FPSO concept mentioned earlier in this chapter is 0.015 fatalities per year, with the following contributions:

- Immediate fatalities: 8.4%
- Escape fatalities: 6.2%
- Evacuation/rescue fatalities: 33.4%
- Occupational fatalities: 52.1%.

The high contribution from evacuation and rescue indicates that there are possibilities for risk reduction associated with this phase. In a real case, these contributions should be discussed more thoroughly, but this is omitted here. Figure 16.4 presents an overview of average FAR values in different areas and rooms. Also the contributions from different phases of the accident are shown.

The diagram shows that immediate and occupational fatalities are dominating in the process areas and turret. Fatalities during escape are dominating for the areas furthest away from the accommodation.

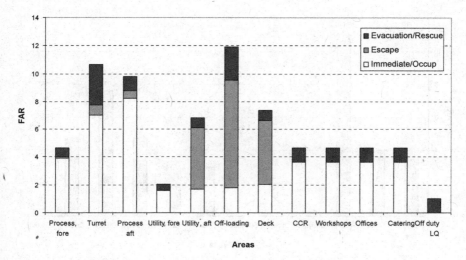

Fig. 16.4 FAR values per area, averaged over day and night shifts

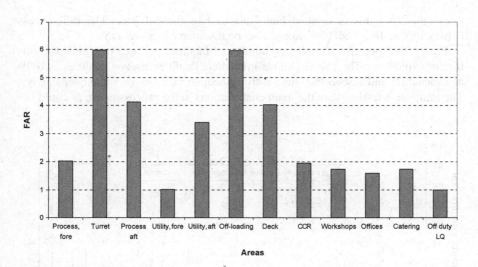

Fig. 16.5 FAR values per area, day shift only

Rarely are there distinctions made between day shift and night shift, in spite of manning levels often being very different. Figure 16.5 shows the FAR levels during day shift, whereas Fig. 16.6 presents the FAR levels during night shift.

The contributions from the different types of accidents i.e., process leaks, riser leaks, fires in cargo tanks, engine room fire, collisions and structural failure were presented in Fig. 16.1. In a real case, these contributions should be discussed more

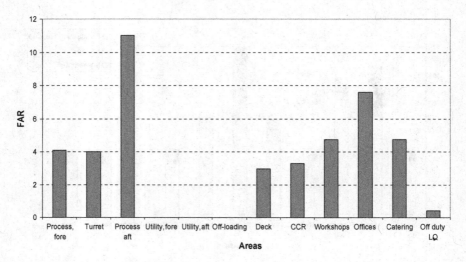

Fig. 16.6 FAR values per area, night shift only

thoroughly, but this is omitted here, due to hypothetical data. The differences between Figs. 16.5 and 16.6 should also be discussed in a real case.

Figure 16.7 presents the average number of fatalities per area per ignited event (fire or explosion). The values include immediate fatalities, escape fatalities as well as evacuation and rescue fatalities, with immediate fatalities presented separately. The immediate fatalities are the main contributions in the process areas, as could be

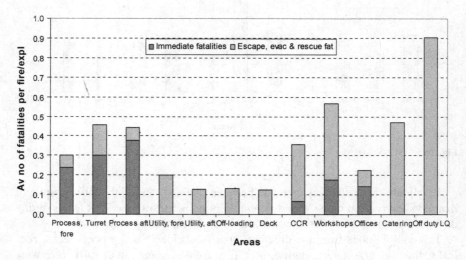

Fig. 16.7 Average number of fatalities per area per fire/explosion

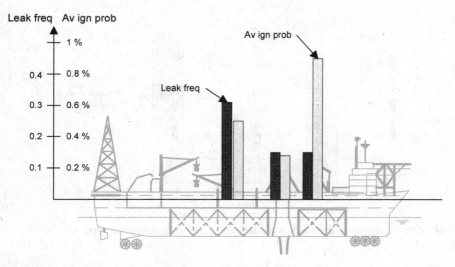

Fig. 16.8 Leak frequency and average ignition probability for process areas

expected. The basis for the diagram is fatalities due to fire and explosion only, other fatalities, such as occupational fatalities are not included. Some areas do not have significant contributions from immediate fatalities.

16.4.2 Contributions for Leak Frequencies

Figure 16.8 presents an illustration of the total leak frequency and average ignition probability for each process area. In a real case, these contributions should be discussed more thoroughly.

Figure 16.9 presents the contributions to overall leak frequencies in the main areas, from technical faults and leaks due to human intervention. In a real case, these contributions should be discussed more thoroughly. Such results are usually not presented, but should be included in the future.

16.4.3 Fire and Explosion Characteristics

Figure 16.10 presents the duration of controlled and uncontrolled fire in different areas on the installation. The following interpretations have been used: A controlled fire is a fire in which all safety systems perform their functions as barrier systems,

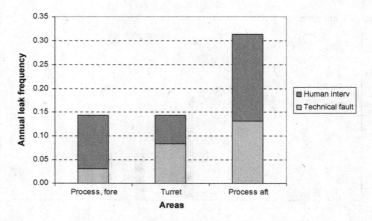

Fig. 16.9 Contributions from technical faults and leaks during human interventions

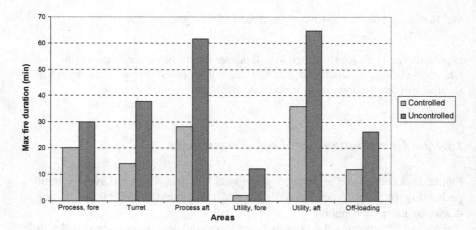

Fig. 16.10 Fire durations in different areas for controlled and uncontrolled fires

except that active fire protection is disregarded. The durations shown are the maximum durations that may occur in the area with a leak rate of 1 kg/s.

The uncontrolled fire is a fire in which depressurisation and/or drain systems are not functional. Likewise, the durations shown are the maximum durations that may occur in the area with a leak rate of 1 kg/s.

Figure 16.11 presents a summary of the explosion load results, indicating the frequency of loads on the blast walls which imply loads that exceed the design basis.

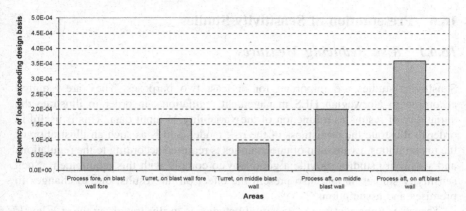

Fig. 16.11 Frequency of explosion exceeding design basis

16.5   Presentation of Significant Improvements

One special kind of sensitivity studies is to present the changes that are required in order to reduce the risk level significantly, say at least by 50%. Figure 16.12 presents overall risk results that illustrate the effect of the following significant improvements:

- Escape tunnel alongside entire length of the vessel
- 'RRM2'.

Only one improvement has been named, whereas the other improvement shown is a hypothetical case denoted 'RRM2'. In a real case, these contributions should be discussed more thoroughly.

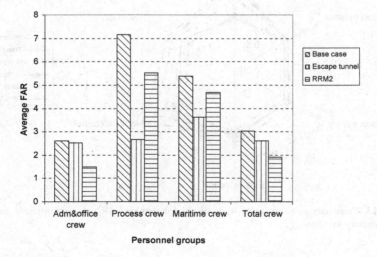

Fig. 16.12 Influence on overall risk results from significant improvements

16.6   Presentation of Sensitivity Studies

16.6.1   *Risk Reducing Measures*

Sensitivity studies are important for at least two purposes. They are required according to Norwegian HES management regulations, in order to illustrate the variability of results as a function of the variability of input data. This is useful in order to illustrate the robustness of the risk model, and is as such an illustration of the uncertainties. Evaluation of uncertainties is required according to the regulations and sensitivity studies provide a good way to comply with this requirement. Such sensitivity studies should also present the effect on the results, due to changes in premises and assumptions.

The other important use of sensitivity studies is to illustrate how the risk levels may change if improvements are introduced with respect to safety systems and/or emergency preparedness actions and plans.

Figure 16.13 presents one way to illustrate sensitivity study results, for different areas and essential safety systems. The variation in input data is an increase of the unavailability of the safety systems with a factor of 2.0. The basis for Fig. 16.13 is average FAR values on the installation, covering the following safety systems:

- Gas detection
- Isolation (including blowdown)
- Ignition control
- Fire detection

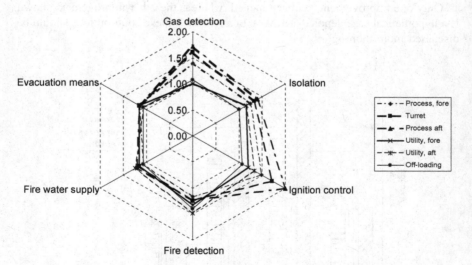

Fig. 16.13 Presentation of sensitivity study results for average FAR values in different areas for essential safety systems

- Fire water
- Evacuation system.

There are six different sensitivities presented in these diagrams, as mentioned above. The effects are shown for six different areas on the installation for each of the sensitivities. For example, for the process 'aft' module, a reduction of the gas detection unavailability by a factor of 5, implies that the frequency of uncontrolled fire is reduced by a factor of 1.75, according to Fig. 16.13. With a similar reduction in the frequency of failure of ignition control, the reduction of the frequency of uncontrolled fire in the 'process aft' area is 2.0, whereas the reduction of the frequency of uncontrolled fire in the off-loading area is the lowest, around 1.1.

The values in Fig. 16.13 are hypothetical, but the general tendencies are as would be expected, for instance that variations in the availability of ignition control have the greatest effect, whereas the effect of changes to some of the other systems have relatively low effect. In a real case, these contributions should be discussed more thoroughly.

Figure 16.14 present sensitivity study results for the frequency of uncontrolled fires, for an increase of the unavailability of the safety systems with a factor of 4.0. It is shown that there is generally a higher influence of most of the safety systems when uncontrolled fires are considered, compared to when fatality risk is used.

In a real case, the contributions in Fig. 16.14 should be discussed thoroughly, but this is omitted here, due to hypothetical data.

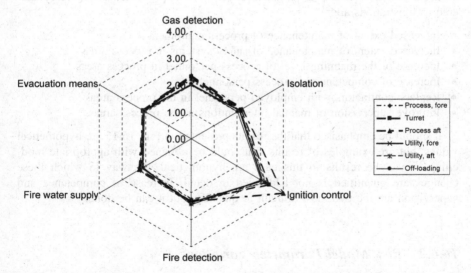

Fig. 16.14 Presentation of sensitivity study results for the frequency of uncontrolled fire in different areas for essential safety systems

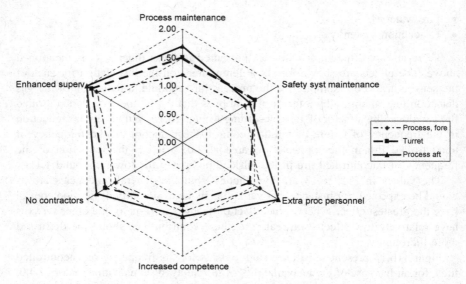

Fig. 16.15 Presentation of sensitivity study results for total leak frequency in different areas for different operational improvements

Figure 16.15 presents sensitivity study results for operational improvements in an MTO-perspective, and the assumed effects. The improvements that are considered as illustrations are:

- Increased extent of maintenance of process equipment
- Increased extent of maintenance of safety systems in process areas
- Increase of the manning level of process personnel in process areas
- Increase of competence of process personnel
- Replace contractors with employed personnel in the process areas
- Improve supervision of manual interventions in the process areas.

It should be emphasised that the results presented in Fig. 16.15 are hypothetical values, but are examples of results that could be obtained with appropriate modelling. Since the results in this case are hypothetical, the way in which these changes are quantified is not addressed in detail. Increase of competence and supervision are the two most difficult to quantify, but it can be done.

16.6.2   Risk Model Parameter Variations

The last illustration is different; it presents the variability of leak frequency (maximum and minimum values) according to variations in the parameters of the risk model, say with a factor of 5 up and down. The hypothetical results are shown

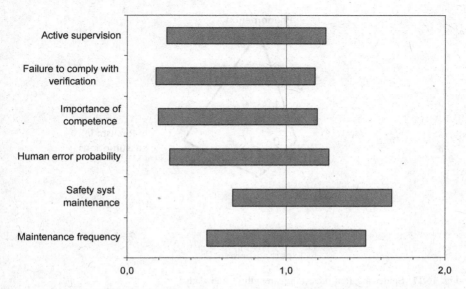

Fig. 16.16 Presentation of variability of study results according to parameter variations in the risk model

in Fig. 16.16. The diagram shows the relative variations around 1.0, for separate variations of each parameter. Similar presentations could be made for other risk results.

16.7 Evaluation of Uncertainty

Evaluation of uncertainty is a required aspect according to the regulatory requirements presented from the management regulations, and as discussed by PSA (2018) (see Sect. 1.5.2). One practical way to evaluate uncertainty is to describe and categorize the strength of knowledge qualitatively as elaborated on in Chap. 2. Several suggestions have been made the last years on how strength of knowledge evaluations can be included in risk result presentations. One example is provided in Fig. 16.17, where a spider diagram is used to present four dimensions relevant for strength of knowledge evaluations, as suggested by Flage and Aven [5]:

- justification of assumptions,
- access to reliable data and information,
- agreement among experts, and
- understanding of the phenomena involved.

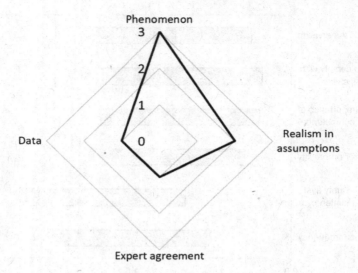

Fig. 16.17 Spider diagram presenting strength of knowledge

The spider diagram example is based on Ref. [6]. If, for example, one spider diagram is presented for each critical assumption, the result is a visual "profile" indicating the strength of knowledge that the evaluations are based upon.

Sensitivity studies as shown in Sect. 16.6 may also be used in order to illustrate parameters are influenced most extensively by variations in premises and assumptions in the quantification of risk, for instance from Figs. 16.13, 16.14, 16.15 and 16.16. Consider the following as an illustration.

Figure 16.13 shows that the probability of failure of ignition control has a strong effect on the results. At the same time, experience data relating to ignition control are few in number and represent usually a small volume of data. The ignition control barrier function is therefore one of the main sources of uncertainty of how likely it is that hydrocarbon leaks will cause uncontrolled fire in the future.

16.8 Presentation Format for Easy Understanding

There have been efforts made during the last years in order to develop result presentation formats that are easy to understand for non-experts. Several formats have been presented, consisting of simplified textual and graphical presentations.

Figure 16.18 is adapted from Safetec Nordic. In this particular example, only risk to personnel is shown, although risk for main safety functions, assets and environment could be summarised in the same manner.

Short description of installation

The sketch shows the installation seen from the side, with main areas outlined. There are two process areas and the turret. Two utility areas, and the accommodation as well as the offloading in the stern of the vessel. This is next to the flare tower.

The lifeboats are installed in the bow of the vessel, next to the accommodation. Forward and above the accommodation is the helideck installed.

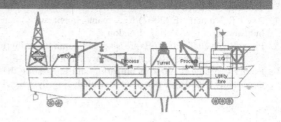

Important assumptions that are used in the QRA

Manning level: Total manning is assumed to be 56 persons, of which the night shift is 10 persons, and the rest is day shift personnel. 12 persons will during day shift have their workplace outside the accommodation, during night there is only two persons outside accommodation.

Evacuation prior to ignition: It is assumed that personnel in many circumstances will be able to escape from their place of work prior to ignition of a hydrocarbon leak.

Shipping traffic: It is assumed an average of 4 supply vessel visits per week. Off-loading will take place with the supply vessel alongside the FPSO, on the leeward side.

Hot work: A total of 8 hours of hot work per year Class A is assumed to take place on the installation, with 600 hours of hot work Class B.

Risk to personnel on the installation

The diagram presents the FAR values for different areas on the installation. The average FAR value is 3.0. The diagram shows that the turret, the process area aft of the turret and the off-loading area have the highest values. The areas furthest away from the accommodation have significant contributions from escape to the shelter area in the accommodation over exposed deck areas.

For the personnel with work place inside the accommodation, occupational accidents are the main threats.

Fig. 16.18 Typical format for simplified presentation of results

Also fire and explosion characteristics, as summarised in Fig. 16.10, could be presented in the same simplified manner as in Fig. 16.18.

References

1. PSA (2017) The management regulations. Petroleum Safety Authority, Norwegian Environment Agency, Norwegian Directorate of Health, Norwegian Food Safety Authority and Norwegian Radiation Protection Authority
2. PSA (2018) Integrated and unified risk management in the petroleum industry. Publication issued by PSA in June 2018

3. Standard Norway (2010) Risk and emergency preparedness analysis. NORSOK Standard Z–013, Rev. 3, 2010
4. Aven T, Vinnem JE (2007) Risk management, with applications from the offshore petroleum industry. Springer Verlag, London
5. Flage R, Aven T (2009) Expressing and communicating uncertainty in relation to quantitative risk analysis. Reliab Risk Anal Theory Appl 2(2):9–18
6. Berner C, Flage R (2016) Comparing and integrating the NUSAP notational scheme with an uncertainty based risk perspective. Reliab Eng Syst Saf 156:185–194

Chapter 17
Evaluation of Personnel Risk Levels

17.1 Current Fatality Risk Levels

17.1.1 FAR in Offshore Operations

This section presents the evaluations of fatality risk levels for Norwegian offshore operations, based on an assessment of previous fatal accidents. The data sources were presented in Sect. 2.2. The scope is limited to production installations and mobile units for drilling and accommodation purposes. Although the scope is limited to the Norwegian sector, some values for the UK sector are also presented as a supplementary source of information.

Separate assessments are performed for occupational accidents, major accidents and helicopter transportation accidents. Total offshore hours are used as exposure hours in order to make the contributions suitable for addition. This is twice the number of manhours as that reported by IOGP and that used by authorities.

17.1.1.1 Occupational Accidents

Occupational accidents have occurred quite regularly, although the trend has clearly been falling, if we consider production installations and mobile units. Accident statistics and trends were presented in Sect. 2.2.4.

If we calculate FAR, taking a 10-year average based on offshore exposure hours (i.e. on-duty as well as off-duty hours), we arrive at the following average values for the period 2002–2011:

Production installations: 0.35 fatalities per 100 million exposure hours
Mobile installations: 1.53 fatalities per 100 million exposure hours

It may be argued, especially for production installations, that a 10-year average does not reflect properly what seems to be a falling trend in the past 10 years. Therefore, we have also calculated values with 20-year averaging:

Production installations: 0.74 fatalities per 100 million exposure hours
Mobile installations: 1.48 fatalities per 100 million exposure hours.

The values for mobile installations are almost identical with 10 or 20 years averaging. For production installations the value with 10-year averaging is considerably lower, reflecting the fact that in the last ten years there has been only one occupational fatality on production installations, whereas the preceding ten years had three fatal accidents.

We have also considered the frequency of severe personnel injury on production and mobile installations, as reported in the Risk Level project, [1]. This frequency was around 1 fatality per million working hours in 1996, which increased to a peak of 2.3 in 2000, dropped back to 1 per million working hours in 2006, and reduced further to 0.48 in 2013, and has increased during the last few years, reaching 0.81 per million working hours in 2017, which is virtually equal to the value ten years ago, 0.88 per million working hours. This implies that the improvement until 2013 has been cancelled during the past five years, at least with respect to the prevention of severe injuries. This supports the argument that a 10-year average would be most appropriate.

The final aspect to be addressed is the number of fatalities per fatal accident, for occupational accidents. According to Appendix B 28 occupational fatal accidents on production and mobile installations in the period 1980–2017 resulted in 27 fatalities in 26 fatal accidents, which correspond to 1.04 fatalities/fatal accident.

17.1.1.2 Major Accidents—Norwegian Sector

Major accidents occurred quite frequently during the first 20 years of operations in the Norwegian sector, but have only occurred once (or never, according to interpretation, see below) during the past 25 years. This does not include the capsize of the mobile accommodation unit West Gamma in August 1990, while being towed from the Norwegian sector to the German sector. This capsize occurred in the Danish sector in the southern North Sea, and the installation during the tow was actually operating under marine jurisdiction, not as a petroleum installation. The major accidents that have occurred in the Norwegian sector include the following events:

- Production installations in the Norwegian sector (see also Chap. 2):

  - Riser rupture, explosion and fire, Ekofisk Alpha, 1975, 3 fatalities
  - Unignited blowout, Ekofisk Bravo, 1977, no fatalities
  - Fire in an enclosed space in a concrete shaft during commissioning work, Statfjord Alpha, 1978, 5 fatalities

- Mobile units in the Norwegian sector:

  - Capsize of mobile drilling unit Deep Sea Driller, 1976, 6 fatalities
  - Capsize of mobile accommodation unit Alexander Kielland, 1980, 123 fatalities
  - Ignited shallow gas blowout, West Vanguard, 1985, 1 fatality.

The interpretation of 'major accidents' in this context is accidents that have the potential to cause at least five fatalities, although in some cases none resulted, such as in some of the cases mentioned above. Two near-misses have occurred in the Norwegian sector in recent years, namely the unignited subsea blowout on Snorre Alpha in November 2004, the massive unignited gas leak on the Visund FPU in January 2006 and the gas leak on the Gullfaks Beta platform in December 2010 [2]. The Snorre Alpha subsea gas blowout in 2004 (see Sect. 4.9) could in theory be classified as a major accident, but it is noted that the official investigation by PSA [3] states that a major accident was avoided. These two events definitely had major accident potential, but they are considered to be near misses, because they did not ignite. The same applies to two unignited shallow gas blowouts of short duration around 1980.

If the subsea blowout on Snorre Alpha is conservatively considered as a major accident, we have had one major accident in the Norwegian sector since 1985 (no fatalities). The last ignited hydrocarbon leak on an installation in the Norwegian sector was in November 1992 [1]. We could also use the entire period from around 1970, then we would have three of four (counting Snorre Alpha) major accidents, but we know that the standards have improved since then, so that would result in another type of uncertainty.

However, there have been some major accidents in different sectors in the recent years (see Chaps. 4 and 5). These include the explosion, fire and capsize of the floating production installation P-36 in Brazil in 2001, the ignited blowout on the Temsah field in Egypt in 2004 (without fatalities), the ignited riser rupture due to collision impact in the Mumbai High field in India in 2005, the blowout from the Kab-101 platform offshore Mexico due to collision with mobile unit Usumacinto in 2007, the burning blowout on the Australian Montara field in 2009 (without fatalities), the Macondo burning blowout in the US Gulf of Mexico in 2010, the Frade underground blowout offshore Brazil in 2011 and fires and explosions in Mexico (Abkatun) and Brazil (São Mateus) in 2015.

17.1.1.3 Major Accidents—UK Sector

With such limited data sources as those presented above, it is also worthwhile considering the number of major accidents in the UK sector of the North Sea, although there are significant differences with respect to operating conditions as well as applied standards, at least prior to 1992. The following are identified as major accidents in the UK sector (not including the Sea Gem capsize in 1965, which occurred during rig movement with 13 fatalities), based on WOAD [4], Appendix B and a study of explosion incidents offshore in the North Sea [5]:

- UK Production installations:

  – Ignited shallow gas blowout, Forties Delta, 1983, 13 fatalities
  – Gas leak, explosion and fire, Piper Alpha, 1984, 4 fatalities
  – Gas leak, explosion and fire, Brent Alpha, 1988, 0 fatalities
  – Gas leak, explosion and fire, Piper Alpha, 1988, 167 fatalities
  – Gas leak, explosion and fire, Cormorant South Alpha, 1989, 0 fatalities
  – Gas leak, explosion and fire, Rough Bravo, 2006, 0 fatalities

- Mobile installations operating in the UK:

  – Capsize of mobile drilling unit Transocean 3, 1974, 0 fatalities
  – Ignited gas blowout, Ocean Oddesey, 1988, 1 fatality.

Figure 2.4, shows the overview of major accidents in the Norwegian sector. The occurrence of major accidents in the UK and Norwegian sectors is illustrated in Fig. 17.1, including the number of fatalities per accident.

17.1.1.4 Frequency of Fatal Accidents—UK Sector

One way to utilise the occurrence of major accidents is to calculate an occurrence frequency for major accidents, based on history. This can also be calculated for the UK sector for the period 1988–2017 as an illustration. The number of manhours in the UK sector for the period was calculated based on the number of employees published by HSE [6]. The split between production and mobile units has been assumed, resulting in 1, 360 million manhours for production installations and 289 million manhours for mobile units. The following frequencies may then be calculated for the period in question, using the incidents as noted above and offshore exposure hours, i.e. double the working hours:

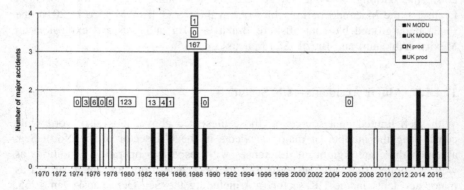

Fig. 17.1 Fatal accidents on UK and Norwegian installations, 1970–2011

Production installations: 0.147 major accidents per 100 million exposure hours
                            (1988–2017)
Mobile installations: 0.173 major accidents per 100 million exposure hours
                            (1988–2017)

17.1.1.5 FAR—Norwegian Sector

Owing to the absence of major accidents (or one) since 1985, it might be argued
that the likelihood of major accidents in the Norwegian sector has been reduced
extensively. However, there have been recent near misses in the Norwegian sector,
and major accidents in the UK sector as well as in more distant areas. A balanced
evaluation is therefore needed.

The crudest approach to the assessment of FAR is to calculate an average FAR
value because of these accidents, normalised against exposure hours in the periods
1974–2017 and 1967–2017, respectively. These values are:

Production installations: 0.43 fatalities per 100 million exposure hours (1974–
                            2017)
Mobile installations: 23.6 fatalities per 100 million exposure hours (1967–
                            2017)

It is obvious that the value for mobile units is strongly dependent on the
Alexander Kielland accident (please note that this accident is classified as 'production' in Fig. 2.4. The values calculated here do not account for the improvement
since the first half of the period. Although the data are sparse, a downward trend is
quite likely, not the least because of the improvements in the regulations and
standards. We therefore need to find ways to include these effects.

An alternative way to utilise the occurrence of major accidents is to calculate an
occurrence frequency for major accidents, based on history. Such incidence rates
may be stated as follows, as an average for the entire operational period:

Production installations: 0.22 major accidents per 100 million exposure hours
                            (1974–2017)
Mobile installations: 0.35 major accidents per 100 million exposure hours
                            (1967–2017)

We may then predict what would be the expected number of fatalities per major
accident in the future, to arrive at a FAR value. This manner gives predictions that
are not influenced by the very high number of fatalities in the Alexander Kielland
accident.

As long as we average over the entire production period for the Norwegian
sector, we are still unable to reflect the improvements that have been made.
However, since no major accidents have occurred for many years, the actual extent
of these improvements is in fact unknown. As a simple alternative, let us calculate

the frequency of major accidents if we only consider the period 1980–2017. The following values then result:

Production installations: 0.057 major accidents per 100 million exposure hours (1980–2017)
Mobile installations: 0.19 major accidents per 100 million exposure hours (1980–2017)

However, it could be argued that major improvements have taken place since 1980, to the extent that averaging over the period 1980–2017 also gives too high values. More representative values may apply to the period 1990–2017. There have been no major accidents on mobile installations in the Norwegian sector in this period, and possibly the subsea blowout on Snorre Alpha in 2004 for production installations. In such circumstances, predictions can be made by assuming that half an event could apply. If this is done for the period 1990–2017, we arrive at the following values:

Production installations: 0.070 major accidents per 100 million exposure hours (1990–2017)
Mobile installations: 0.105 major accidents per 100 million exposure hours (1990–2017)

If we assume that the expected number of fatalities in the future is eight fatalities per major accident for production installations and six fatalities for mobile units, we arrive at the following values (using the period 1990–2017):

Production installations: 0.56 fatalities per 100 million exposure hours
Mobile installations: 0.63 fatalities per 100 million exposure hours.

An alternative source is to look at what is calculated in QRA studies, which consider the predicted frequency and fatalities for individual installations in the future. We have over the years had access to a number of such studies, which are confidential and cannot be named. From these studies, the following typical values may be stated:

Production installations: 3.0 fatalities per 100 million exposure hours
Mobile installations: 1.20 fatalities per 100 million exposure hours.

However, it should be noted that FAR values predicted in QRA studies are often quite conservative. This was shown in [4] for explosions, but it is applicable in general. The author has argued that if you added predicted fatalities from QRA studies for all installations in the Norwegian sector, you would probably end up with up to five predicted fatalities from major accidents each year, and there has been none since 1985. These last values are therefore most likely too high.

A third type of input comes from the Risk Level project [1]. The main presentations of risk are relevant in this study. We can observe that the indicators of major accident risk suggest a stable level during the past 10 years for production installations and a falling trend for mobile units. However, some additional input is available from this source. Expert evaluations were made (and have been refined over the years) in order to determine weights for major hazard precursory incidents (such as unignited hydrocarbon leaks) in the Risk Level project [1]. These precursor events and weights imply the following FAR values:

Production installations: 0.46 fatalities per 100 million exposure hours
Mobile installations: 0.40 fatalities per 100 million exposure hours.

When all these predictions and the suggestions for the trends from the Risk Level project are considered, we can take an average between the average assumed for the period 1990–2017 and the basis for the values in the Risk Level project [1], and end up with the following values:

Production installations: 0.60 fatalities per 100 million exposure hours
Mobile installations: 0.62 fatalities per 100 million exposure hours.

It can be observed that the various predictions of the fatality numbers for production and mobile installations are reasonably consistent; implying that the uncertainty associated with these predictions should be limited.

17.1.1.6   Number of Fatalities Per Accident

If we calculate the average number of fatalities per major accident based on the events shown above, the average value is 23 if Alexander Kielland is included, and three if it is excluded, for the Norwegian sector. If we include the UK and Norwegian sectors, the average number of fatalities per accident becomes 21 if Alexander Kielland and Piper Alpha (1988) are included, and three if these two catastrophes are excluded. Details are presented in Table 17.1.

17.1.2   FAR in Worldwide Offshore Operations

The most extensive data source for accidents in the worldwide offshore industry (as well as onshore operations) is the IOGP annual statistical summaries, which have been

Table 17.1   Average number of fatalities in major accidents

Area	Production installations	Mobile installations	All installations
Norwegian sector	2.7	43	23
UK sector	31	0.5	23
Average both sectors	21	26	23
Average both sectors without Alexander Kielland and Piper Alpha	3.1	2.0	2.8

published since 1998. Figure 17.2 presents the 20-year trend of offshore FAR from these IOGP annual reports [7]. The generally falling trend in the period is obvious.

The report covers occupational accidents, major accidents and transportation accidents, for company personnel as well as for contractors. The number of man-hours included increased by more than 150% during the period, and it corresponds in 2011 to nearly 780 million manhours (offshore), from 45 companies in 98 countries. The highest number of fatalities in one year is 32 lives lost in 2004, while the lowest is 8 lives lost in 2005. Two helicopter accidents occurred in 2004, which contributed significantly to this high value. This underlines that fatalities on installations are mixed with helicopter transportation fatalities in IOGP statistics, whereas all other presentations in this book consider installation and helicopter fatalities separately.

One critical limitation of IOGP data is clear from the values in 2001, 2005, 2007 and 2015. The capsize of P-36 offshore Brazil in 2001 following an explosion and fire which resulted in 11 fatalities, is not included in the IOGP values. If included, it would increase the number of deaths from 19 to 30. In 2005 the total number of deaths in the IOGP analysis is 10. There was however, one major fire and explosion on the Mumbai High North field, operated by ONGC in India. This accident resulted in 22 fatalities, which if included, would affect the value for 2005 considerably. Similarly, the burning blowout on the Usumacinta installation in 2007 caused 22 fatalities during evacuation, which are not included in the 21 fatalities reported for that year. One major accident in 2015 occurred in Brazil (Petrobras) with 9 fatalities, this accident is not included in the number of fatalities reported by IOGP, 21. It is not known what other accidents are excluded from the IOGP statistics.

A burning blowout occurred in the Mediterranean Sea offshore Egypt in 2004, without fatalities, but with the total loss of the Temsah field wellhead platform, which is also excluded from the IOGP reporting. If further companies were included, a higher volume of exposure hours would also be included, such that FAR values might not be significantly affected. The lack of completeness may not imply under-prediction in the case of occupational accidents. Major accidents are so rare however, that significant effects may result when these are not included.

Figure 17.2 illustrates what the effect of the inclusion of these four major accidents in 2001, 2005, 2007 and 2015 would be if fatalities and assumed man-hours for the three companies in question were included. The number of employees was found from the companies' web-pages, from which offshore manhours could be 'guestimated'. Figure 17.2 is based on offshore exposure hours, including on-duty as well as off-duty hours.

It should be noted that the effect of Petrobras, ONGC and Pemex data are only considered in the four years 2001, 2005, 2007 and 2015. In fact we do not know if these operators had other fatal accidents in the years in question. If this were the case, the values in Fig. 17.2 would be under-predicted. The values in Fig. 17.2 are therefore not precise, but are the best predictions that can be made with the available data.

Finally, we had hoped to be able to separate the IOGP data into categories of occupational fatalities and major accident fatalities for offshore operations.

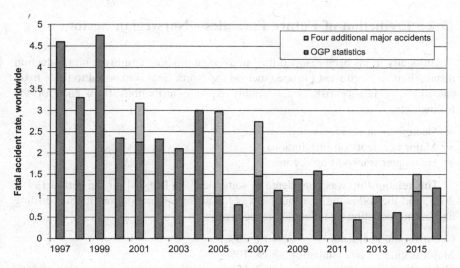

Fig. 17.2 Worldwide trend, offshore FAR, based on total offshore exposure hours and IOGP reporting

However, the reporting by IOGP was not sufficiently detailed in order to allow this categorization.

Another source of international statistics is IRF (see Sect. 6.5). Figure 6.11 presented a comparison of FAR values for most of the members of IRF. Figure 17.3 presents the trend of average FAR values for IRF member countries in the period 2007–2016. The contributions from known major accidents in 2008 (helicopter accident), 2010 (blowout) and 2015 (explosion and fire) are marked separately in the diagram. When the major accidents are excluded, there is a falling trend for the remaining fatalities.

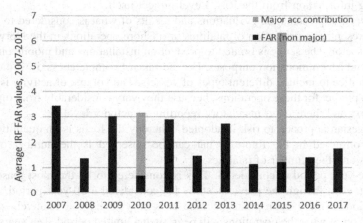

Fig. 17.3 Average FAR values for IRF members, 2007–2016

17.2   Prediction of Future Fatalities—Norwegian Sector

Traditionally it has been claimed that an average offshore employee on production installations in Northwest Europe (such as the North Sea) is exposed to three main categories of fatality risk, with roughly equal contributions from each of the following:

- Occupational accidents
- Major accidents on installations
- Helicopter transport accidents.

This relationship was considered in some detail in Ref. [8] for the period 1998–2007, and the following distribution was found for production and mobile drilling and accommodation installations, respectively:

Occupational accidents   40% 75%
Major accidents on installation   30% 10%
Helicopter transport accidents   30% 15%

However, this assessment was performed more than 10 years ago, and these relationships may have changed. Therefore it is interesting to consider what the distribution would be with the revised values from recent operations. The basis for the assessment is offshore operations in the Norwegian sector, including the North Sea, the Norwegian Sea and the Barents Sea.

The approach to the assessment is as follows: the predicted number of fatalities in a future five year period in the entire Norwegian sector is considered, separately for production installations and mobile drilling and accommodation units. These predicted values are used in order to establish the relative distributions. Another important principle adopted is that of triangulation, corresponding to the approach taken in the Norwegian Risk Level project [1]. This is implemented through input from various sources and perspectives, quantitatively as well as qualitatively, including information from the Risk Level project itself.

This section presents the evaluations and results of what is considered to be the most representative prediction of fatalities in offshore operations in the Norwegian offshore sector. The scope is limited to production installations and mobile units for drilling and accommodation purposes. For the sake of completeness, it would have also been nice to include different types of vessels. The volume of activity is much harder to predict for these operations, because they vary considerably from one year to the next and there is no mandatory reporting of activity level.

A Bayesian approach to risk is adopted, whereby the focus is on quantities that may be observed in the future. What can be observed is the number of fatal accidents and the number of fatalities.

A five-year period is considered. This is considered to be the most reasonable approach, even though the previous study [7] considered a 10-year period. A five year period is more appropriate because it is easy to have a relatively precise appreciation of what the operations will be in such a limited period. Ten years is too long, because premises may change radically during such a period, for instance if the oil price level changes radically as experienced recently.

We have therefore made predictions for a future five-year period, namely 2019–2023. Further, the expected time between accident types is such that the most likely number of occurrences for these accident types during the five years is zero occurrences.

Separate assessments are performed for occupational accidents, major accidents and helicopter transportation accidents. Total offshore hours are used as exposure hours, in order to make the contributions suitable for addition. This is twice the number of manhours as reported by IOGP above as well as that used by authorities.

17.2.1   Important Assumptions and Evaluations

17.2.1.1   Occupational Accidents

The different sources and interpretations discussed for occupational accidents in the recent years are quite consistent and point to a slowly decreasing level, which has resulted in the longest period ever without occupational fatalities on production installations and mobile units in the Norwegian sector.

The assumption is therefore made that the average FAR during the past 10 years is most applicable for the future five years:

Production installations: 0.35 fatalities per 100 million exposure hours
Mobile installations: 1.53 fatalities per 100 million exposure hours.

17.2.1.2   Major Accidents on Installations

The assessment of risk levels for major hazards on the installation is much more complex, due to the fact that more than 25 years have passed since the last major accident with fatalities in the Norwegian sector. We have shown quite different fatal accident rates according to how they are considered. Our approach is based on no occurrences in the period 1990–2011, supplemented with data from QRA studies and the Risk Level Project, as discussed above. It may be observed that the various predictions of the fatality numbers for production installations are reasonably consistent; implying that the uncertainty associated with this prediction should be limited. The predictions for mobile installations are not quite as consistent. When all the predictions are considered, we ended up with the following values:

Production installations: 0.60 fatalities per 100 million exposure hours
Mobile installations: 0.62 fatalities per 100 million exposure hours.

17.2.1.3 Helicopter Transportation Accidents

The assessment of risk associated with helicopter transportation risk is somewhat similar to major hazard risk on installations, although many more statistics are available. The challenge is to evaluate the effect of recent improvements. Some assumptions have been made in order to reflect this, as discussed in Chap. 13. The discussion concluded that a representative average value for the period 2008–2017 may be the following for the Norwegian sector:

Helicopter transport: 80 fatalities per 100 million person-flight hours

It should be noted that this value refers to modern helicopters, such as Sikorsky S92, operated with the highest standards as do the two major helicopter operators in the Norwegian sector.

17.2.2 Occupational Accidents

The assumptions on which the predictions are based are presented above. The predicted numbers of fatalities because of occupational accidents in the period 2019–2023 are as follows:

Production installations: 0.43 fatalities
Mobile installations: 0.77 fatalities

This implies that 1.2 fatalities is the predicted number of fatalities during the period 2019–2023 because of occupational accidents. We have shown that the average number of fatalities per fatal accident was 1.04 in the period 1980–2017. We have therefore predicted 1 fatality per fatal occupational accident.

17.2.3 Major Accidents on Installations

Similarly, the predicted number of fatalities because of major accidents in the period 2019–2023 is as follows:

Production installations: 1.50 fatalities
Mobile installations: 0.62 fatalities

This implies that 2.1 fatalities is the predicted number of fatalities during the period 2019–2023 because of major accidents. The predicted number of major accidents for this period can be calculated if assumptions on the average number of fatalities per major accident are made. Table 17.1 demonstrated that this is very sensitive to whether the Alexander Kielland and Piper Alpha accidents are included or not. If weighted averages are calculated, in which these two large accidents are weighted by 10%, the average comes out as 4.9 and 4.2 fatalities per accident for production and mobile installations. With these weighted average values, the number of major accidents is 0.29 for production installations as well as mobile installations for this five-year period.

17.2.4 Helicopter Transportation Accidents

Based on the statistical basis and assumptions, the predicted number of fatalities because of helicopter accidents in the period 2019–2023 is as follows:

Flights to production installations: 1.50 fatalities
Flights to mobile installations: 0.60 fatalities

This implies that 2.1 fatalities is the predicted number of fatalities during the period 2019–2023 because of helicopter accidents.

17.2.5 Summary of Predicted Fatalities

The preceding sections can now be used in order to predict the number of fatal accidents in the period 2019–2023 for the entire NCS. Table 17.2 presents the expected values. The following are the expected number of fatal accidents:

Occupational accidents: 1.2 fatal accidents
Helicopter accidents: 0.19 fatal accidents
Major accidents: 0.44 fatal accidents

This implies the following probabilities for fatal accidents:

Probability of occupational accident(s) (2019–2023): 70%
Probability of helicopter accident(s) (2019–2023):   17.5%
Probability of major accident(s) (2019–2023):   36%

This implies that there is a 84% probability that there will be at least one fatal accident in the Norwegian sector in the five-year period 2019–2023. Prediction intervals are also important, in order to illustrate the uncertainty of these predicttions. The following prediction intervals apply for the number of fatal accidents:

Table 17.2 Overview of predicted fatalities in the Norwegian sector during 2019–2023

Hazard type	Production installations	Mobile installations	Sum
Occupational accidents	0.43	0.77	1.2
Helicopter accidents	1.50	0.62	2.1
Major accidents	1.50	0.60	2.1
Total	3.4	2.0	5.4

Occupational accidents: 0–3 fatal accidents
Helicopter accidents: 0–2 fatal accidents
Major accidents: 0–2 fatal accidents

The final step is to present the values as relative contributions. Since the values are uncertain predictions, the values are rounded off somewhat, but the exact values are shown here in parentheses:

- Production installations

  – Occupational accidents: 10% (13%)
  – Major accidents on installation: 45% (44%)
  – Helicopter transport accidents: 45% (44%)

- Mobile drilling and accommodation units

  – Occupational accidents: 40% (39%)
  – Major accidents on installation: 30% (30%)
  – Helicopter transport accidents: 30% (31%)

There are significant changes in these relative contributions, especially for production installations, which is mainly the result of a significant reduction in the frequency of occupational accidents on production installations. The contribution from occupational accidents on production installations is only 10% of the overall risk exposure. There are less changes for mobile units, which in are close to the 'one third' all distribution.

Some of these probabilities are high, and suggest strongly that accidents are quite likely to occur during the five-year period considered. There is a 84% probability that a fatal accident will occur in the operation of production and mobile installations in the period 2019–2023. There is about a 70% probability that an occupational accident will occur. By contrast, there is about a 53% probability that a major accident will occur, either on installations or in the helicopter transport. These data strongly indicate that fatal accidents are quite likely, unless further significant improvements are implemented over and above what was discussed in Chap. 13.

The values in this section have demonstrated that the efforts made by the industry and authorities during the past 15–20 years have paid off. The risk levels associated with operation of production installations and mobile units in the Norwegian sector today are less than 60% lower than that predicted 20 years ago, as illustrated by Fig. 17.4. This reduction is more extensive for mobile units (more than 70% reduction), while the contribution from occupational accidents has been reduced by more than 85%.

This study has further suggested that the relative contributions from the main sources of fatality risk offshore on production installations are largely unchanged from those predicted in 1997, except for occupational accidents on production installations. Risk levels are more than 60% lower compared with 1997. There has also been a reduction for mobile installations of, almost 70%. This implies that

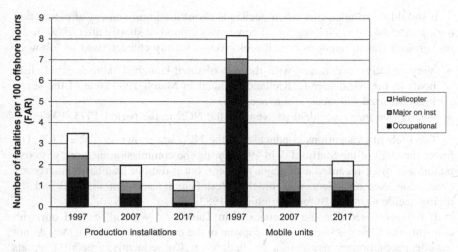

Fig. 17.4 Comparison of predicted FAR values for production installations and mobile units in 2017 with values in 1997 and 2007

considerable improvements have been made in all three main categories. However, it also implies that continuous improvement towards the 'zero vision' adopted across the industry, will need further work in all three areas.

17.3   Major Accident and Evacuation Frequencies

17.3.1   Life Boat Evacuations on the NCS

Around 1980 several initiatives were taken in the Norwegian petroleum industry to improve lifeboats; a very significant factor was the negative experience from the Alexander Kielland accident. This process produced the first freefall lifeboats, which were installed on the first installations around 1983–1984. More than 200 freefall lifeboats have been installed in the subsequent 30 years, mainly on Norwegian production installations, but also on some Norwegian mobile installations. Other countries have not made freefall lifeboats mandatory as Norway has for production facilities, but some installations also have this type of lifeboats, especially on installations in hostile environments. Although freefall lifeboats have never been tested in a real evacuation-to-sea scenario, tests carried out offshore on some Norwegian installations in 2005–2006 revealed that they had severe weaknesses. There has thus been considerable focus on lifeboats and evacuation procedures in recent years. It has therefore considered to be important to reflect on the frequency of the future use of evacuation means on the NCS as a background to the evaluation of possible improvements.

It should be recalled that when decisions about the improvement of evacuation means were taken by Norwegian oil companies around or shortly after 1980, it the background was different to how it looks today, mainly characterised as follows:

- Very negative experience with the use of davit-launched (conventional) life-boats in the Alexander L. Kielland accident in March 1980 (one of the seven lifeboats used for successful evacuation)
- Nearly one major accident per year on the NCS in the period 1975–1980.

The lifeboat evacuations conducted on the NCS are shown in Table 17.3. The fire in the shaft of the Statfjord A in 1978 during the commissioning work with five fatalities is often regarded as a major accident, but it did not lead to an emergency evacuation. See also Sect. 4.24 on the West Gamma accident in 1990 and the diving accident on the Byford Dolphin in 1983.

It is emphasised that the evacuations in Table 17.3 were all carried out with conventional lifeboats (and a rescue capsule in the evacuation in 1976), they do not include precautionary evacuation by helicopter. Consequently, freefall lifeboats have never been tested in a 'real' case, as noted above.

One can speculate whether this was a development that could have been imagined when the freefall lifeboats were introduced in the early 1980s on the mobile installation Dyvi Delta (now known as Deepsea Delta) and Heimdal production facility. There is reason to believe that advocates for freefall lifeboats expected that they would be used to some extent to evacuate personnel from these

Table 17.3 Overview of evacuation cases by conventional lifeboats in the Norwegian sector, 1966–2017

Year	Type of accident	Number of fatalities		Number of persons involved	Weather and wave conditions
		On installation	During evacuation		
1966	Evacuation following Ocean Traveller's collision with supply vessel	0	0	51	Good weather
1975	Evacuation following ignited riser rupture, causing explosion and fire	0	3	71	Wind: 5 m/s 0.5 m waves
1976	Evacuation following grounding of Deep Sea Driller during tow to shore	0	6		Beaufort 11–12
1977	Evacuation following unignited blowout	0	0	112	Beaufort 7
1980	Evacuation following capsize of Alexander L. Kielland flotel	(unknown)	123 (total)	212	Strong gale
1985	Evacuation following ignited shallow gass blowout causing strong explosion	1	0	79	Low wind 2–3 m waves

facilities. More than 30 years after the first installation, however, this has not occurred so far.

17.3.2   Experience Data from Freefall Lifeboat Tests

Freefall lifeboats have never been used in offshore accidents, not on the NCS nor in foreign sectors. Further, there has never been an emergency evacuation ships that have freefall lifeboats (skid mounted boats) installed. Empirical data on freefall lifeboats must therefore come from training activities.

There have been few reported injuries in training centres, while there seems to be an unknown number of injuries and back problems related to pre-existing back problems. Such damage is an issue in relation to the scope of training in the use of freefall lifeboats. However, no injuries are considered to be decisive for judging whether lifeboat evacuation by freefall lifeboats is safe in emergencies.

This leaves us with no other option than to evaluate miscellaneous experience data from various sources, in order to establish as relevant frequencies as possible. The major accidents we are concerned with here are those on installations; helicopter accidents can also be considered to be major accidents, but they are not relevant in the present context.

17.3.3   Major Accident Frequency—Norwegian Sector

If we conservatively classify the Snorre Alpha in 2004 (see Sect. 4.9) as a major accident, although there were no fatalities, there has been one major accident in the period after 1985, i.e. one event in 27 years, whereas there were six cases in the period 1975–1985. There have several near-misses during the past 15 years, which could under different circumstances have developed into major accidents. This certainly underlines that the frequency is not zero, although only one event occurred in 35 years.

Figure 17.5 shows the curves for the NCS, expressed as rolling 30-year average values, for production and mobile installations together:

- Number of major accidents (NCS) per 100 million working hours
- Number of major accident fatalities (NCS) per 100 million working hours.

Figure 17.5 shows that the major accident frequency is falling in the entire period, from a value around one major accident per year for the first few years. There is a peak in 1980 when the number of fatalities is included. This peak is because of the Alexander Kielland accident, after which the curve falls rapidly, and is 0.12 fatalities per 100 million working hours as an average for the period 1982–2017.

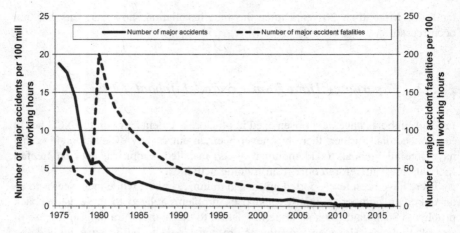

Fig. 17.5 Frequency of major accidents and major accident fatalities in the Norwegian sector, 30-years rolling average

Figure 17.5 also shows that the frequency of major accidents in 2017 is only 2% of what it was in 1980 (i.e. 5.9, reduced to 0.10 in 2017). The 40-year average becomes 0.25 per 100 million working hours in 2017.

The number of fatalities in major accidents shows an even more extensive reduction from the peak value in 1980 when 30-year rolling average is considered. The 40-year average value of the number of fatalities per 100 million working hours is 14.5, corresponding to a FAR value. The number of fatalities averaged over the past 30 years multiplied by the total number of working hours on the NCS for production and mobile installations corresponds to 19.4 years between major accidents that required a lifeboat evacuation.

However, even though Fig. 17.5 is based on all emergency evacuation s on the NCS, it is still impossible, based on experience data, to predict the future risk contribution from emergency evacuations by lifeboats on Norwegian installations. Luckily, there are too few major accidents, and the accidents reflected in Fig. 17.5 are too old to give a good basis for future risk prediction. The next section therefore looks at statistics from other sectors outside Norway.

17.3.4   Major Accidents Worldwide

A number of major accidents worldwide have called for evacuation-to-sea procedures, several of which have occurred without fatalities, but also several with multiple fatalities. The most well known are presented in Table 17.4. The table does not include the loss of jack-up installations during towing.

Only four of the accidents in Table 17.4 occurred on mobile installations: Ocean Ranger (1982), Ocean Odyssey (1989), Deepwater Horizon (2010) and Aban Pearl

(2010). Some ignited blowouts on mobile installations may be missing from Table 17.4 when no investigation report is available in the public domain.

There are many major accidents worldwide in the period in which there have been only two major accidents on the NCS. Data from these offshore areas do not provide a good basis for us to calculate risk levels on the NCS. Freefall lifeboats are not generally used on any other shelf, and the regulations are not as strict as they are in the Norwegian sector, except on the UK continental shelf. Nevertheless, the data in Table 17.4 provide an insight into emergency evacuations. The average is 13.7 fatalities per major accident, for the emergency evacuation phase alone, if Piper Alpha is excluded, and 16.9 fatalities per evacuation if all fatalities (except Piper Alpha) are included.

17.3.5   Major Accident Probability—Norwegian Sector

The values stated in this section tend to suggest that a return period for major accidents on the NCS may be as long as 30–40 years, when all types of installations are taken together. This should be regarded as a statistical expected value, which

Table 17.4 Overview of some of the major accidents with evacuations to sea in worldwide operations worldwide, 1980–2017

Year	Country	Installation	Type of accident	Number of fatalities (total/ evacuation)
1982	Canada	Ocean Ranger	Capsize	84/84
1984	Brazil	Enchova	Ignited blowouts	42/42
1988	UK	Piper A	Explosion and fire	167/?
1989	UK	Ocean Odyssey	Ignited blowout	1/0
2001	Brazil	P-36	Explosion and fire	11/0
2002	Brazil	P-34	Serious listing	0/0
2004	Egypt	Temsah	Ignited blowout	0/0
2005	India	Bombay North High	Ignited riser rupture caused by supply vessel collision	22/22
2007	Mexico	Usumacinta	Unignited blowout[a]	22/22
2009	Australia	Montara/West Atlas	Unignited blowout[b]	0/0
2010	US	Macondo/ Deepwater Horizon	Ignited blowout	11/0
2010	Venezuela	Aban Pearl	Capsize	0/0
2015	Mexico	Abkatun	Explosion and fire	7/0

[a]The blowout ignited at a later stage, probably by a vessel involved in the combatment
[b]The blowout ignited during the well killing operations, but was unignited when the evacuation took place

has limited applicability for a single installation and for a short-term period of, say, the next five to 10 years.

Philosophically, such a long return period may imply challenges that may render the return period an invalid projection. A return period of 30–40 years is longer than the average field lifetime offshore. This implies that new installations may have significantly changed characteristics during such a return period. Such a long period also implies that most employees offshore would never have experienced a major accident or been indirectly affected by a major accident on another installation during their employment offshore.

If the return period were 40 years, and there would be roughly 100 installations in the Norwegian sector, and thus the annual frequency of major accidents per installation would be 2.5×10^{-4} (0.025%). This is quite a low annual frequency.

However, this shall not be interpreted to imply that emergency evacuation does not need to be included in the design basis. The predicted frequency is low, but not negligible.

17.4   Risk Tolerance Criteria

17.4.1 Definition

The management of HES regulations [9] do not define risk tolerance criteria at all. The third revision of the NORSOK standard on risk and emergency preparedness analysis was issued in 2010 [10], and this defines risk tolerance criteria (actually 'risk acceptance criteria' is the term used) as 'criteria that are used to express a risk level that is considered as the upper limit for the activity in question to be tolerable'.

The NORSOK definition of risk tolerance (acceptance) criteria is adopted in the following. In addition, the requirements stated by PSA and HSE are presented in the following sections.

17.4.2 Philosophical Dilemma

The first risk tolerance criteria were implicitly introduced into Norwegian offshore legislation around 1980, and were extensively used until 2000. The new regulatory regime introduced from January 2002, referred to risk tolerance criteria as well as the ALARP (As Low As Reasonably Practicable) principle. Risk tolerance criteria continued to be the main instrument for risk acceptance, but since 2004 there has been an increased focus on the ALARP principle as a supplement or substitution for risk tolerance criteria. The ALARP principle implies that all risk reduction proposals that are well founded should be implemented unless it can be shown that costs and/or other negative effects are in gross disproportion to the benefits.

Aven and Vinnem [11] argued that a mechanical approach to risk tolerability based on risk tolerance criteria should be replaced by an approach based on the generation and consideration of alternatives. This approach is similar to an ALARP evaluation. The authors believed that in the very mature Norwegian offshore industry risk tolerance criteria could be removed as a requirement from the regulations. The regulations, however, still have requirements for the formulation and use of risk tolerance criteria. It is thus important to address the use of such criteria. Their practical use is further discussed in Sect. 17.6. Since 2007 however, the majority of the long time players in the Norwegian sector have left the region or have sold their assets to new organisations, and the petroleum industry is not as mature as it was some 10–15 years ago.

17.4.3 Norwegian Regulatory Requirements

PSA's requirements regarding risk tolerance criteria and their use are presented explicitly in HES management regulations [9]. The regulations consist of normative requirements and guidelines. The wording of Section 9 of the regulations was described in Sect. 1.5.2. The following is the text of the Guidelines to Section 9:

Re Section 9

Acceptance criteria for major accident risk and environmental risk

The acceptance criteria that the party responsible sets for the design of a facility, has great significance for that the acceptance criteria can be met in the operational phase. Hence, both the party responsible for operating a mobile facility and the operator shall set acceptance criteria in areas under their responsibility.

Acceptance criteria as mentioned in the first subsection, shall express and represent an upper limit for what is considered an acceptable risk level for the various categories mentioned in literas a to d. As ensues from Section 11 of the Framework Regulations, complying with health, safety and environmental legislation constitutes an important parameter for this upper limit and it is accordingly not permitted to set aside specific requirements in the health, safety and environmental legislation in respect of risk calculation. Additional risk reduction shall always be considered, even if the results of risk analyses or risk assessments indicate a level of risk that is within the acceptance criteria, cf. Section 11 of the Framework Regulations.

The acceptance criteria shall be formulated so that they are in accordance with the requirement for suitable risk and preparedness analyses, cf. Section 17, and are suitable for providing decision-making support in relation to the risk analyses and risk assessments carried out.

Major accident means an acute incident such as a major spill, fire or explosion that immediately or subsequently entails multiple serious personal injuries and/or loss of human lives, serious harm to the environment and/or loss of major financial assets.

Acceptance criteria for acute pollution shall include the risk of acute pollution to occur (the area of authority of the Petroleum Safety Authority Norway) as well as the risk of harm to the external environment/environmental risk (the area of authority of the Norwegian Environment Agency).

Offshore petroleum activities

See Annex A of the NORSOK Z-013 standard for a description of different types of acceptance criteria that may be used for major accident risk and environmental risk as mentioned in subsection 2 literas a, c and d. See Annex B Chapter 4 of the standard for a complementary description of the acceptance criteria for loss of main safety functions as mentioned in subsection 2 litera b, cf. Section 11 of the Facilities Regulations. For information, see also Section 7 of the Facilities Regulations.

The operators that have facilities and activities in the same area, should cooperate on principles for establishing acceptance criteria, so that they are in a comparable form among operators, and so that they form a suitable basis for e.g. establishing joint emergency preparedness, cf. Section 21 of the Framework Regulations.

The main regulatory requirements for risk tolerance criteria can be summarised as follows:

- Risk tolerance criteria shall address all personnel as well as groups that have particular risk exposure.
- Risk tolerance criteria shall be defined for the personnel and environment.
- The risk exposure of third parties is only relevant for petroleum installations onshore.

These requirements imply a conscious approach to risk assessment, in the sense that it is needed to enhance competence and promote communication between different disciplines. They also establish the limitations of risk assessment and its importance in safety management. It should be noted that PSA does not formulate any requirements or guidelines for how risk tolerance criteria shall be established.

17.4.4 Risk Tolerance Criteria Requirements According to UK Regulations

The criteria used by the UK petroleum industry are mainly those that have been formulated by HSE and are embodied in statutory legislation. These are introduced briefly below.

17.4.4.1 ALARP Principle

The focus in the Safety Case Regulations (SCR, [12] is on the ALARP principle. This was expressed in the first version of the regulations, Schedule 2, 'Particulars to be included in the Safety Case for a Fixed Installation' where it was stated that the safety case must include:

'A demonstration, by reference to the results of quantitative risk assessment, that the performance standards….are adequate to reduce risks to the health and safety of persons to the lowest level that is reasonably practicable…'.

In this extract, the ALARP principle is directly tied to the safety of persons. However, in the SCR [11] the same principle is used in relation to the loss of the integrity of the temporary refuge (TR). Although these are not conflicting, it should be pointed out that risk to persons can also be expressed in more direct ways than through the impairment of the TR.

As mentioned in Sect. 1.6.1, this wording was removed from the regulation, in the 2005 version. Nevertheless, HSE expects no changes to the practice because of similar requirements elsewhere in the regulations. This quote may therefore still be considered to be appropriate. The SCR were updated in 2015 to reflect the EU 2013/30 Directive, but the essence of the requirements has not changed.

17.4.4.2 TR Impairment

The SCR require that the frequency of loss of integrity of the TR is less than 10^{-3}/ year. This criterion is often regarded as the main criterion for UK installations. It is worthwhile noting that this criterion is not detailed in the SCR, although it is mentioned in the HSE Guidance to the regulations.

The Schedules detail the content of safety cases for fixed and mobile installations, which is further described in the General Guidance on Content of Safety Cases. The TR is part of the arrangements for protecting personnel from the hazards of explosion, fire, heat, smoke, toxic gas or fumes, together with escape routes to and from TR, evacuation means, and facilities for monitoring and controlling the incident. The following aspects relating to the TR are detailed in a safety case:

- The functions to be performed by the TR.
- The range of possible accidental events which may threaten the integrity of the TR.
- The criteria as regards the conditions that constitute a loss of integrity (impairment) of the TR. Four main types of failure need to be considered:

  - Loss of structural support
  - Loss of life support
  - Loss of command support
  - Loss of availability of evacuation.

- The minimum period for which the TR must maintain its integrity in the face of 'design events' (endurance time).

The quantitative risk tolerance criterion in the SCR is tied to the TR, but the regulations also require the operator to provide escape routes to and from the TR, as well as sufficient evacuation means. The operator shall also specify criteria for these functions.

It is interesting to note how closely this can be compared with the Norwegian Main Safety Functions, previously given as shelter area, escape ways and evacuation station. This shows clearly the strong parallels between the two different sets of legislations.

17.4.4.3 Personnel Risk Criteria

The only quantitative criterion given in the SCR is the maximum allowable frequency for the impairment of the TR. This is in line with the Cullen report [13], which stated that operators should specify other criteria.

Because of this, the focus in the initial phases of the Safety Case work was on the performance of the TR and on ensuring compliance with the maximum upper limit impairment frequency. Later, however, operators looked at personnel risk more directly, as expressed through expected fatalities, individual fatality risk, or other risk measures.

Several operators clearly favour individual fatality risk as a measure of risk. Typically, risk is calculated by considering a group of personnel, such as production operators and calculating the annual expected fatality risk for the individuals in the group.

This is a very direct measure of risk, but it also has its weaknesses, primarily because it does not address the whole risk picture for an installation. Measures that are efficient at reducing risk for the group that has highest individual risk are not necessarily the most efficient at reducing total risk for the entire platform. For this purpose, the total expected number of fatalities for the entire platform would be a more useful risk measure.

In the 1990s in one of its supporting documents, HSE indicated that the average individual risk, arising from major hazard s, was in the order of 10^{-3} per year. The 10^{-3}/year value was calculated for a period that included the Piper Alpha accident, which had a very large effect on the calculated risk level. The UK and Norwegian statistics now show fatality frequencies that are substantially lower.

17.4.5 General Requirements

In order that risk tolerance criteria can support HES management decisions, they need to have the following qualities:

- Be suitable for decisions regarding risk reducing measures.
- Be suitable for communication.
- Be unambiguous in their formulation.
- Be independent of concepts, namely not favour any particular concept.

17.4.5.1 Suitability for Risk Reduction Decision-Making

Risk tolerance criteria that are simple to use in a decision-making process are often precise and associated with particular features of an installation or an activity. Such risk tolerance criteria are suitable for measuring the effect of risk reducing actions and other changes to design or operations.

17.4.5.2 Suitability for Communication

In order for risk tolerance criteria to be suitable for communication, they should:

- Be easy to understand for non experts
- Allow for the comparison of risk with other activities.

The communication of risk acceptance relates to how criteria are interpreted and understood among all involved parties, such as those that are exposed to the risk, the management of the company, authorities and the public. However, risk tolerance criteria that are easy to understand may represent an oversimplification if the problem being addressed is complicated or difficult to understand.

Risk tolerance criteria that express a societal dimension will often enable comparison with other activities in society. Such criteria are often related to the parameters that are associated with the latter stages in the sequence of events (such as fatalities).

17.4.5.3 Suitability with Respect to Unambiguity

In the present context, unambiguity means that risk tolerance criteria should be formulated in such a way that unreasonable or unintentional effects do not arise when expressing or evaluating risk levels. Possible problems with ambiguity may be associated with:

- Imprecise formulation of risk tolerance criteria,
- Unclear definition of system limits for the analysis, or
- Alternative ways of averaging the risk.

17.4.5.4 Suitability with Respect to the Independency of Concepts

Another possible problem is that different risk tolerance criteria may be aimed at the same type of risk (e.g. risk to personnel expressed by means of FAR versus impairment risk for main safety functions). In this event, different ranking of risk may occur for alternative solutions when considering different risk tolerance criteria.

17.5 Criteria Used for Personnel Risk by the Petroleum Industry

The discussion below demonstrates some important aspects relating to how Norwegian companies have developed their risk tolerance criteria. It must be noted that the values shown are illustrative only, and they may vary significantly from one operator to another. The following types of risk tolerance criteria are considered:

- PLL
- FAR
- AIR
- f–N
- Risk Matrix.

17.5.1 Group Average Risk

Two main approaches are used by Norwegian oil and gas offshore operators, with respect to the use of risk tolerance criteria. Statoil (now Equinor) and other companies have stipulated FAR = 10 as a risk tolerance limit expressed as an average of the total offshore installation workforce.

The regulations also require a risk acceptance limit for the groups/individuals exposed to the highest risk levels. Some companies have stipulated FAR = 20 as the tolerance limit for most exposed groups.

The other main approach used by Norwegian operators is to use AIR = 0.001 per year as a risk tolerance limit. Some companies have explicitly stated that this is the risk tolerance limit for most exposed groups, whereas other companies are less clear about this.

It was shown in Chap. 2 that AIR = 0.001 per year may correspond to a FAR of just above 30. Accordingly, there is a factor of about three between these limits. Whereas FAR = 10 may be considered to be a high limit, AIR = 0.001 per year is a very high and very relaxed limit that has no risk reduction potential. Even when AIR = 0.001 per year is used as a limit for most exposed groups, it has little risk reduction potential.

FAR = 10 is to some extent relaxed for large integrated installations, but quite challenging for a small installation. If the value FAR = 10 was claimed as the risk acceptance limit for most exposed groups, it would be a challenge for all installations. This is discussed further in Sect. 17.6.

17.5.2 Risk Distribution

One company in Norway has used an f–N curve as the tolerance limit. For an installation with POB = 100 persons, the curve is expressed as follows:

$$f(N) = \frac{0.01}{N} \qquad (17.1)$$

Using Eq. 2.16 this corresponds to an average FAR = 5.9, which signifies that this limit is more challenging than an average FAR = 10. In addition, the fact that a distribution is used instead of a single value, makes it more challenging to satisfy this limit. The values are adjusted according to the max POB onboard, as follows. For an installation with max POB = M, the values according to Eq. 17.1 are adjusted with the following factor:

$$\sqrt{\frac{M}{100}}$$

It may be noted that also one of the international oil companies uses an f-N curve as the risk tolerance limit, in fact they also use a relatively high aversion factor (see Sect. 2.1.4.6).

17.5.3 Potential Loss of Life (PLL)

It is uncommon to define risk tolerance limits for PLL, but PLL can be used for decision-making in some circumstances. PLL is well suited to comparing alternative solutions for the same development objective, if it is calculated across all life cycle phases. The objective will then be to minimise the PLL value.

PLL is relatively easy to understand for non-experts because of its calculation of an absolute level of fatalities. However, this risk tolerance limit does not consider the number of individuals in the population. This has to be observed carefully when comparing with other activities, especially if the number of individuals exposed is different.

By contrast, PLL is difficult to express in physical terms, as the value is usually below 1.0 (PLL = 0.1 does not imply that 10% of a person is dead!). Such a value may be illustrated by expressing it as a 10% probability of one fatality, although this is slightly imprecise, because there is also a small probability of more than one fatality. It is usually possible to define PLL unambiguously, however, it is not well suited for averaging differences between personnel groups.

The PLL value will often favour the development concept that has the lowest manning level resulting from a lower number of individuals exposed to risk. This underlines the fact that one way to reduce societal risk is to limit the number of personnel exposed to risk.

17.6   Use of Risk Tolerance Criteria in Personnel Risk Evaluation

Risk tolerance criteria are used by oil companies in the evaluation of risk, i.e. to assess the results of QRA studies and to decide on the need for risk reduction. Oil companies have the responsibility to develop their own risk tolerance criteria under Norwegian legislation.

The corresponding UK approach is that authorities have formulated the upper tolerability limit, namely AIR less than 10^{-3} per year. However, they have also explicitly stated that the value should be a factor of 10 lower for new installations, i.e. AIR less than 10^{-4} per year. These values correspond to FAR = 34 and 3.4, respectively.

Most Norwegian companies use a risk tolerance limit of FAR equal to 10, as an average for all employees onboard, with total offshore hours as exposure, including major accident risk on installations, risk during helicopter transportation and occupational accident risk. This value has been unchanged for many years; the last revision performed by Equinor was about 20 years ago, when the limit was revised to include the full helicopter transportation phase. Most other companies have never revised their risk tolerance criteria. For the most exposed group, a value of FAR = 25 is often used.

Some Norwegian companies use AIR less than 10^{-3} per year as an upper tolerance limit in an ALARP context, but without any stricter limit for new installations.

The preceding subsections demonstrated clearly that current statistical levels are well below FAR = 10, for the Norwegian sector in particular, as well as for worldwide operations. The current statistical values are now so far below FAR = 10 that they are starting to be meaningless as a risk tolerance criterion for new installations.

The use of such a very relaxed limit no longer has any impact on design and operation, and it is surprising that PSA does not require that the petroleum industry in Norway revise its risk tolerance criteria. Even though there is a requirement for an additional ALARP evaluation, the level of risk tolerance criteria gives some indications of an ambitious level, and these indications are nowadays extremely weak.

This should be seen in the context of what Abrahamsen and Aven [14] found, namely that authorities should define risk tolerance criteria rather than industry.

Another factor should also be considered. Norwegian regulations require the operators to adopt a continuous improvement attitude in their implementation of HES work. The lack of revision of risk tolerance criteria over so long periods does not signal a continuous improvement attitude at all. The impression created is that the industry favours flexibility, also in the exposure of employees to risk levels.

The discussion of these aspects focuses on risk to personnel, but there are some parallel evaluations with environmental risk (see also Chap. 18). Oil companies in Norway also use risk tolerance criteria for environmental risk that are so relaxed that they have no impact on the design of installations and equipment as well as the planning and execution of operational activities.

In an ideal world, ALARP evaluations should provide additional risk reduction compared with the use of risk tolerance criteria. In practice however, the author has come to realise that the industry needs to have absolute limits to relate to and that it is unable to adopt reasonable risk tolerance criteria that affect facilities and operations. This is consistent with the theoretical findings of Abrahamsen and Aven [13].

Therefore, what recommended FAR tolerance limits would affect the planning of new facilities and operations? The values predicted in Sect. 17.2 correspond to FAR values of 1.3 and 2.2 for production and mobile installations, respectively.

On this basis it may seem to be a reasonable 'stretch target' to define a FAR limit of 1.0. Such a value would be considered as a challenge to meet. However, it may be somewhat less of a challenge if helicopter transportation was omitted, and the values limited to risk on installations. This is actually more natural, as the owners and operators of installations have no or little influence on the risk level during helicopter transportation, except in the choice of helicopter models used. If this were done, the predicted values in Sect. 17.2 would correspond to FAR values of 0.8 and 1.4 for production and mobile installations, respectively. This suggests that such a FAR limit would be a real 'stretch target' for mobile installations, but not for production installations.

17.7   Risk Tolerance Criteria for Environmental Spill Risk

The most recent documentation of how risk tolerance criteria for the environment are derived is found in the MIRA report [15]. These guidelines are being updated at the end of 2018, but only general aspects of the updating have been available. The development of tolerance criteria for environmental spills is based on the following main principles:

- Environmental damage is mainly classified according to the quantities of pollutant that will reach the shoreline.

- The duration of environmental damage (i.e. until restoration has been completed) is the main expression of environmental damage.
- The duration of environmental damage shall be insignificant in relation to the expected time between such damage occurrences.

The first principle should work satisfactorily in open sea, but if the activity is undertaken in waters close to shore or in environmentally sensitive areas, it may have to be omitted, because the drift models are too coarse under these circumstances.

The second principle is implicitly based on the assumption that a finite restoration period can be established, namely that the damage to the environment is not beyond repair. There is some resistance to this principle by environmental specialists, based on claims that sometimes the extent of damage is so severe that restoration is impossible.

The third principle is subjectively considered to be sound, based on the fact that apart from the spill caused by the 'Braer' tanker in 1996 no spill has occurred in the North Sea in the past 40 years (for which the restoration period has been more than one year). Some may be inclined to connect the 'Braer' spill with the offshore industry, but it should really be considered to be an 'ordinary' tanker accident. The third principle is thus more difficult to implement in practice. This was thoroughly discussed in Vinnem [16], and it is summarised below.

17.7.1 Initial Approach

The first Norwegian approach to risk tolerance for the environment was based on the decision to split environmental consequences into three categories, according to the effects on marine life, with the frequency limits as shown in Table 17.5. The

Table 17.5 Initial approach to risk tolerance criteria for the environment

Effect category	Typical effects	Acceptable annual frequency of exceedance
Major, long term effect	Environmental effects that are not restored within 5 years Effect on more than 100 km of coastline Spills with a major effect on wildlife	1×10^{-5}
Significant effect	Environmental effects that are restored within 2–5 years Effect on more than 10 km of coastline Spills with a significant effect on wildlife	1×10^{-3}
Minor effect	Environmental effects that are restored within 2 yrs Effect on more than 1 km of coastline Spills with a limited effect on wildlife	1×10^{-2}

likely effects of spills at inshore locations should be assessed based on an evaluation of wind, wave, current and topological effects. A rough estimation would suffice unless this risk aspect was critical, in which case an oil dispersion simulation might be necessary.

The environmental spill risk should be considered based on the average risk level for all field installations (on a "per field" basis) over the field life.

17.7.2   Current Approach

The so-called 'current' approach in the Norwegian offshore industry is based on the same overall principles as stated above. The relationship between recurrence and duration should be made in relation to the most vulnerable resources (often referred to as 'Valued Ecological Component'—VEC) that may be affected. The criteria can be applied at three levels:

- Field specific values
- Installation specific values
- Operation specific values.

Installation-related criteria should be used for all activities considered together in relation to an installation. The use of criteria related to specific operations should be for isolated operations, such as the drilling of an exploration well.

Risk tolerance criteria cannot be considered to be definite limits. Risk reduction measures should always be considered when the risk results are in the ALARP interval. Risk tolerance criteria should thus be satisfied for all environmental damage categories for the risk level to be acceptable.

The quantification of risk to the environment should be carried out with respect to both frequency and consequences. The presentation of risk is perhaps best undertaken using a risk matrix which is consistent with the accuracy in modelling. Nevertheless, a continuous function is sometimes used, which often gives a false impression of accuracy that is not warranted.

The tolerance limits are quantitatively different depending on whether a field, an installation, or an operation is considered. It may not always be obvious at which level acceptance should be considered; this has to be decided explicitly in each case. If in doubt, it is recommended that the highest level is used, namely field-related before installation-related and installation-related before operational-related. The categories of environmental damage are defined as follows:

- Minor—environmental damage with recovery between 1 and 12 months
- Moderate—environmental damage with recovery between 1 and 3 years
- Significant—environmental damage with recovery between 3 and 10 years
- Serious—environmental damage with recovery in excess of 10 years.

Table 17.6 Current tolerance limits for environmental damage [17]

Environmental damage	Duration of damage (restoration time)	Operational specific risk tolerance limits
Minor	1 month – 1 year	$<1 \times 10^{-3}$
Moderate	1–3 years	$<2.5 \times 10^{-4}$
Significant	3–10 years	$<1 \times 10^{-4}$
Serious	>10 years	$<2.5 \times 10^{-5}$

The approach takes as its starting point the resources (VEC s) and their recovery. Which installation is the origin of the spill is not important from a resources point of view. The principal basis is that a particular resource in total should not be exposed to accidental spills for a longer proportion of time than can be considered to be 'insignificant'.

The interpretation of this principle in practice has resulted in relaxed risk tolerance limits, and these limits are therefore not recommended. Table 17.6 presents a typical example from a recent exploration drilling project. The results from the analysis confirm that the limits are relaxed.

The first impression of these risk tolerance criteria is that they form a rational set of criteria that should address the overall environmental risk limits by including both field-level values and requirements for both installations and operations. However, several aspects of the criteria need further attention:

- The manner in which these criteria have been derived implies that it is assumed that such values may be based on observations, which is impossible with rare events. As an illustration, only one blowout with a significant oil spill has occurred during 50 years of exploration and production in the Norwegian sector, and this spill did not reach the shore.
- The level implicit in the interpretation of 'insignificant' may correspond to a too high risk level.
- The criteria are sometimes used to consider if a new region or area should be opened for petroleum activities or not. This may lead to a narrowminded evaluation, without taking the full spectrum of potential consequences into account. If this is the case, the criteria are not suitable for such decision-making.

In summary, the derivation presented above has some weaknesses and thus it should be further developed. There are also a number of complicating aspects in relation to the use of these criteria that have not been fully considered:

- What should be done when there are more (or fewer) installations in a field than the average used in the derivations? Should the average or the real value be used?
- The risk quantification is calculated for either the most vulnerable resource or some of the most vulnerable resources. Risk quantification is thus a representative value, rather than a full summation of all contributions.

The main challenges are on the other hand associated with the application of the environmental risk analysis approach in risk evaluations, not the criteria as such. This is discussed in Chap. 18.

Alternative approaches are presented in Refs. [16] and [18]. The approach presented in Ref. [16] is old and could be considered as outdated, since it has never been adopted by anybody. There is at the same time the fact that risk acceptance principles or tolerance limits have never been convincing to environmentalists. The approach in Ref. [16] could be reformulated to the following:

- Consider a new region which is not currently opened for petroleum activity (for instance Lofoten, Vesterålen and Senja in the Norwegian sector). The petroleum sector should be willing to state a challenging goal for environmental risk for the next 30–50 years, if petroleum activity were to be allowed in this area during such a period.
- Such target value could for instance be 1%, or even 0.1%, as the total probability of significant environmental damage (restoration time 1 year or more) in the entire region during the full period.
- The target value for the full region would then have to be translated into values for a number of well drilling activities and production activities for individual fields, to split it up into allowable risk contributions for activities and facilities.
- These new levels would be far lower than the current tolerance limits and would represent considerable challenges for the industry. This could be a way for the petroleum industry to convince the public that they 'mean business' with respect to environmental protection.
- This discussion attempts to separate environmental risk from climate warming issues. Such separation is not commonly found but might sometimes be beneficial to clarify what are the most important concerns.
- This illustration does not take into account that petroleum activities sometime into the distant future may not be allowed to expand further (or even continue), due to climate concerns. This would require some form of international cooperation and ratification, which at the moment is challenging to foresee. In any case, such considerations are outside the scope of the present discussion.

17.8 Risk of Material Damage/Production Delay

Production delay is often used as the means to express material damage risk. The frequency of events that cause delays (of equivalent full production) are typically determined for the following delay categories:

- 1–7 days
- 1 week–3 months

- 3 months–1 year
- 1–2 years
- above 2 years.

The contributions of accidental delays to production unavailability should be low; namely less than one 10th of the total production unavailability. A risk aversive approach should be taken to require that the most serious accidents make the lowest contributions to total risk. It is not normal practice to define explicit risk tolerance criteria for the risk to assets. The common approach is to analyse this risk, and implement risk reduction measures based on cost benefit analysis. Approach to cost benefit analysis is outside the scope of this book (see [11]).

17.9 Risk Tolerance Criteria for Temporary Phases

Traditionally, the probability of structural failure has been the safety factor considered in the temporary phases of large constructions. This is not natural in the context of offshore safety, where (according to the Norwegian legislation) the responsibility for safety covers:

- Personnel
- Environment
- Material assets/investment.

This implies that objectives and risk tolerance criteria need to be focused on all three dimensions of risk. The overall objectives for safety management in the construction and installation phases, are:

- Protection of personnel involved in the construction and installation activities.
- Protection of the investment in the structures and equipment.

Although the protection of the environment also falls within the responsibility of the operator, accidents that may cause significant pollution are rarely possible during the construction and installation phases. This is therefore not seen as a main objective, but may need to be addressed for the sake of completeness.

There have been several attempts to formulate risk tolerance criteria for temporary phases, such as the installation phase. Experience, however, suggests that quantitative criteria are impractical and should not be developed. A risk matrix approach is often used, instead, in conjunction with an ALARP based approach to risk reduction.

Normally unmanned installations represent another type of temporary phases (see further discussion in Chap. 22).

References

1. PSA (2012) Trends in risk level in the petroleum activity. RNNP 2011 report. April 2012 (Summary report in English, other reports only in Norwegian)
2. PSA (2011) Investigation—gas leak on Gullfaks B 4.12.2010, PSA. www.psa.no
3. PSA (2005) Investigation of gas blowout on Snorre A, Well 34/7-P31A (in Norwegian only), Petroleum Safety Authority [Norway], undated
4. Woad (1994) WOAD database. DNV GL, Høvik, Oslo
5. Vinnem JE (1998) Blast load frequency distribution, assessment of historical frequencies in the North Sea. Preventor, Bryne, Norway, report no 19816–04
6. HSE (2012) Offshore safety statistics bulletin 2011/12. http://www.hse.gov.uk/offshore/statistics/stat1112.pdf. Accessed 31 Aug 2012
7. IOGP (2012) Safety performance indicators—2011 data, report no 2011s, May 2012
8. Vinnem JE, Vinnem JE (1998) Risk levels on the Norwegian Continental shelf. Preventor, Bryne, Norway, Aug 1998, report no 19708–03
9. PSA (2011) The Management Regulations; Petroleum Safety Authority, Norwegian Pollution Control Authority and the Norwegian Social and Health Directorate, Stavanger
10. Norway Standard (2010) Risk and emergency preparedness analysis, Z–013. Standard Norway, Oslo
11. Aven T, Vinnem JE (2007) Risk management, with applications from the offshore petroleum industry. Springer, London
12. HSE (2015) The offshore installations (Offshore Safety Directive) (Safety Case etc.) Regulations 2015. Health and Safety Executive, HMSO, London
13. Cullen L (The Hon) (1990) The public inquiry into the piper alpha disaster. HMSO, London
14. Abrahamsen EB, Aven T (2012) Why risk acceptance criteria need to be defined by the authorities and not the industry? Reliab Eng Syst Safety 105:47–50
15. Norwegian Oil and Gas (2007) Recommended method for environmental risk analysis—MIRA (in Norwegian only), Norwegian Oil and Gas, April 2007
16. Vinnem JE (1997) Environmental risk analysis of near–shore wildcat well, approach to rational risk acceptance criteria, SPE/UKOOA. In: SPE/UKOOA European environmental conference, 1997, 15–16 Apr, Aberdeen, Scotland
17. DNV GL (2018) Environmental Risk and Emergency Preparedness Analysis (MRABA) for exploration wells 16/1-31 S&A in PL338 and 16/5-8 A in PL 815 in the North Sea for Lundin Norway AS, report 2018–1122
18. Vinnem JE, Wagnild BR, Heide B (2011) New approach to risk monitoring for acute environmental spill to sea on the Norwegian Continental Shelf, presented at SPE European Health, Safety and Environmental Conference in Oil and Gas Exploration and Production held in Vienna, Austria, 22–24 Feb 2011

Chapter 18
Environmental Risk Analysis

18.1 Overview of Environmental Risk—Norway

18.1.1 Acute Spill Statistics for the Offshore Industry

Table 18.1 presents the largest oil spills on the NCS since the beginning of offshore petroleum activities in the late 1960s.

If tankers spill were included, only one or two of the spills on offshore installations would be included on a 'top 10 list', namely the Macondo oil spill (see also Chap. 5), and possibly the Ixtoc oil spill in 1979.

The only offshore oil spill in the North Sea or Norwegian Sea exceeding 10,000 tons (with certainty) was the Ekfisk B blowout in 1978, where the spill is stated as 12,000–22,000 tons in the literature. No oil spill reached shorelines in this case.

18.1.2 RNNP Presentation of Environmental Risk in Norway

The RNNP work by PSA (see also Sects. 6.3 and 22.3) was extended in 2009 to include risk for acute environmental spills in addition to risk to personnel health and safety. This section summarises the work in the RNNP extension to the environment.

18.1.2.1 Main Spill Mechanisms and Sources

In order to monitor risk related to the full range of spills, all potential acute leak mechanisms and sources need to be addressed. The two main spill types are limited releases and potential major accidents that have oil leak potential (i.e. blowouts,

© Springer-Verlag London Ltd., part of Springer Nature 2020
J.-E. Vinnem and W. Røed, *Offshore Risk Assessment Vol. 2*, Springer Series
in Reliability Engineering, https://doi.org/10.1007/978-1-4471-7448-6_18

Table 18.1 Overview of the largest oil spills on the NCS [3]

Year	Spill (m³)	Installation	Event summary
1977	12,700	Ekofisk Bravo	Blowout during well intervention
1989	1,400	Statfjord C	Oil leak due to fracture in concrete wall in storage cell in platform structure
1992	900	Statfjord field	Oil spill due to valve on hose from loading buoy left in open position
2003	784	Draugen field	Oil spill due to rupture of connection to subsea inst
2005	340	Norne field	Oil spill due to manual valve in produced water system left in open position
2007	4,400	Statfjord A	Oil spill from loading hose rupture during offloading from Statfjord A to shuttle tanker

catastrophic events such as fire and explosion, etc.). The split between sources with limited volume and those that have large volume is for convenience, not as a matter of principal difference. The sources with limited volume are various volumes in the hydrocarbon production and processing systems on the installation for which there is a statistical database available.

The potential sources of large spills are subsea reservoirs, storage tanks and storage cells (condeep structures) and pipeline systems that have large volumes. The coupling to major accident precursor categories is discussed in a subsequent subsection. Note that blowouts, process leakages and leakages from sub-sea production systems, pipelines, risers and so on are included in the term 'hydrocarbon releases'.

18.1.2.2   Reported Acute Environmental Spills

Limited acute crude oil releases occur typically in the order of up to once per installation per year on average, and there is sufficient statistical data available at a national level to produce risk indicators that are capable of showing trends. Environment Web is a database in which all operators on the NCS report acute spill incidents and this database has been used to evaluate the acute releases that have occurred of crude oil, other oils and chemicals.

According to the Petroleum Act's definition of petroleum activity, data for the following offshore facilities/systems are reported in the Environment Web data-base until 2013, replaced by the EPIM Environmental HUB (EEH) from 2014 [1]:

- Permanently located installations. This also includes the installation phase
- Mobile facilities when on location, including drilling installations, well intervention installations (subsea) and so on
- Multiservice vessel flotel/crane if the main function is flotel

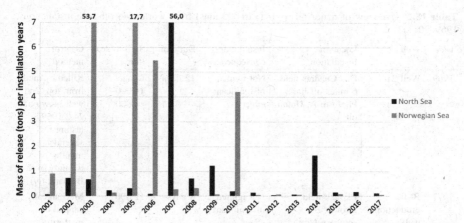

Fig. 18.1 Mass of oil release to sea per installation years because of occurred releases North Sea and Norwegian Sea, 2001–2017

- Well stimulation/well processing installations
- Transport systems.

Figure 18.1 presents a summary of the mass of released oil in occurred releases, 2001–2017, for the North Sea and the Norwegian Sea, expressed in per installation years.

The years 2003, 2005 and 2007 are dominated by one large spill each of the years, Table 18.1. For the other years, there is a great variation throughout the period for both areas.

18.1.2.3 Potential Spills from Major Accidents

Blowouts are not rare events; in fact there were 15 blowouts in the Norwegian sector in the period 1999–2009, all of which were gas blowouts, and all but one were shallow gas blowouts. By contrast, blowouts and other events that cause major oil spills to sea rare events, as can be seen from Table 18.2. Such occurrence data are therefore not suitable for providing indicators that can be used in a meaningful manner; they would virtually never change, even on a worldwide basis. Perhaps more importantly, the rare occurrences would give a false impression that these events are not important, which would be severely misleading.

This is the same situation as for major accident indicators relating to personnel safety. The adopted solution is also parallel, in the sense that major accident precursors have been applied, and almost the same set of major accident precursors is used for personnel safety indicators. It can be noted that the following list of categories of major hazard precursors is used:

Table 18.2 Overview of major oil spills (>10,000 tons) from worldwide offshore installations, 1969–2018)

Date	Well	Location/Installation	Installation consequences	Duration	Spill to sea (tons)	Control method
1969	Well 21	Dos Cuadras field, 6 miles offshore, Platform A, Union oil. U.S.	Not ignited, insignificant effect	12 days	Range: 10,000–14,000	Killed by mud from top. 2nd well blew out 24.2.1969, lasting for several months (County of St. Barbara, 69)
1977	B–14 production well intervention	Ekofisk field, Ekofisk B, southern North Sea, Norway	Not ignited, insignificant effect	8 days	12,000	Mechanically capped on platform (see Sect. 4.2)
1979	Ixtoc 1 exploration well	Gulf of Mexico, Mexico (800 km south of Texas)	Fire, total loss	294 days	Range: 350,000–450,000	BOP closed, had to be reopened 2 relief wells (see Sect 4.3)
1988	Secondary blowouts due to total loss	Piper Alpha platform, Piper field, UK	Explosion, fire, total loss caused loss of well control	22 days	<10,000 (uncertain)	Mechanically capped on wreck (see Sect. 4.13)
2009	Montara H1 ST1	Montara field, Australia	Ignited during killing, total loss	74 days	Range: 4–30,000	Relief well, 5 killing attempts (IOSH, 2009, see Sect. 4.11)
2010	MC252–1	Macondo field, Deepwater Horizon semi-submersible, U.S.	Explosion, fire, sank after 2 days	87 days	Official prediction: 655,000	Static kill (bullheading, BP, 2010, see Sect. 4.5)

- DFU1 Non-ignited hydrocarbon leaks
- DFU2 Ignited hydrocarbon leaks
- DFU3 Well kicks/loss of well control
- DFU5 Vessel on collision course
- DFU6 Drifting object
- DFU7 Collision with field-related vessel/installation/shuttle tanker
- DFU8 Structural damage to platform/stability/anchoring/positioning failure

- DFU9 Leaking from and damage to subsea production systems/pipelines/risers/flowlines/loading buoys/loading hoses.

The categories of precursor events are called 'DFUs', which is a Norwegian abbreviation corresponding to scenarios for which emergency arrangements must exist. The industry has used the same categories in their data collection schemes. The full list of DFUs is presented in Ref. [2].

DFU4, Fire/explosion in other areas (non-hydrocarbon fuelled) has been omitted, as it is considered to be incapable of causing escalation to the extent a secondary spill could occur. It can further be noted that in the period these precursor data have been collected, starting in 1996, DFU2 has never occurred (see also Sect. 6.4.3). The mechanisms of these types of precursors are as follows:

- DFU3 may lead to uncontrolled blowouts that would cause acute spills, irrespective of whether ignition occurs or not (ignition may reduce the spilled amount, but this is disregarded). Ignited blowouts may lead to 'secondary spills' if the wellheads and/or X-mas tree fails in addition to the failure of the downhole safety valve (DHSV).
- Hydrocarbon leaks (DFU1, DFU9) may cause fire and explosion which may escalate to wells, risers or storage if several barriers fail, thereby causing 'secondary' spills. Releases that have occurred are included in the reported acute environmental spills.
- Damage to subsea production systems, pipelines, risers, flowlines, loading buoys and loading hoses (DFU9) may lead to hydrocarbon leakages.
- Construction failures owing to impact or internal failure (DFU5, DFU6, DFU7 and DFU8) may also cause 'secondary' spills if several barriers fail.

It should be noted that some of these mechanisms are very unlikely to contribute to acute spills, reflecting the fact that many barriers have to fail simultaneously in order for a spill to occur. Their contributions are nevertheless quantified as a conservative approach, in order not to omit any mechanism that may cause acute environmental spills, as well as to reflect the fact that some of these mechanisms, even though they are unlikely, have occurred in the past. An overview of risk indicators based on the potential amount of oil spilled is presented in Fig. 18.2. This figure presents an index based on considering potential alternative scenarios/outcomes resulting in oil spills and the associated probabilities. The figure presents accumulated data for all the above DFUs. The source of the information is the 2018 version of the RNNP report [3].

For the Norwegian Sea there were three years with significantly higher values: 2002, 2005 and 2010. This is mainly due to well control incidents in those years. For the other years, there have been variations from year to year with no particular trend, for both the Norwegian Sea and the North Sea.

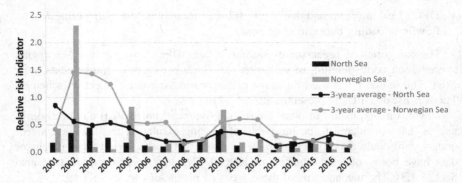

Fig. 18.2 Risk indicator for potential releases based on all DFUs, NCS, North Sea and Norwegian Sea, 2001–2017

18.1.2.4 Relationship Between Operational Spills and Major Spill Events

We referred to the ratio between the average fatality risk for offshore employees from occupational accidents, major accidents at the facility and helicopter accidents in Sect. 17.2.5, In a similar way one can use the values in the RNNP study to predict the contribution to the amount released from spills, including releases that have occurred, potential releases from precursor events, and potential releases in connection with the transport of crude oil by tankers. This is calculated as the average over the period 2001–2010 for the Norwegian sector as a whole. The chart below shows the ratio for the entire shelf.

All spills that have occurred are included, except those from injection and drill cuttings. Several of the largest spills that have occurred relate to the discharge of crude oil onto shuttle tankers. This is not covered by precursor events.

The RNNP work has not developed models for the risk of accidental release associated with oil transport by tankers to shore, a simplified approach is rather used in terms of an activity indicator for the volume of crude oil transport by tankers. The contribution from such transport is therefore calculated based on values from a risk assessment for a development project [4]. Since models for precursor events are based on risk analysis, there are no significant fundamental differences in the methodological approach. However, there is greater uncertainty in this contribution, since not all the conditions and assumptions of the risk analysis are fully known. Moreover, the contribution limited to shuttle tankers and navigation to an onshore terminal. This may be part of the explanation for the low contribution from such accidents. Accidental releases related to offloading are included among the releases that have occurred (Fig. 18.3).

As regards the contribution of precursor events that can cause acute releases resulting from major accidents if barriers fail, the contribution from well control failure events is 73% of the contribution from precursor events in total. They therefore constitute 73% of 78%, i.e. 57% of the grand total. Other contributions are

Fig. 18.3 Relationship between releases that have occurred and potential releases from precursor events, calculated as contributions to released average amount for the NCS, 2001–2010

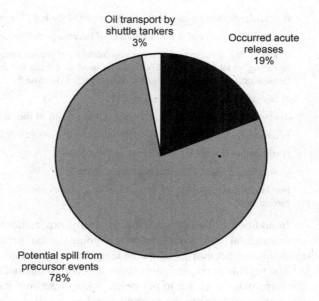

Oil transport by shuttle tankers 3%

Occurred acute releases 19%

Potential spill from precursor events 78%

from fires and explosions as well as from structural damage that may cause spills in the case of the failure of multiple barriers.

18.2   Regulatory Requirements etc.

18.2.1   Norway

The following is quoted from the Management regulations:

Section 17

The responsible party shall carry out risk analyses that provide a balanced and most comprehensive possible picture of the risk associated with the activities. The analyses shall be appropriate as regards providing support for decisions related to the upcoming processes, operations or phases. Risk analyses shall be carried out to identify and assess what can contribute to, i.e., major accident risk and environmental risk associated with acute pollution, as well as ascertain the effects various processes, operations and modifications will have on major accident and environmental risk. Necessary assessments shall be carried out of sensitivity and uncertainty.

The risk analyses shall

a) identify hazard and accident situations,

b) identify initiating incidents and ascertain the causes of such incidents,

c) analyse accident sequences and potential consequences, and

d) identify and analyse risk-reducing measures, cf. Section 11 of the Framework Regulations and Sections 4 and 5 of these regulations.

Risk analyses shall be carried out and form part of the basis for making decisions when e.g.:

a) identifying the need for and function of necessary barriers, cf. Sections 4 and 5,

b) identifying specific performance requirements of barrier functions and barrier elements, including which accident loads are to be used as a basis for designing and operating the installation/facility, systems and/or equipment, cf. Section 5,

c) designing and positioning areas,

d) classifying systems and equipment, cf. Section 46 of the Activities Regulations,

e) demonstrating that the main safety functions are safeguarded,

f) stipulating operational conditions and restrictions,

g) selecting defined hazard and accident situations.

For larger discharges of oil or condensate, simulations of drift and dispersion shall be carried out.

In addition to the above, there are many expectations with regards to how the environmental risk assessments are going to be performed, expressed in the guidelines to Section 17 of the Management regulations.

The regulative requirements stress that environmental risk is to be considered in the same manner as risk to personnel, taking reservoir and fluid characteristics into account, in addition to the spreading and evaporation of oil on the water, for acute spills to the environment.

Acute releases into the atmosphere and the marine environment are in principle of equal importance. The practice however, is that acute releases into the marine environment are analysed, whereas releases to the atmosphere are not.

The guidelines presented above are applicable to offshore petroleum facilities in the Norwegian sector. The same requirements also apply to major onshore petroleum facilities, but the guidelines are somewhat different.

18.2.2 IOGP Oil Spill Risk Assessment Standard

IPIECA and IOGP has published a guideline for oil spill risk assessment and response planning for offshore installations [5]. Part 1 of the document presents how oil spill risk assessments can be performed in accordance with the ISO31000 standard [6]. Part 2 of the document presents oil spill response planning including determination of oil spill response resources.

For those countries that adopt this IOGP standard, the requirements will be more or less parallel with the Norwegian legislative requirements to environmental risk analysis. This implies that also some of the challenges that are discussed in Sects. 18.6, 18.7 and 18.8 may be of interest to more people than just those who are faced with Norwegian legislative requirements for environmental risk assessment.

18.3 Modelling of Environmental Risk Analysis

18.3.1 General Aspects Relating to Environmental Risk Analysis

Some requirements to environmental risk analysis are stated in the PSA management regulations and associated guidelines, such as those quoted above. Additional requirements are also stated in NORSOK Z–013 [7]. The following are the requirements of Section 7.5.12:

(a) the analysis shall include all relevant scenarios as identified in the HAZID. Relevant scenarios are those that can contribute to the risk level of the system;
(b) for the identified release scenarios, discharge rate and duration distributions shall be established. The analysis shall reflect the variation in release rates and release durations;
(c) the analysis shall include modelling of drift and dispersion of the relevant harmful substance(s) on the sea surface and in the water column, and exposure of coastline, if relevant;
(d) the analysis shall consider the effect of relevant barriers such as detection, drain systems etc.;
(e) the analysis shall include modelling of the exposure of sensitive environmental resources, at least for the period for the planned activity and subsequent month;
(f) the analysis shall include calculation of environmental consequences. The consequences shall be a function of the relationship between amount of harmful substance and environmental sensitivity. Consequences shall be calculated for the identified risk indicators, e.g. populations or habitats;
(g) if calculations of environmental effects are performed, these results shall be documented;
(h) it shall be possible to compare the environmental risk contributions from different facilities in an unambiguous way, i.e. the calculation of environmental consequences must be comparable;
(i) the analysis shall also reflect planned emergency preparedness measures.

An informative Annex G of Z–013 also provides more details. There are few published papers on environmental risk analysis for drilling and production projects for the offshore sector. For Norwegian offshore exploration drilling and production projects, a specific method of environment risk analysis (i.e. MIRA) was developed in the mid 1990s (further description below). Even that approach has not been published in any recognised academic journals. However, the former US authority Minerals Management Service published a paper on OSRA (Oil Spill Risk Analysis) [8], which appear to have been the main emphasis from its point of view.

18.3.2 Event Trees

The event trees usually used in the analysis of environmental risk are often relatively simple and focus on aspects that may determine the duration of the uncontrolled flow. The factors that determine the duration of a blowout are usually the following:

- Immediate well 'killing' before developing into full blowout
- Mechanical isolation of the flow ('capping')
- Self stopping of flow in the reservoir ('bridging')
- Drilling of relief well(s).

Ignition of the blowout is also an important indirect factor, because an ignited blowout will place severe restrictions on the movement of personnel on the installation. In this event mechanical isolation activities may be prevented or take a longer time. The spill will also be less extensive because of the amount of oil that burns off.

Another important factor for the likely success of isolation activities is whether the well is a so-called 'dry completion' or a 'wet completion', namely whether the wellhead and Xmas tree are on a platform deck ('dry') or on the seabed ('wet'). The installation of mechanical devices in the well will be more complicated for a subsea completed well, which implies that a higher fraction of the blowouts may require drilling a relief well. A typical event tree for the environmental consequence analysis of oil spills is shown in Fig. 18.4.

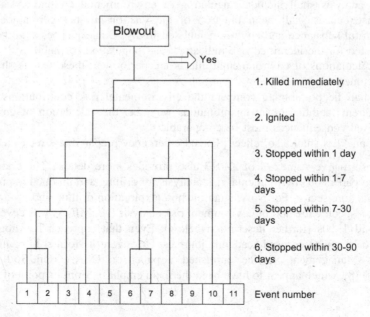

Fig. 18.4 Event tree often used in oil spill analysis

18.4 Overview of the MIRA Approach

18.4.1 General Principles

The MIRA approach [9–11], was developed in the mid 1990s. It is limited to one installation, and does not aggregate over ecological components. It is therefore not suitable for the presentation of an overall picture.

MIRA has been used for virtually all new drilling projects on the NCS since the late 1990s as well as new production fields. The basis is usually in a distribution from a QRA study, giving the frequency, leak rate and duration of possible leak scenarios. The main bulk of the work in the environmental risk analysis relates to simulating the possible behaviour of the spills under the influence of weather, waves and streams, as well as the consequences to valued ecological components (VECs) if the spill is stranded and possible local consequences in the water column.

MIRA [12], can be carried out according to the resources and time available for the analysis and the level of prior knowledge from comparable conditions. The three levels are called:

- **Source based analysis**: The simplest approach, based on the duration and rate of release, as well as distance to shore.
- **Exposure based analysis**: More extensive approach, based on the duration, rate and amount of release as well as oil drift simulation. The resources and effects of releases are considered in separate grid quadrants, typically 15 km by 15 km.
- **Damage based analysis**: Most extensive approach, based on the duration, rate, and effect potential of release, as well as oil drift simulation.

Consequences are related to the most vulnerable populations, including beach habitats. The source based analysis is the most conservative, and it has been indicated that the overprediction of frequencies by almost one order of magnitude is possible with this approach. The damage based analysis is the least conservative, but there is still distinct conservatism in the approach. The source based calculation should be used as a quick first round to determine whether a closer examination is warranted or not. Otherwise this approach can be used in order to provide detailed results a closely related project.

After extensive evaluation, the team involved in MIRA development focused on **recovery time** as the single parameter for the quantification of consequences. This is in principle the case irrespective of which analysis level that has been chosen, however only in the damage based analysis is recovery time calculated quantitatively. In the other two approaches, more qualitative and indirect assessment is performed.

Recovery time as a measure of environmental damage can be illustrated by considering actual data from some large spills [13]. This is shown in Table 18.3. It can be observed that more than half of the accidents shown were caused by tankers or other types of vessels. Further, all the impacts with the longest durations were caused by vessels.

Table 18.3 Overview of recovery times of some large oil spills

Source of spill	Year of spill	Calculated total spill (bbls)	Observed recovery time (years)
Exxon Valdez	1989	375–500,000	Around 10
Mercantile Marcia	1989	?	4
Oil pipeline, Louisiana	1985	?	1
Oil pipeline, Texas	1984	?	2
Amoco Cadiz	1978	20,000	5–10
Esso Bernica	1978	–	9
Ekofisk B platform	1977	22,000	1
Tsesis	1977	–	5–10
Arrow	1970	7,000	5–10
Santa Barbara	1967	>8,500	1
Torrey Canyon	1967	30,000	5–10

It can also be observed that there is no direct relationship between the amount of oil spilled and the resulting recovery time. The longest recovery times that have been recorded are in the order of 10 years. No spill with a recovery time shorter than one year is shown in the table because we only selected some of the largest spills as the basis for the presentation.

Irrespective of what level of analysis is selected, it will never be a calculation of the total environmental risk. First, the calculation is performed only for the most vulnerable populations. Second, if the activities are carried out in several seasons, then these are not added, and frequencies are predicted for one consequence category and each blowout scenario at a time. Four consequence categories are thus stated [14]:

Minor: recovery time 1 month to 1 year
Moderate: recovery time 1 to 3 years
Significant: recovery time 3 to 10 years
Serious: recovery time above 10 years

The derivation of allowable frequencies for each consequence group is based on the following:

- The overall goal for environmental protection is that the environment shall be as undisturbed as is practicably possible.
- This is taken to imply that the recovery time after environmental damage shall be insignificant relative to the frequency of occurrence of such damage (or the 'return periods'). It is suggested that 'insignificant' in that context is taken as 5% of the time, per damage category and per offshore area (see discussion below).
- This principle is applied to each category of recovery times.

- The NCS is subdivided into 5–10 areas, which are assumed not to overlap significantly with respect to exposure of vulnerable populations along the coastline.
- Average activity levels are defined for these areas, in order to compute the relationship between allowable frequencies for the area in total and for the installations and operations that take place in an area.

It is worth noting that the approach presented by Klovning and Nilsen [14], is the first that attempts to develop a rational basis for determination of tolerability limits for the environment. The approach therefore represents an important step forward. However, some aspects of the approach need in-depth consideration. It is difficult to understand the basis (according to [14]) for transforming the requirement about an insignificant extent of disturbance into an assumed fraction of time of 5% per area and for each consequence category. If the risk contribution from accidents in each of the four categories has the same level, then there will be disturbance nearly 20% of the time, rather than 5%, from each of the 5–10 offshore areas.

The division in areas may also be questionable from an acceptability point of view. It is likely that society's perception of the risk level will be independent of whether different areas are affected by accidental spills. The perception is likely to be based on evaluating the entire coast as a whole. Seen from a public perception point of view, there could be a virtually unbroken effect on one strip of coastline at all times if these principles were taken to their limit. This aspect is also to some extent related to so-called risk aversion. The risk aversion implied by the criteria is discussed in a later section.

Another aspect indicated in Ref. [14] is that risk acceptance would be applied even to operations, because the overall values are broken down into field specific values, installation specific values and operation specific values. This breakdown is to a level lower than what is considered to be justifiable in the case of risk to personnel.

MIRA is conducted through the execution of a series of steps that may be summarised as follows:

- Identify release scenarios into the environment

  - Based on an environmental HAZID
  - Scenarios from QRAs

- Analyse barriers on the installations that may prevent spills or reduce amount of spilled volume into the environment

  - Including the reduction of discharge rates and duration of spills

- Establish release scenarios

  - Leak location (geographical, topside and subsea)
  - Contaminant characteristics
  - Discharge rates and durations

- Simulation of the drift and dispersion of the oil for relevant scenarios

  - Spread on the sea surface, contamination of water column
  - Drift time, evaporation, emulsification
  - Stranding of oil on shore

- Establish the occurrence of environmental resources within the influence area and their vulnerability/sensitivity towards the contaminant
- Calculate drift time to and the exposure of these environmental resources to the contaminant

  - Overlap between the contaminant and scenarios for distribution of biological recourses

- Assess (qualitatively or quantitatively) the short and long term effects on these environmental resources

  - On individuals and populations (establish relevant, reliable and valid consequence categories)

- Assessments shall be based on updated science and the monitoring and mapping of biological recourses
- Calculate the risk as a combination of the probability of a certain event causing environmental damage and the degree of the seriousness of the damage
- Indicators for sensitivity can be

  - Vulnerability of the resources (individual and/or population/habitat level) to the contaminant
  - Scientific value or administrative protection value of the resources.

The use of the MIRA approach requires that detailed meteorological data are available as well as software for the simulation of the drifting and spreading of the oil over time with currents, wind and waves. These simulations will also require considerable IT resources in order to perform a large number of simulations within reasonable time limits.

Damage to ecological components are in MIRA given as 'restoration time', which is the period after the population of a component has been affected by a spill, until it has reached the previous level, or at least close to it, taking natural variations into account.

The use of restoration time implicitly assumes that it will be possible for resources and habitats to restore their levels after serious oil spills. Some experts will argue that the most severe oil spills will have such dramatic effects that such restoration is impossible. One of the most severe oil spills in a vulnerable environment was the Exxon Valdez spill in Prince William Sound in Alaska in March, 1989. It is commonly accepted that most of the species affected in this accident have been restored or are close to being restored. Restoration time in this accident is thus over 20 years. The shortest restoration time considered in MIRA is one month. The four categories considered for restoration time were stated above.

18.4.2 Environmental Damage Distribution

The environmental risk will be expressed as the frequencies of environmental damage in the categories outlined above. The following would be the complete calculation of frequencies:

$$\lambda_{damage,i} = \sum_T \sum_J \lambda_{end,j} \cdot P_{A,j}(t) \cdot P_{B,jL}(t) \cdot P_{damage,i,j}(t) \tag{18.1}$$

where

$\lambda_{damage,\ i}$ frequency of damage for damage category i

$\lambda_{end,\ j}$ frequency of end event in Fig. 18.4 i.e., a release with specified duration according to the categories stated above and valued component j

$P_{A,\ j}(t)$ probability of exposure of an area with component j present at time t

$P_{B,\ j}(t)$ probability of the presence of the valued component j at time t

$P_{damage,i,\ j}(t)$ probability of damage in category i and valued component j at time t

T   total time over which damage frequencies are considered

J   total number of valued components.

The common approach to the implementation of MIRA [10] is that a few of the most vulnerable VECs are selected for analysis. These VECs are then considered individually, such that Eq. 6.9 is implemented as follows:

$$\lambda_{damage,i,j} = \lambda_{end,j} P_{A,j}(t) \cdot P_{B,jL}(t) \cdot P_{damage,i,j}(t) \tag{18.2}$$

where

$\lambda_{damage,\ i,\ j}$ frequency of damage for damage category i and valued component j.

18.5 Presentation of MIRA Results

The presentation of MIRA results is possibly the weakest aspect of this approach. A typical presentation of results is shown in Fig. 18.5 for the four seasons spring, summer, autumn and winter. Figure 18.5 shows the consequences in terms of restoration times in these four categories for various vulnerable species in the influence area, i.e. where spill scenarios may hit the coast under various weather conditions.

Figure 18.6 is to some extent special in the sense that is actually shows a frequency that exceeds the relevant tolerance limit. It appears from Fig. 18.6 that this applies to Grey seal for scenarios during autumn with a restoration time exceeding 10 years.

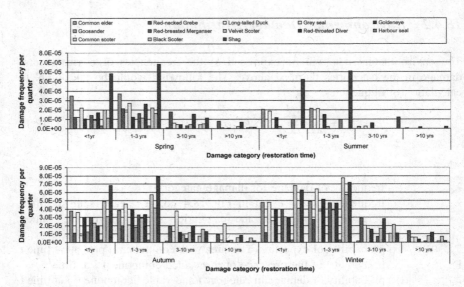

Fig. 18.5 Typical results from MIRA, presented separately for the four seasons

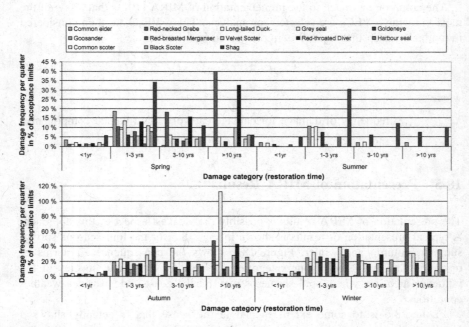

Fig. 18.6 Typical results from MIRA, presented as a percentage of the tolerance limit separately for the four seasons

However, this is not representative. In most cases, the values are far below the tolerance limits, and the environmental risk analysis is used to demonstrate that the risk to the environment is trivial.

A slightly modified version is presented in Fig. 18.6, which does not show frequencies of damage in the four categories, but rather the frequency value as a percentage of the tolerance limits that apply to the damage categories.

18.6 Discussion of the Current Practice

The implication of the presentations in Figs. 18.5 and 18.6 is that no overall risk can be presented, because the results are presented for each VEC and each quarter of the year.

It may be meaningful to distinguish between parts of the year if an exploration drilling project is considered, but for production this is irrelevant. Nevertheless, consultants continue to present results for each quarter. It would be more meaningful to present annual values, which may be easily calculated by summing the values per quarter (see Fig. 18.7). Tolerance limits for production are usually expressed as annual values, such that the values in Fig. 18.7 are needed. It is emphasised that the activity has to be continuous throughout the year in order to add the values for each the season of the year.

The tolerance limits can be included in the results presentation, as illustrated in Fig. 18.8. Logarithmic scale is usually required because the tolerance limits span more than one order of magnitude. Tolerance limits are shown as red lines, indicating the tolerance frequency for each damage category.

However, no overall risk level presentation is available, even with the presentation in Fig. 18.8. The following premises and assumptions may thus be utilised in order to develop an overall presentation of risk.

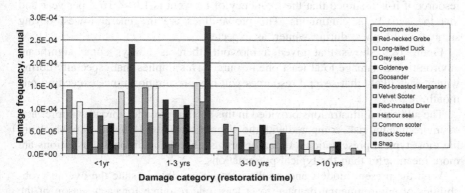

Fig. 18.7 Typical results from MIRA, with quarterly values summed to annual values

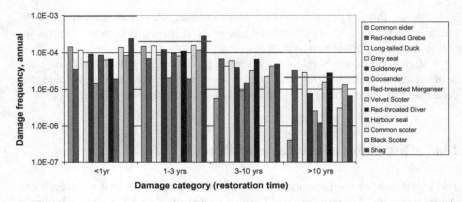

Fig. 18.8 Typical results from MIRA, with annual values in relation to tolerance limits (red lines)

From a public perception and evaluation point of view, it may be virtually irrelevant how long the restoration time will be (it is unknown at the time of the accident in any case). From the public debate, it may seem that environmental damage is intolerable in any case. This may be taken as an argument to sum all damage categories, as long as there is a significant damage, i.e. over one month restoration time.

If we sum all the damage categories in Fig. 18.8, the following annual frequency of environmental damage results:

$$1.32 \times 10^{-3} \text{per year}$$

This value corresponds to a 0.13% probability of significant environmental damage during a year, while over 10 years the value is approximately 1.3%. If the field lifetime is 30 years, the corresponding probability of significant environmental damage is 4%.

Figure 18.9 shows the variations in the probability of significant damage to each resource if it is assumed that the frequency of blowout is 1.0×10^{-3} per year and that the activity is continuous. The probabilities are in general lowest during summer and highest during winter, as expected.

Figure 18.9 shows that given a blowout, there is always some significant environmental damage to at least one resource. This implies that especially during winter, it is likely that several resources will have restoration times exceeding one month.

The additional illustrations provided in this section provide a broader impression of environmental risk, compared with the diagrams in Figs. 18.5 and 18.6 which are the most typical presentations. It can be argued that the additional illustrations are more meaningful than the typical presentations.

With the present models and results it is not possible to state the overall probabilities of environmental damage to at least one resource for each season of the

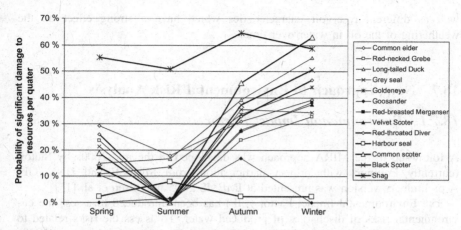

Fig. 18.9 Typical probabilities of significant damage by season given that a blowout occurs

year, but it should be easy to modify the modelling in order to provide these results. The results may look like Fig. 18.10.

Figure 18.10 shows the conditional probability of significant damage to at least one resource given that a blowout occurs in one of the periods throughout the year. A blowout during autumn and winter is certain to lead to a restoration time above one month for at least one resource, whereas the probability during the summer season is just over 50%. For the year as a whole, there is a 82% probability that at least one resource will have restoration time above one month given that a blowout occurs.

It is expected that such values would vary quite considerably between offshore fields, because of differences in location and distance to vulnerable resources as

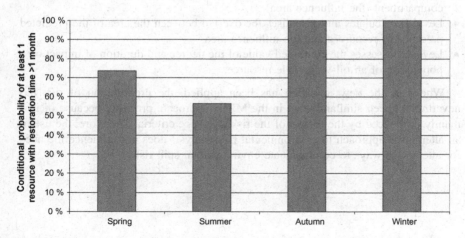

Fig. 18.10 Typical probabilities of significant damage by season given that a blowout occurs

well as different reservoir characteristics which have a strong effect on the weathering of the oil in water over time.

18.7   New Approach to Environmental Risk Analysis

18.7.1   Environmental Impact Factor

A follow-up to the MIRA approach was developed in the mid-2000s by Statoil (currently Equinor) with project participation from SINTEF and DNV GL. A preliminary version was presented at IMEMS 2005 by Nilsen et al. [15].

The Environmental Impact Factor (EIF) has been introduced to assess the environmental risks of discharges of produced water. To assess the risks related to accidental oil discharges, the EIF model for the risk assessment of acute oil spills (EIF Acute) is currently under development. The EIF Acute concept aims to:

- Be comparable, at some level, to the EIF concept implemented for produced water as well as account for the presence of resources vulnerable to oil
- Include a comparable measure of the environmental damage risk in different environmental compartments (water masses, sea surface and shoreline)
- Be able to comprise the expected probability or frequency of the incident, as well as the extent and duration of the damage
- Be applicable to single installations, oil fields and regions worldwide
- Serve as an environmental management tool, allowing analyses of risk reducing efforts.

EIF Acute is planned to have three levels:

- Level I calculates the area where oil amounts exceed the threshold level for each compartment (the influence area)
- Level II calculates and visualises the overlap between the area of the predicted presence of resources and the influence area
- Level III assesses the combined value of the degree and duration of impact on a population of an oil-vulnerable resource.

Whenever the new approach has been applied, the presentation of risk has nevertheless been similar to that in the MIRA approach, probably because of it is mainly influenced by the format of the risk tolerance criteria. Therefore, the use of an alternative approach to environmental risk analysis does not influence the need for alternative ways to communicate environmental spill risk.

18.7.2 The ERA Acute Model

The status in 2017 is that the MIRA approach will most likely be replaced by a new approach, referred to as ERA Acute. The main change from MIRA is the application of consequence assessments, where more sophisticated models than in MIRA are being implemented. In general, where MIRA used historical data to predict the consequences, the ERA Acute model is to a larger extent based on understanding and modelling physical mechanisms.

Consequence assessments in ERA Acute are based on continuous functions, and are thus more able to respond to changes in input information, compared with the use of discrete categories in MIRA.

Where recovery time played a key role in the MIRA approach, the ERA Acute model in addition includes a resource damage factor (RDF) as a combination of impact (for example, loss of a population) and recovery time.

In general, the risk calculations in the ERA Acute model are more sensitive to the effect of consequence-reducing measures, such as the effect of oil recovery strategies and other emergency preparedness measures, than the MIRA approach. This makes it possible to quantify the effect of consequence-reducing measures in terms of updated risk calculations.

18.8 Need for Alternative Ways to Assess and Communicate Risk

Section 18.6 presented several additional illustrations of environmental risk based on the results generated with the MIRA methodology or easily derived from these results or intermediate results.

When environmental risk owing to several installations and activities in a larger area is being considered, this is usually carried out by running MIRA results for each installation and activity, and presenting all results. This is virtually meaningless, or least entirely misleading.

The environmental risk assessment for an area should aim at providing an overview which cannot be achieved by the presentation of detailed results per location/activity.

Such evaluations must also consider the oil spill over the field lifetime, not only year by year.

Such evaluations must also consider that it is extremely unlikely that two blowouts will occur within a period of say several years. During more than 45 years of the exploration and production of oil and gas offshore, only once has a blowout with a significant oil spill occurred., thus underlining that two blowouts close in time are very unlikely.

A realistic environmental objective of the development of a new field could therefore be that the total national probability of significant environmental damage over the field lifetime should not increase significantly from that of the new field alone, calculated as an additional contribution to the total probability of significant environmental damage owing to offshore oil and gas alone. An interpretation of 'a significant increase' could be an upper limit of a 1% (possibly a 0.1%) increase in probability. This implies that the probability of significant spills because of shipping is ignored.

Another parameter may also be interesting for the overall environmental classification of new drilling and production permits. Figure 18.10 suggests that some locations in certain seasons may have very high conditional probabilities of causing significant environmental damage. The average conditional probability of causing significant environmental damage over the year in Fig. 18.10 is 83%. The highest probability in any season could be used as an alternative. These probabilities reflect the reservoir conditions, location of the facilities in relation to the valuable resources and expected performance of barriers. The reservoir conditions and location are not possible to influence, but the performance of barriers is. Unfavourable reservoir conditions (such as if little of the oil is expected to evaporate) and location could be compensated for by higher performance requirements (PRs) for the barriers. This is currently not carried out in practice, because it is always argued that risk tolerance criteria are satisfied. With such an approach, more focus could be placed on performance of barriers. A review of the results from a number of environmental risk analysis studies (published by the Norwegian Environment Agency) demonstrates large variations with respect to the resource that has the highest conditional probability of significant damage (i.e. over one month restoration time):

- Barents Sea:

  - 45%

- Norwegian Sea:

  - 11%
  - 2%

- North Sea:

  - 72%
  - 71%
  - 19%
  - 12%

It may seem somewhat surprising that the highest values are from the North Sea, but it should be noted that the two locations with over a 70% probability are among those with the shortest distance to the coast.

All these cases have been presented with only a few percentages of the relevant risk tolerance limits, which leaves the impression of low or negligible risk. Conditional probabilities of significant environmental damage given that a blow-out occurs provide a much more severe impression, and therefore suggest much more incentive to reduce such values, which would be good in an ALARP context.

References

1. Norwegian Oil and Gas (2018) Recommended guidelines for emission and discharge reporting, rev. 16
2. Vinnem JE et al (2006) Major hazard risk indicator for monitoring of trends in the Norwegian offshore petroleum sector. Reliab Eng Syst Safety 91(7):778–791
3. PSA (2018) Trends in risk levels, report on acute spills (in Norwegian only). Petroleum Safety Authority
4. Lilleaker (2007) Volume and frequency analysis of oil spills, LA-2007-16
5. IPIECA and IOGP (2013) Oil spill risk assessment and response planning for offshore installations. Oil spill response joint industry project
6. ISO (2018) ISO/IEC 31000 risk management—guidelines
7. Standard Norway (2010) Risk and emergency preparedness analysis, Z–013. Standard Norway, Oslo.
8. Price JM et al (2003) Overview of the oil spill risk analysis (OSRA) Model for environmental impact assessment. Spill Sci Technol Bull 8(5–6):529–533
9. Vinnem JE (1997) Environmental risk analysis of near-shore wildcat well; approach to rational risk acceptance criteria. Presented at UKOOA/SPE European environmental conference, Aberdeen, 15–16 Apr 1997
10. Sørgård E, Jødestøl K, Hoell E, Fredheim B (1997) A stepwise methodology for quantitative risk analysis of offshore petroleum activities. In: SPE/UKOOA European environmental conference, 15–16 Apr 1997, Aberdeen, Scotland
11. Norwegian Oil and Gas (2007) Method for environmental risk analysis—MIRA (in Norwegian only), Revision 2007. DNV report 2007-0063
12. Jødestøl KA, et al (1995) Method of environmental risk analysis (MIRA) (in Norwegian only) DNV report 95-3562, Høvik
13. Vinnem JE, Vinnem JE (1998) Risk levels on the Norwegian Continental shelf. Preventor, Bryne, Norway, Aug 1998. Report no.: 19708–03
14. Klovning J, Nilsen EF (1995) Quantitative environmental risk analysis. SPE paper 30686. Presented at SPE annual technical conference & exhibition, Dallas, 22–25 Oct 1995
15. Nilsen H, Brude OW, Hoel E, Johansen Ø et al (2005) Development of the environmental impact factor model for risk assessment of acute oil spills (EIF Acute). In: International marine environmental modeling seminar (MEMS) conference, 23–35 Aug 2005, Helsinki, Finland
16. EU (2013) The safety of offshore oil and gas operations directive 2013/30/EU and amending directive 2004/35/EC. http://eur-lex.europa.eu/homepage.html
17. Vinnem JE, Wagnild BR, Heide B (2011) New approach to risk monitoring for acute environmental spill to sea on the Norwegian Continental shelf. Presented at SPE European health, safety and environment in oil and gas exploration and production, Vienna, 22–24 Feb 2011

Chapter 19
Approach to Risk Based Design

19.1 Overview

19.1.1 About the Need for Risk Based Design

Risk-based design has been the focus of attention for several years, although in the last few years it has frequently been referred to as risk-informed rather than risk-based, to emphasize that decisions should not only be made based on risk assessment results. In real applications, however, the difference between risk-based and risk-informed is not always easy to see. In this book we have used the term 'risk-based'.

The idea of risk-based design is that protective and mitigative measures are chosen and designed according to the hazards and the risks that are present, instead of applying deterministic design solutions, given by authority requirements, standards or previous projects. This gives an opportunity for optimisation in terms of the solutions adopted in each particular case.

One consequence of risk-based design is the extensive use of risk assessment. Since the criteria used in risk-based design are related to aggregated risk (as in the 10^{-4} criterion), the risk calculations may change along the design of the offshore installation. For example, as more details are included in the 3D models, explosion pressures tend to increase, resulting in a need for more robust blast walls in order to meet the 10^{-4} criterion. This means that risk-based design implies an inherent risk of design changes in late project phases in order to meet the criteria, and such late changes may be very expensive. To prevent this effect, there has been discussion in recent years on developing an alternative to risk-based design, where the intention is to use experience from previous projects more directly. This method is referred to as the RISP method, and is elaborated on in Chap. 14. By using the RISP method, the risk of late changes is reduced. However, the result may also be additional robustness and hence additional costs compared to the use of risk-based design.

© Springer-Verlag London Ltd., part of Springer Nature 2020

225

J.-E. Vinnem and W. Røed, *Offshore Risk Assessment Vol. 2*, Springer Series in Reliability Engineering, https://doi.org/10.1007/978-1-4471-7448-6_19

The current chapter focuses on risk-based design only. The following questions will be the focus of the discussion in this chapter, which addresses the risk-based approach in general:

- Do we not trust the design solutions chosen on this basis?
- Do we not trust risk analysis as a tool?
- Is the data not good enough?
- Do we not trust the risk assessors who use the approach?

The above will be discussed for the main types of risk:

- Fire
- Explosion
- Collision impact
- Impact by falling loads.

As a further illustration of the approach to risk based design, some comments may be made regarding design against blast loads. The approach taken before 1990 could be characterised as follows:

- Worst case conditions for explosion were defined.
- Blast loads were simulated or calculated for the worst cases, using an approach based on empirical data or CFD (Computational Fluid Dynamics).
- Design solutions were established and implemented for the worst case conditions.

Design solutions could be implemented for these conditions in a cost effective way, therefore no-one saw the need for a more advanced approach. In QRA studies, however, simple probabilistic assessments of blast load distributions were made.

In 1998/99 the situation regarding design against blast loads was dramatically changed. This followed the completion of the large scale test programme at the British Gas Spade Adam test centre, in the BFETS test programme—Blast and Fire Engineering for Topside Structures—SCI [1]. The results from these tests suggest that:

- It is not possible to design for the worst case conditions or if it is, then cost-wise it becomes impossible.
- Some alternative approach is needed.
- The probabilistic approach has become an industry standard.
- An extensive approach to probabilistic modelling of blast loads was developed by the Norwegian oil companies (at that time) Norsk Hydro, Statoil, and Saga Petroleum. This procedure has been further developed following its proposition in 1998.

Thus for explosions it is not a question of who wants to adopt a probabilistic approach, but rather if anyone can afford not to do so.

19.1.2 Scope for Risk Based Design

Risk based design is in accordance with the normal offshore design practices used in both structural and topside regimes. There are differences in the regulatory requirements in these two regimes as discussed in the following section. The systems and equipment that may be considered using a risk based design approach are detailed in Table 19.1.

In the following sections of this chapter reference will be made to three different classes of study, according to the three categories in Table 19.1. The design of structural systems and passive safety systems is essentially a structural design, although only the former is according to the dedicated standards for structural design. The other class in Table 19.1 applies to safety and emergency systems or systems that are vital in a safety context.

19.1.3 Challenges for Design

A number of aspects may pose a challenge to risk based design, either because typical solutions may be expensive, or because there is considerable uncertainty about analytical results. The subject itself is controversial and this may in itself be a problem. The brief listing below is split into what is called 'standard installations' and >minimum installations=.

19.1.3.1 Standard Installations

This term includes integrated installations (or bridge linked) with permanent manning. Both fixed platforms and floating production units of various types are considered.

Table 19.1 Overview of the scope for risk based design

Structural systems	Active safety and emergency systems	Passive safety systems
Support structure	Gas detection	Passive fire protection
Hull structure	Fire detection	Blast protection
Deck structure	ESD valves	Dropped load protection of topside
Buoyancy	Process safety valves	systems
compartments	Process safety instrumentation	
Subsea installations	Fire water supply	
	Deluge systems	
	Sprinkler systems	
	Ventilation system	

The design challenges may be related to the importance of design accidental loads and premises. Experience has shown that these are particularly important for:

- Passive structural fire protection

  - Particularly for fire on sea, but also for protection of topside equipment.

- Design to blast loads

  - Blast and fire barriers.
  - Blast resistance of equipment.
  - Blast resistance of structures.

- Dimensioning against collision impact

  - Particularly steel platforms.
  - Floating concrete structures.
  - Also other floating installations, if reserve buoyancy in the damaged condition is not provided within the deck structure.

- Dimensioning against impact from falling objects.

  - Particularly for buoyancy elements.
  - Special criticality for TLP, if not designed for dynamic failure of one tether or the filling of one compartment.

In the case of FPSOs, the frequency of impact by shuttle tankers has been quite high, although not in relation to scenarios which have implications outside local damage. Nevertheless it is an important issue which was discussed in Sect. 10.4.

19.1.3.2   Minimum Installations

These installations are primarily the 'not normally manned' platforms. They are virtually without exception fixed installations, typically wellhead platforms with or without minimum level processing.

Nevertheless the size of the platform and the extent of the systems installed can vary considerably. Sometimes mobile drilling units are coupled with other installations, such that they together form a production installation. We have chosen to regard the combination as a 'standard' installation, even though each of the installations may in isolation be considered as 'minimum'.

All the aspects that may be important for the standard installations are also applicable to minimum installations and in many cases may have higher importance. In addition other aspects may also be critical, such as:

- Extent of use of active safety and emergency systems.
- Availability requirements for active safety systems.
- Protection against external impact for minimum structures.

19.1.3.3 Challenges for Operational Safety

In some cases the use of floating production installations has introduced new hazards that need particular attention because they fall in the border area between operation and design. The use of active systems for marine control of installations is one such aspect.

The most obvious historical example of a severe accident relating to a design solution which relied heavily on operational control, is the accident on the Ocean Ranger in Canadian waters in 1982. Ocean Ranger had a port light in the ballast control room in one of the columns smashed during storm conditions. The ingress of sea water caused short circuits in the ballast control system thereby leading to uncontrolled operation of the ballast valves.

The operators were not able to rectify the situation sufficiently quickly and believed they had closed the valves when they actually had opened the valves. The heel angle therefore increased, and the rig soon came to an inclination where it could not be righted. The unfortunate detail of the design was to have ballast pumps only at one end of the pontoons. The result of this was that under some conditions of inclination, the suction head for the pumps was such that certain tanks could not be emptied. A summary of these events is given in Sect. 4.21.1. It should be noted that all onboard the rig were lost.

Another example of the interaction between design and operation is the turret arrangement on some FPSO designs with a so-called drag-chain system. Active manual control of the turning arrangement employing a complex turning/locking system is required. This necessitates extensive coordination between two people at different locations who do not both have the same information.

19.2 Authority Regulations and Requirements

19.2.1 Norwegian Installations

19.2.1.1 Dimensioning Accidental Events

Since 1990 the Norwegian offshore regulations have contained provisions for conducting risk based design of structures as well as equipment. In the previous set of regulations, the fundamental events were called 'Dimensioning Accidental Events', which according to the Risk Analysis regulations [2], were defined as follows:

• Dimensioning Accidental Event	'Accidental event which according to the defined acceptance criteria represents an unacceptable risk, and which consequently serves as a basis for design and operation of installations and otherwise for implementation of the activities'

The industry had problems with this definition since it was launched and the NORSOK Risk and Emergency Preparedness guidelines [3], formulated an alternative definition. In the revised definition the dimensioning accidental load is the crucial definition, from which the definition of the dimensioning accidental event follows.

• Dimensioning Accidental Load	'Most severe accidental load that the function or system shall be able to withstand during a required period of time, in order to meet the defined risk tolerance criteria'
• Dimensioning Accidental Event	'Accidental events that serve as the basis for layout, dimensioning and use of installations and the activity at large'

There were two main problems with the original definition, the first is the obvious reference to 'an unacceptable risk'. What are the implications if the risk level as assessed in the concept phase is acceptable? Shall this be taken to imply that no accidental events shall be the basis of design? Probably not, but there is no apparent answer. The second problem relates to the interpretation of the wording 'an unacceptable risk'.

The NORSOK standards for structural design NORSOK N-001 [4]; N-003 [5] and N-004 [6] do not refer to anything but 'accidental actions' and 'Accidental Limit State' (ALS).

The facilities regulations [7], define 'Dimensioning accidental load' as follows: 'An accidental load that the facility or a function shall be able to withstand for a defined period of time.' There is no specification of how these loads shall be determined. The need to distinguish between DiAL and DeAL was discussed in Sect. 7.7.

19.2.1.2 Use of Dimensioning Accidental Loads

The regulations still refer to the expression 'dimensioning fire and explosion loads' and use of risk analysis for various design purposes, such as in the following instances:

- Design of passive fire protection
- Design of fire divisions
- Decision on communication equipment
- Combat of acute environmental pollution
- Use of standby vessels
- Selection of survival suits for personnel protection in an emergency
- Specification of emergency unit for personnel treatment in an emergency
- Use of chemical for emergency preparedness.

One additional aspect should be noted. Throughout the regulations there are some aspects where the Norwegian regulations stipulate minimum or deterministic solutions, which cannot be altered through any kind of analysis. Examples of such minimum requirements are:

- The need to disregard the cooling effect provided by active fire protection systems, when designing the passive fire protection.
- The need for the fire water supply to be able to provide fire water to the largest area and the largest of its adjacent areas.
- The need to provide free fall type lifeboats with sufficient capacity.

19.2.1.3 Structural Design

The design of structures according to NORSOK N-001 [3] requires an accidental limit state (ALS) design check which shall consist of:

(a) Resistance to accidental actions. The structure should be checked to maintain the prescribed load-carrying function for the defined accidental actions.
(b) Resistance in damaged condition. Following local damage which may have been demonstrated under (a), or following more specifically defined local damage, the structure shall continue to resist defined environmental conditions without suffering extensive failure, free drifting, capsizing, sinking or extensive damage to the external environment.

The approach refers back to risk analysis in order to determine what shall be the accidental actions.

19.2.2 UK Regulations

A brief overview of UK regulations was given in Sect. 1.6. It should be noted that the UK regulations do not have similar requirements for the use of risk analysis in order to define dimensioning accidental loads.

19.3 Relationship with Risk Analysis

The risk based design approach has an obvious close relationship with risk analysis. This is obvious and is not controversial in a general sense, but it becomes more of a problem when considering in detail how it may be implemented.

This section addresses the use of risk analysis for risk based design of topside systems and equipment. Some of the aspects which need clarification are:

- Which type of risk analysis is most suitable?
- How to select the dimensioning accidental events in an event tree?
- Which are the important aspects to model in an event tree?
- What type of consequence models to use in the risk analysis?
- Is the risk analysis sufficiently sensitive to changes in safety systems?

19.3.1 Suitable Risk Analysis

The requirement that the risk analysis is suitable also has implications in the determination of what are suitable risk tolerance criteria. This discussion will cover both the type of criteria and analysis, because these are quite close considerations. These topics are strongly dependent on the type of systems or elements addressed in the risk analysis.

Another aspect of suitability is that the risk analysis is carried out in accordance with high ethical standards, i.e. without willingly trying to manipulate the risk picture. If this is done the risk based design will fail dramatically, as in the case study presented in Sect. 3.12. This aspect will to a large extent depend on how assumptions are made and verified. If unrepresentative or unrealistic assumptions are made in order to eliminate some hazards or accident mechanisms, then the risk based design will fail catastrophically.

19.3.1.1 Structural Systems

Under Norwegian legislation structural systems must be designed in accordance with the approach prescribed by the relevant regulations. The acceptance limits are stated in the regulations, defining functional requirements with respect to structural consequences, and the upper limit of exceedance frequency, 10^{-4} per load type per year. Both the statement of the maximum consequences and the frequency of occurrence are related to structural aspects explicitly, as follows:

- Consequences: Functional capability of the structural system in the damaged condition.
- Frequency: Occurrence frequency for accidental loads on the structure.

This implies that the analysis of consequences need not be taken very far and therefore it is sufficient to consider structural aspects alone. This limits the need for extensive consequence analysis, and implies that the uncertainties will be limited, as discussed in Sect. 2.4.3.

19.3.1.2 Passive Safety Systems

The passive safety systems are to a certain extent coupled with the structural events considered in the preceding section, in that the dimensioning aspect for these systems is usually related to a load specification, such as:

- Heat load capacity for passive fire protection
- Blast load capacity for blast protection
- Impact load capacity for dropped load protection.

This should determine the type of risk tolerance criteria and risk analysis to be used in this context. They should resemble the criteria and approach for the structural systems as much as possible. The facilities regulations [6] have defined similar requirements for the escalation prevention as for the structural systems. This type of criterion is often referred to as 'escalation criteria', in the sense that the function of the passive barrier is often to prevent escalation from one area to the neighbouring areas.

The advantages noted for the structural design approach, with respect to limited extent of consequence modelling and reduced extent of uncertainties, are also applicable for the design of passive barriers.

19.3.1.3   Active Safety and Emergency Systems

The active safety systems are the most difficult class to handle in relation to risk based design, because the aspects that need to be addressed are so diversified. Table 19.2 has attempted to capture parts of this by listing some of the design parameters that will be of interest in relation to these systems.

Table 19.2   Overview of parameters for design of active safety systems

System	Parameters	System	Parameters
Gas detection	Type of detectors Location of detectors Detection logic Detection availability Test interval	Fire water supply	Number of pumps Type of pumps Reliability of power supply Availability Independence from operational systems
Fire detection	Type of detectors Location of detectors Detection logic Detection availability Test interval	Fire water distribution	Routing of supply lines Redundancy Sectionalisation Inspection and test intervals
ESD valves	Location of valves Sectionalisation Type of valves Closing time Availability Fire protection	Deluge systems	Location of valves Sectionalisation Location of deluge heads Availability Test interval
Process safety valves	Location of valves Sectionalisation Type of valves Closing time Depressurisation time Fire protection	Sprinkler systems	Location of valves Sectionalisation Location of deluge heads Availability
Process safety instrumentation	Need for additional instruments	Ventilation system	Natural/mechanical systems Location of ducts Separate systems Availability

The variety of these parameters implies that no universal type of analysis is particularly suitable for all of these purposes. It must also be considered which of these parameters is actually suitable for risk based design.

Availability is a parameter which is usually included explicitly in the event trees used in the studies. The aspects of the systems in Table 19.2 that are related to availability are therefore the most obvious candidates for risk based design, including the following:

- Gas detection
- Fire detection
- ESD valves
- Fire water supply
- Deluge system
- Sprinkler systems
- Ventilation systems.

It may be argued that availability is just about the only parameter that is well suitable for risk based design in relation to the active safety systems. However, the need for sectionalisation of the process volumes may also be addressed in a risk base context. Also requirements for test of safety systems may be determined with input from risk based design.

Other parameters such as type of valves, location of different equipment etc., are not possible to address in the context of a risk based approach. This is not a question of whether the studies are sufficiently sensitive to variations in these issues, but just the fact that current risk modelling is not able to address such aspects in a meaningful way.

A comment is also needed on what the analysis shall be used for in relation to these systems. The results from the analysis will in many cases be used to identify design **premises**, rather than design loads.

There may be several types of risk analysis and risk tolerance criteria used, as long as the risk results can be grouped into different scenario and severity categories. Further, the risk aspect needs to be quantified on a continuous scale, which reflects variations in accident frequencies and consequences. This implies that fatality risk, impairment risk and escalation risk parameters may be used.

19.3.2   Use of Event Trees

19.3.2.1   Design of Active Safety Systems

Event trees are mostly used in the analysis for the risk based design of active safety systems, and the question is then whether relatively coarse event trees or much more detailed trees should be used.

If the simple event tree shown in Fig. 15.13 is considered, then it will be seen that the following aspects are included:

- ESD system for isolation of segments
- Ignition
- Strong explosion
- Escalation to other equipment and area.

This is a rather limited selection of systems. The simple event tree is aimed primarily at aspects that dominate the risk to personnel. Therefore it may be defensible to have a simple tree, but this tree structure implies that the following systems are not addressed:

- Gas detection
- Fire (flame/heat/smoke) detection
- Fire water supply
- Deluge system
- Sprinkler systems
- Ventilation systems.

A sophisticated event tree is therefore required, in order to use risk analysis to select design premises for active safety systems. A tree structure as shown in Figs. 15.14 and 15.15 should be more appropriate.

The following is a list of issues that need careful consideration when planning a risk analysis study in a design context:

- Release rate (should be considered as time dependent function) for flow inside/-outside tubing, with/without unrestricted flow, etc.
- Gas leak geometry i.e., direction, obstructions, etc.
- Calculation of flammable cloud size as a basis for calculation of ignition probability and explosion overpressure.
- Performance of drain systems in order to limit the duration of pool fires.
- Possible ignition sources, their time dependency, strength, etc.
- The distinction between ignitions that will cause a fire versus explosion.
- Estimation of gas explosion overpressure.
- Calculation of fire sizes for alternative scenarios and the impact of radiation at points which may be critical in respect to trapping of personnel or impairment of safety functions.
- Calculation of the smoke impact at points which may be critical with respect to trapping of personnel or safety functions.

Some Norwegian oil companies have established through cooperation target values for safety system availability.

19.3.2.2 Design of Structures and Passive Safety Systems

The choice of the type of event trees to use is mainly determined by the choice of consequence models as discussed in the following section. Detailed consequence models will require detailed event trees, but if coarse consequence models are selected, then coarser event trees may also be used.

The event trees will often be different from those that are required in relation to active safety systems.

19.3.3 Use of Consequence Models

The question of whether coarse or sophisticated consequence models should be used, apply to design of structures and passive safety systems to resist accidental loads. This appears to be an easy choice, because obviously sophisticated consequence models should be used, but the dilemma is **when** to use these models. The following options are available:

- Use of sophisticated consequence models in risk analysis. The approach implies that:
    - All the sophisticated consequence models are integrated into the risk analysis.
    - Design accidental loads may be extracted directly from the analysis.
- Use of sophisticated consequence models when designing. The approach implies that:
    - Risk analysis is performed with 'order of magnitude' consequence models.
    - Selection of preliminary DiALs and DAEs is done on this basis.
    - Advanced consequence modelling is performed for the design of structures and/or passive protection in order to confirm or revise the preliminary design accidental loads.

Both fire and explosion are complex physical processes, which ideally should call for sophisticated modelling. But with the practical constraints outlined here, this becomes a dilemma without any obvious solution. This is illustrated in the table below. Table 19.3 very clearly demonstrates that the modelling of fire loads is considerably more complex than explosion and collision modelling.

It should also be pointed out that coarse modelling usually is more conservative, thus allowing more margin vis-à-vis minimum requirements. This implies that coarse modelling will imply more robustness in the design, which is an advantage.

Table 19.3 Comparison of characteristics of fire, explosion and collision loads

	Fire	Explosion	Collision
Characteristics	Heat load Duration Fluctuations Direction Susceptibility for different elements	Peak overpressure Impulse	Impact energy (Impact energy)

The preferred solution for use of sophisticated consequence modelling is the second option listed above, that is the use of the sophisticated consequence models when the design is being undertaken. This has the following advantages:

- The risk analysis may be performed with consequence models that are quick to carry out.
- It avoids the need to integrate advanced consequence models into the risk analysis, which is problematical in that many scenarios are needed, which may require very extensive resources and time if advanced consequence models are used.
- Advanced consequence modelling may be limited to those scenarios that have been selected as DAEs.

Even though this option may be the easiest to implement in practice, there may be phenomena which are so complex in a probabilistic sense, that the only defensible option is to chose the first option, namely to integrate the advanced consequence modelling into the risk analysis.

19.3.4 Sensitivity to Changes in Active Safety Systems

The sensitivities considered in Vinnem et al. [8] are reductions of the following failure probabilities (i.e. safety improvement), relating to nodes in the event trees:

- Leak detection (Leak Det)
- Operator intervention (Man SD)
- Number of hot work hours (Hot Work)
- Maximum likely overpressure (Expl ovpr)
- Fire detection (Fire Det)
- Manual combatment of small fire (Man fifi)
- ESD operation (ESDfail)
- Fire water unavailability (Fwfail)
- Escalation by neighbouring equipment (Eqm esc)
- Time to structural collapse (Str colps).

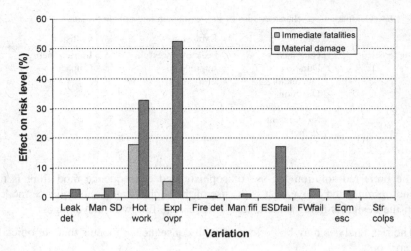

Fig. 19.1 Overall sensitivity results

Reduced number of hot work hours implies reduced ignition probability and reduced maximum explosion over-pressure implies reduced probability of strong explosions. It should be noted that the blowdown system was also part of the study, but was eliminated from the graphical presentation.

Figure 19.1 presents the overall results from the sensitivity studies conducted. The blow-down sensitivity showed no effect for risk to personnel, and an insignificant effect, less than 2% reduction of the frequency for total loss of platform as the highest contribution.

It should be noted that all sensitivities, except hot work and time to structural failure, were based on 50% reduction of failure probability. A 90% reduction of hot work duration was used, and the time to structural failure was reduced from 30 to 20 min (i.e. increase of risk).

It may be questionable whether 50% reduction of the failure probability is a representative improvement, or whether the insensitivity shown for most of the systems is due to too low improvement of each parameter. The 50% reduction of the failure probability was chosen because it is generally achievable with limited upgrading of an existing system without needing to install additional redundancy. Such improvement may be achieved by changes to maintenance, inspection or test procedures or intervals. This is really the main scope of possible improvement to an existing installation.

Later, an additional set of sensitivity studies was undertaken, to demonstrate the effects of more extensive changes of the availabilities. These later studies confirmed the findings in initial overall studies.

19:4   Approach to Risk Based Design of Topside Systems

19.4.1   Basis for Approach

During the search for an approach that might be sufficiently sensitive for the stated purpose, the situation demonstrated in Fig. 19.2 was noted. This figure shows the fatality risk contributions (to PLL) from different leak categories in the process systems.

One of the reasons why QRA results appear to be insensitive to variations in safety system standards, performance and functions is that the main contributions to process systems risk are often medium and large gas and liquid leaks. Safety systems and functions are less effective in providing protection against these, because the resulting conditions are too extreme for the systems to be fully effective.

On the other hand, perhaps the safety systems do not need to be capable of handling these larger leaks because they are relatively rare. For the referenced platform, the expected frequency of **ignited medium or large leaks** is:

$$0.007 \text{ per platform year}$$

If this is adopted, then the basis for design of safety systems and functions should be limited to small leaks only, which are obviously more frequent. The importance of these systems and functions could then be expected to be much more dominating. Figure 19.3, shows how the three leak sizes contribute to the frequencies of three different scenarios:

- Ignited and unignited leaks
- Ignited leaks only
- Significant material damage risk.

It should be noted that the proposed approach addresses the concern often expressed by authorities, that often only overall risk is considered and that local variations are 'washed out'. The approach suggested here, must surely be considered to consider 'local effects'.

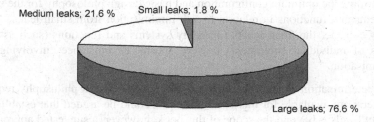

Medium leaks; 21.6 %   Small leaks; 1.8 %

Large leaks; 76.6 %

Fig. 19.2 Contributions to fatality risk from process system leak categories

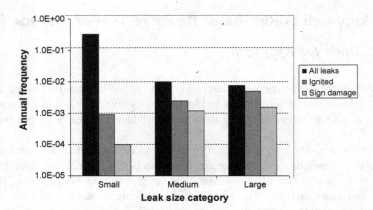

Fig. 19.3 Illustration of small leak frequency, ignited leaks, significant material damage frequency

The diagram shows that the small leaks are highest in number, but have the lowest contribution to risk (as could be expected). As an illustration, the conditional probability of significant material damage resulting from a leak is considered. For small leaks this is 0.03% whereas for large leaks it is 20%.

The possibility that safety systems could be important in guarding against the effects of small leaks is considered in detail in the following.

19.4.2   Fundamentals of Proposed Approach

The proposed approach may be described as follows:

- Calculate the risk from small gas and oil leaks separately.
- Divide the resulting scenarios into the following sub-categories:

  – Strong explosion
  – Spreading fire outside area (i.e. escalating on platform)
  – Spreading fire within area
  – Non-spreading fire

- Use this basis for sensitivity studies for safety systems and functions.
- Determine the optimum configuration and basic design philosophy for the safety systems and functions in relation to risk minimisation from small leaks.
- Only consider the finer details for safety systems and functions (such as locations of individual equipment) in very special circumstances involving risk optimisation.

The determination of an optimum configuration and design philosophy requires that criteria are available for this optimisation. It could be argued that establishing the target levels is beyond the scope of this book, however, a suggested approach is presented later in this chapter.

19.4.3   Overview of Sensitivities

This section summarises the individual sensitivities of the different safety systems and functions. Only an overview is presented, and the main systems and functions are discussed separately in the following section.

When discussing the sensitivities of the individual systems and functions, a parameter referred to as **relative sensitivity** is used. The relative sensitivity of a safety system is calculated as follows:

$$S_{rel} = \frac{Change\ of\ scenario\ frequency}{Change\ of\ safety\ system\ parameter\ unavailability} \qquad (19.1)$$

where

S_{rel} relative sensitivity of a system or function.

Figure 19.4 gives an overview of the relative sensitivities, according to the definition above, implication being that a relative sensitivity of 1.0 corresponds to a situation where 50% reduction of a certain parameter value leads to 50% reduction in the risk value related to small leaks.

The relative sensitivity of the explosion over-pressure exceeds 1.0, which is due to the definition of this sensitivity. The definition is somewhat indirect, as is further described in Ref. [9].

With one exception, all those parameters that could be labelled safety systems or functions have relative sensitivities that are close to or above 0.5 i.e., 50% reduction of a certain parameter value results in at least 25% reduction of risk associated with small leaks.

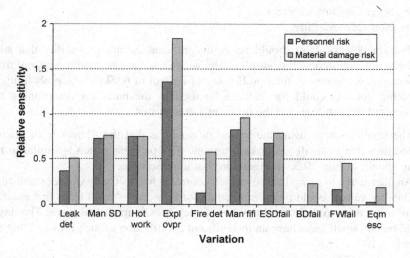

Fig. 19.4 Overview of relative sensitivities for personnel risk and material damage risk

19.4.4   What Should Be the Target Protection Level

It has been demonstrated that risk analysis, if carefully planned and executed, may be used as the basis for design and operational planning of safety systems and functions. Parallel conclusions could also be drawn for design of emergency preparedness systems. The next problem is how to establish the availability requirements for these systems.

A stepwise approach is proposed below, starting with an evaluation of whether the overall risk can be used. This will rarely be the case. The solution is then to use the minor leaks as the basis for establishing requirements, realising that the overall risk will be virtually unchanged irrespective of what values are chosen.

When small leaks are chosen it should be realised that the number of expected leaks per platform per year will usually be in the order of 0.1–1 per year. Therefore, the conditional probability that control is lost, needs to be quite low. Now the ignition probability for such a scenario is usually quite low, say a maximum of 1%. The overall annual frequency of a small ignited leak should therefore be less than 0.01 per platform per year. The proposed approach is:

1. First check to see if the overall risk is after all sufficiently sensitive to variations in some or all of the availabilities of the safety systems. If this is the case, then Step 2 and 3 are bypassed, and the availabilities that imply tolerable risk are chosen, possibly after having completed an ALARP evaluation.
2. If overall risk is not sufficiently sensitive to variations in availability of safety systems, then the contribution to risk from small leaks is reversed, and the relative contributions to risk are calculated for the following scenarios:

   - Strong explosion
   - Spreading fire out-side area (i.e. on platform)
   - Spreading fire within area
   - Non-spreading fire.

3. For small leaks there should be an insignificant overall probability that such leaks develop into an uncontrolled fire, which escalates beyond the system from where it originates. A total probability of control of 0.99 (or 1% probability of losing control) could for instance be used as the basis for determining the required availabilities for the different safety systems.

The implication of using the values proposed is that the frequency of uncontrolled fires due to small gas leaks is around 10^{-4} per year which is similar to the value shown in Fig. 19.3, and therefore not unreasonable.

An alternative to using fires for which control is lost as the basis for establishing availability values, could be to use the ratio of fires due to small leaks that escalate out of a fire area, to the overall number of such fires from all leak sizes. The target could be that small leaks have an insignificant contribution to such fires i.e., that the ratio is 1:100 or less.

It should be observed that the proposed approach does not tie in with the risk tolerance criteria, but the approach is constructed in a way which implies that over-all risk results are not significantly affected.

19.5 Risk Based Design of Structural and Passive Safety Systems

The risk based design of structural systems and passive safety systems is easier than the risk based design of topside systems. There are a few crucial issues involved in this:

- Optimisation with respect to design accidental loads may in some cases be possible. A simple example of this is shown in Sect. 9.6.
- The detailed analysis to be used in the determination of the explosion load probability distributions for the different areas on an installation is very critical. This is also discussed in Sect. 9.4.

19.6 Practical Considerations

19.6.1 Design Against Fire Loads

Design against fire loads implies the need for analysis of all relevant hydrocarbon containing systems on the platform. This includes three main categories of leaks; blowouts, riser leaks and process leaks. These areas are discussed separately below in order to point out the essential aspects to be considered. It should be noted that blowouts are not addressed here in relation to possible spill effects, only in relation to their effects on personnel and/or the installation itself.

19.6.1.1 Analysis of Blowout

Blowout is one of the main sources of fire risk, if there are wellheads on the installation (i.e., most fixed installations) or immediately below the installation. The second case applies to some floating installations, but not the majority. About 25% or the floating production installations in the Norwegian sector have this configuration with wellheads on the seabed immediately below the installation. When a semi-submersible type of floating production installation is equipped to perform its own well drilling and other operations, then the wellheads are virtually below the installation. In most cases of floating production, the wells are some distance away and the blowout hazard is not relevant in relation to the risk to the FPU.

Environmental effects are very dependent on the time needed to control a blow-out. This is not so important for the effects on personnel and structures, because the blowout will have a long duration in virtually any case. A short duration blowout in an environmental context has a duration of some few hours (as opposed to days or weeks), whereas an ignited blowout may be catastrophic to either personnel or structures if it has a duration of several hours.

Blowout consequences in the form of fire dimensions and loads are strongly dependent on the rate of the uncontrolled flow from the reservoir. It often appears that the assumptions used in predicting the flow rates in a blowout scenario are rather conservative. It would appear that more differentiation between particular situations is sometimes needed in the analysis of these scenarios in respect of:

- Distinguishing between unrestricted and restricted flow.
- Distinguishing between flow through tubing or through casing.
- Determination of the expected well flow potential as a function of the time.

In relation to a blowout the design basis for fire protection would be to protect personnel for the time needed for an orderly and safe escape and—if required—subsequent evacuation. It would not be realistic to design so extensive fire protection that the structure may survive an ignited blowout for any extended period. External cooling is normally required to provide the required extent of structural integrity if the blowout has a long duration.

19.6.1.2 Analysis of Riser Leaks

The analysis of riser leaks also includes leaks from flexible flowlines, which are critical from a risk point of view for floating production units. Also pipeline leaks close to an installation should be included in the same context, but these are usually of lower importance. The fire dimensions will be determined by the following aspects:

- Hole dimensions
- Pipeline diameter
- Pressure in the pipeline
- Obstructions in the flow path.

The dimensions of the fire may be extreme in the case of a full bore rupture of a riser or flowline. Very sophisticated software may be required in order to predict the time dependent flow rate in such a scenario, where the friction losses will be the determining factor. It is debatable to what extent an exact prediction of the flow rate is required, but sometimes it is important to know quite precisely the duration of the fire. This is the case if it is anticipated that personnel may remain in the shelter area (only relevant if this is on a bridge linked installation) or if structural fire protection is provided for the duration of such a scenario. The duration of the flow (and any fire) will be determined by the following aspects:

- Whether a Subsea Isolation Valve (SSIV) is installed and if so:
  - the response time for activation
  - the location
- Hydrocarbon inventory in the pipeline.

19.6.1.3 Analysis of Process Leaks

This analysis probably requires the most extensive analytical efforts because most of the leaks in this category are minor leaks. Under normal circumstances such leaks will not lead to escalation of the accident if the safety systems provided are adequate. That is the reason why the safety systems need to be considered carefully in relation to a variety of scenarios.

The idealised relationship between time, fire frequency and probability of escalation are shown in Fig. 19.5, assuming that these curves may be derived from an event tree. The frequency of fire is assumed to decrease according to the duration of the fire, such that long duration fires are less likely than short duration. Analysis of different leak scenarios and paths in the event tree should be the basis of establishing this curve. The probability of escalation is at the same time a result of considering the different escalation mechanisms in the paths of the event tree, and modelling this as a time dependent function.

There are a number of aspects that need to be considered in the analysis in order that the models have the required sensitivity to reflect changes in the safety systems, including:

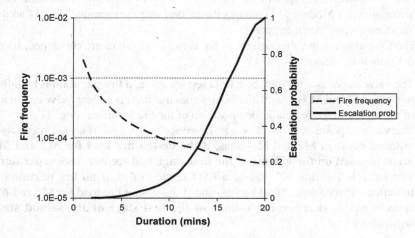

Fig. 19.5 Idealised modelling of duration dependent frequencies

- Application of idealised models to complex geometries involving confinement etc.
- Effect of isolation and blowdown
- Effect of deluge
- Escalation to other systems.

In terms of protection against escalation inside one area or module (i.e. a fire cell), there are several cases which may need separate consideration, as follows:

- If all equipment units inside a fire cell belong to the same segment, then there is in principle no explicit requirements for protection against escalation inside the cell as the entire contents of the segment will have been used as the basis for the design of the fire barriers against other areas.
- If there are several segments in the same cell, then two sub-cases will have to be considered:

  - If the fire barriers with other areas are designed on the basis of the hydro-carbon contents from all segments in the cell considered, then the case is as above.
  - If the fire barriers with other areas are designed only on the basis of the largest hydrocarbon contents from one of the segments in the cell considered, extensive protection against escalation from one segment to the others has to be provided.

A diagram such as in Fig. 19.5 should in principle be developed for each leak scenario in each module, but it would be expected that a discretisation is performed, such that a few representative cases are established for each module.

Figure 19.6 shows the location of isolation valves between a first stage separator and the downstream equipment on the liquid side. Module M2 is the first stage separator area and Module M3 contains the second stage separation. The M4 area is the produced water treatment area.

The following are the alternatives, as far as isolation valves are concerned, based on the principles above:

- The valve in position 'C' will be a PSD service valve, **if fire partition is installed between M2 and M3 areas**. This also assumes that there is a check valve up-stream of the second stage separator, down-stream of the fire partition. (Pos. 'H'.)
- The valve in position 'C' will be a PSD service valve, even if no fire partition is installed between M2 and M3 areas, **if the design fire load for M2 and M3 areas is based on the volume of the first stage and second stage separators**.
- The valve in position 'C' will be a ESD service valve, if **no fire partition is installed** between M2 and M3 areas, and if the design fire load for M2 and M3 areas is based on **either the volume of the first stage or the second stage separators**.

As regards the function of valve 'C', the following scenarios may develop: A fire in M2 will not affect equipment in M3 (and vice versa), if there is a fire partition

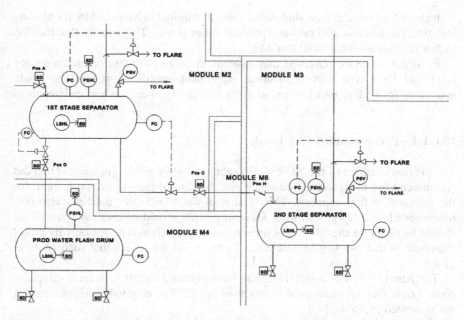

Fig. 19.6 Simplified schematics of three process vessels and possible isolation valve positions

between M2 and M3 areas. This implies that a check valve in M3 will not be affected by a possible fire in M2. The pressure vessel in M3 will accordingly be safely and reliably protected from feeding the fire in M2, by this check valve. The same applies for a fire in M3.

The assessment of the design fire load for M2 will have to be based on the maximum fire potential in M2, namely the contents of the pressure vessel in M2. No particular requirements apply to the valve in M2, when the fire is in the same module. Hence, an ESD valve is **not** required in the case of a fire partition between the modules.

The design fire load is the decisive factor when no fire partition is installed between the two modules. No particular requirements apply to the valve between M2 and M3, if the total contents of hydrocarbons in **both** modules M2 and M3, is the design basis. The valve may hence be of PSD type.

The valve in position 'C' **needs** to be of ESD type if the design fire load corresponds to the volume with the highest loads of the vessels in M2 and M3. An ESD valve gives the highest reliability of the closing function in a fire scenario. This is required to limit the duration of the fire to correspond to the design basis. Similarly isolation of the pressure vessel in M2 may be required, in the case of fire in M3.

Exactly the same requirements apply to the valve in position 'D', between stage separation and the produced water treatment tank.

It should be noted that no shut-down valve is required in Module M6, if only one isolation valve is installed between pressure relief points. This implies that the line in M6 is relieved either in M2 or M3.

Finally it could be observed that whereas there are different requirements for ESD and PSD type valves, identical equipment specifications may be made applicable for both types of valves in order to simplify operation and maintenance.

19.6.1.4   Recommended Fire Loads

The previous version of the NORSOK Technical Safety S-001 guidelines [10], did recommend fire loads to be used in the specification of passive fire protection, for the selection of fire scenarios. This version of the NORSOK guidelines also recommended how average and local maximum loads should be used, and that credit should be given for the effect of active fire water protection. The reason why this is important is that Norwegian authorities currently do not accept that such credit is given.

The new revision of S-001 [11], has recommended slightly changed values, to some extent they are somewhat higher than before. The new values recommended are presented in Table 19.4.

19.6.2   Design Against Explosion Loads

Design in relation to explosion loading usually has to be based on a probabilistic consideration, because design to worst-case conditions will be impossible, either technically or economically. The following main characteristics need to be considered in an advanced explosion load assessment:

- The leak itself i.e., the time dependent flow
- The dispersion of gas
- Ignition characteristics

Table 19.4 NORSOK recommended fire loads

	Jet fire		Pool fire
	For mass leak rates >2 kg/s	For mass leak rates 0.1–2 kg/s	
Local peak heat load (kW/m^2)	350	250	150/250[a]
Global average heat load (kW/m^2)	100	0	100

[a]150 kW/m^2 for burning rates 0.1–2 kg/s. 250 kW/m^2 for burning rates >2 kg/s

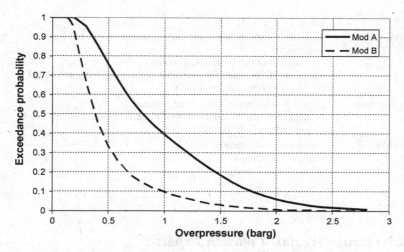

Fig. 19.7 Idealised exceedance probability curves for explosion loads

- The explosion loads, the time dependent (and location specific) pressure transients
- Load responses of equipment and structures
- Possible effects of load reducing actions.

An example of an idealised probability distribution for two modules is shown in Fig. 19.7. Two different curves are shown, with the same shape, which represent rather similar scenarios. The important sections of the curve are in the range 0.5–1.0 bar, where secondary failures may occur. The difference between the two curves is quite distinct in this range, with conditional probabilistic varying by a factor of up to 4.

The following aspects should be addressed in a detailed, probabilistic evaluation of explosion loads and responses:

Time dependent leak	• Flow regime i.e., gas jet or condensate flow
	• Location of leak point
	• Geometrical considerations
	  – Gas: Direction of jet
	  – Condensate: Direction of flow
	• Time dependent flow rate
	• Modelling of segments, explicit or averaged

Dispersion of gas	• Ventilation characteristics, wind direction/force • Dispersion modelling • Choice of idealised cases for consideration
Ignition characteristics	• Interaction between dispersion and ignition • Distinction between types of ignition sources • Location of ignition source s • Time dependent ignition probabilities

(continued)

(continued)

Explosion loads	• Choice of explosion overpressure modelling (i.e. CFD-based or empirical) • Resolution of the geometrical modelling • Simplifications made for cases that are not explicitly calculated • External effects from explosion
Load responses	• Type of load analysis has to be selected • Criteria for failure definition need to be established, in relation to equipment and structures
Load reducing effects	• Increased ventilation • Reduced blockage • Active use of deluge

19.6.3 Design Against Collision Impacts

Collision risk usually consists of two types of collision scenarios:

- Ships/vessels ramming the installation (powered as well as drifting),
- Collision between the installation and floating units located near-by (flotels, crane vessels, shuttle tankers, etc.).

The nearby floating units usually do not give extra design loads, except for floating concrete structures, if a local puncture of the wall may be caused. A traffic survey for the area should usually be carried out for a new platform location as the basis for quantification of the collision probability. This may be used for the probability and consequence analysis.

Speed distributions need to be established for each vessel type to distinguish between 'head-on' and drifting (see Chap. 9) collisions, in the latter case, wind and wave conditions need to be considered. The determination of collision energy will need to differentiate between various collision scenarios, taking geometrical aspects and dimensions into account.

When making the consequence analysis, the loads on the installation will be the primary focus, however, the following effects need also be considered:

- The superstructure of the vessel may impact on the deck structure.
- The vessel may impact risers, conductors or non-structural elements.
- The ship may sink and cause damage to pipelines and other subsea installations.
- The ship may catch fire which may also be a threat to the installation.

The probability of impact by passing vessels needs to consider the aspects outlined in Chap. 9. Also warning and/or assistance as well as emergency procedures for assistance to drifting vessels need to be considered.

19.6.4   Design Against Dropped Load Impact

Impact by dropped objects may cause damage to hydrocarbon systems, thereby causing a leak, or significant structural damage. The following accidental events should be addressed:

- Fall of crane, boom, or load onto topside equipment or structures.
- Fall of crane, boom, or load into sea.

Different dimensions and the loads of possible objects need to be addressed. The probability of damage needs to address:

- The probability of the fall, reflecting the design of the crane, the brake systems, and, if known, procedural restrictions.
- The probability of the object hitting a particular system.
- The loads from the impact on the system.

If the energy from the falling load is sufficient to cause a leak, these scenarios should be transferred to the analysis of HC leaks, including ignition probability to reflect the possibility that the fall itself is an ignition source. In cases where the falling load causes structural damage the possibility of progressive collapse shall be considered.

19.7   Safety Integrity Levels

The IEC standards 61508 [12] and 61511 [13], introduce a categorisation scheme for safety system reliability requirements, which is extensively used in the offshore oil and gas industry. The Norwegian Oil and Gas document 070 [14] presents recommended guidelines for the application of the abovementioned IEC standards into the Norwegian petroleum industry. Whereas the IEC standards are voluminous and will lead to extensive processes as design input, the 070 Guideline presents simplified requirements, which eliminate the need for extensive analyses.

The IEC standards state several requirements to the planning and implementation of the safety systems, including familiarity and definition of equipment under control (EUC), identification of hazards, interaction between EUC, and risk assessments. However, as few details about the installation are available in the early conceptual phase, all these evaluations have to be carried out in a general manner. This means that the decision makers have to rely on knowledge from other similar projects.

In principle the SIL requirements are supposed to be based upon a risk assessment. The 070 Guideline recommends minimum Safety Integrity Level (SIL) requirements for each safety function, and that such minimum SIL requirements are used as input to QRA. The Norwegian Oil and Gas 070 Guidance has defined SIL requirements for a number of essential safety functions, as was shown in Table 15.2.

19.8   Cyber-Resilient Design

As part of the design of offshore installations, there is a need to analyse cyber risk sufficiently and ensure that the offshore installation is designed in a cyber-resilient way. As mentioned in Chap. 1, the cyber security risk has increased recently as more and more of the equipment on offshore installations is 'on-line' and more and more of the control functions are moved away from the offshore facilities.

There is a need to improve the way that the above-mentioned risk is managed. At present, risk and technical safety experts typically do not have sufficient expertise to understand the challenges that cyber threats imply, and to understand what is a sufficiently robust design. On the other hand, IT and cyber risk experts may not have sufficient understanding of the application of their knowledge in technical safety systems. In total, to succeed in improving the way offshore installations are designed in the future, to manage cyber threats, increased focus and attention by the operating companies is needed.

References

1. SCI (1998) Blast and fire engineering for topside systems, Phase 2. Ascot, SCI. Report No.: 253
2. NPD (1990) Regulations relating to implementation and use of risk analysis in the petroleum activities, Norwegian Petroleum Directorate, Stavanger, Dec 1990
3. Standard Norway (2010) Risk and emergency preparedness analysis, NORSOK Standard Z-013, Rev. 3, 2010
4. Standard Norway (2012) Integrity of offshore structures, N-001, Rev. 8, Sept 2012. Standards Norway, Oslo
5. Standard Norway (2007) Action and action effects, N-003, Rev. 2, Sept 2007. Standards Norway, Oslo
6. Standard Norway (2004) Design of steel structures, N-004, Rev. 2. Standards Norway, Oslo
7. PSA (2011) Regulations Relating to design and outfitting of facilities, etc. in the petroleum activities (the Facilities Regulations), last amended 18, 2017 Dec, Petroleum Safety Authority, Stavanger
8. Vinnem JE, Pedersen JI, Rosenthal P (1996) Efficient risk management: use of computerized QRA model for safety improvements to an existing installation. In: 3rd International conference on health, safety and environment in oil and gas exploration and production, New Orleans, USA. SPE paper 35775
9. Vinnem JE (1997) On the sensitivity of offshore QRA studies. In: European safety and reliability conference, 1997, Lisboa, Portugal
10. Standard Norway (2008) Technical safety, Edition 4, Feb 2008. Standards Norway, Oslo
11. Standard Norway (2018) Technical safety, Edition 6, June 2018. Standards Norway, Oslo
12. IEC (2010) IEC 61508—Functional safety of electrical/electronic/programmable electronic safety–related systems. IEC, Geneva
13. IEC (2016) IEC 61511—Functional safety—safety instrumented systems for the process industry sector. IEC, Geneva
14. Norwegian Oil and Gas (2018) Application of IEC 61508 and IEC 61511 in the Norwegian Petroleum Industry, Guideline 070, revised June 2018

Chapter 20
Risk Based Emergency Response Planning

20.1 Establishment of Emergency Response on the NCS

How to plan for emergency response in the Norwegian oil and gas activities is elaborated on in the NORSOK Z-013 standard [1]. With reference to the risk management process described in ISO 31000, as presented in Chap. 3, emergency response is part of the risk treatment phase. The NORSOK Z-013 standard defines emergency preparedness as:

> technical, operational and organisational measures, including necessary equipment that are planned to be used under the management of the emergency organisation in case hazardous or accidental situations occur, in order to protect human and environmental resources and assets.

The above definition indicates that the primary aim of the emergency preparedness measures is to reduce the consequences if a hazard occurs. In other words, the focus of the emergency preparedness measures is to establish consequence-reducing measures. On the other hand, planning for emergency response is self-evidently an activity carried out in advance. This means that even though emergency preparedness measures belong on the right-hand side of the bow-tie figure, planning for such measures is carried out in advance, before an initiating event occurs, and thereby belongs on the left-hand side of the bow-tie figure.

The establishment of emergency preparedness is based on performing four steps (Fig. 20.1):

- the QRA,
- the emergency preparedness analysis,
- the emergency response plan, and
- emergency training and exercises.

In the following, the four steps will be introduced and presented from a practical point of view. For details on the approach, the reader is referred to the NORSOK Z-013 standard [1].

© Springer-Verlag London Ltd., part of Springer Nature 2020
J.-E. Vinnem and W. Røed, *Offshore Risk Assessment Vol. 2*, Springer Series in Reliability Engineering, https://doi.org/10.1007/978-1-4471-7448-6_20

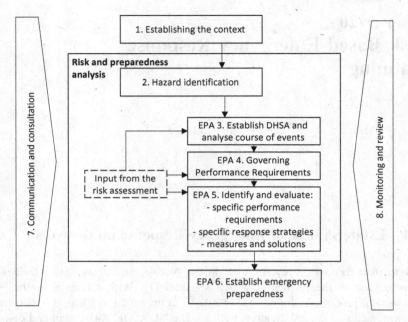

Fig. 20.1 Establishment of emergency preparedness (simplified version of a figure in the NORSOK Z-013 standard [1]

20.1.1   QRA

One of the purposes of a QRA is to provide input to the establishment of the emergency response. The QRA describes potential hazards, scenarios and consequences that may occur at the installation, and this information is relevant when an emergency response is planned for. Understanding the scenarios that may occur includes understanding the barrier functions and the barrier elements in place to achieve the barrier functions. It is also crucial to understand scenarios where barrier elements do not work as intended, for example, if isolation and blowdown is not achieved or if active or passive fire protection measures fail.

The QRA also provides information about assumptions, boundaries and uncertainties related to the risk models and the risk calculation results, which may affect emergency response planning. If, for example, the QRA questions the ability of the structure to withstand a fire after a certain number of minutes, the QRA may end up with an assumption/recommendation that mustering should be finished within a specific time to prevent impairment of safety functions. In such case, the emergency response planning of mustering must ensure that the assumptions made in the QRA are met. Otherwise, we may end up with a situation where the personnel are not taken care of in time.

20.1.2 Emergency Preparedness Analysis

While the QRA elaborates on the hazards and consequences that may occur on an offshore installation, the emergency preparedness analysis elaborates on how these challenges can be dealt with in practice. This includes defining performance measures.

20.1.2.1 Establish the Context

Usually, the emergency preparedness analysis is based on a series of workshops performed by a workgroup. It is essential that the workgroup is supported by key stakeholders, such as the offshore installation manager. The workgroup must have competence to understand how the hazards identified in the QRA may develop and have the authority to decide what is a sufficient level of emergency preparedness. Typically, a workgroup consists of the offshore installation manager, the safety representative, a nurse, a process operator, a crane operator, a drilling/well expert, a technical safety representative and a safety delegate. The workgroup needs to understand both the organisation (roles and responsibilities) and technical safety systems and other systems that influence how scenarios may develop. Examples of relevant background information are as follows:

- regulatory requirements, sector-specific guidelines and standards,
- relevant information from the QRA (Sect. 20.1.1),
- existing emergency preparedness analysis (if present),
- drawings and/or 3D design model,
- cooperation agreements with emergency response resources,
- company-specific governing documents,
- experience from training and exercises, and
- information about technical condition of assets, including barrier element status.

20.1.2.2 Hazard Identification

In many cases, the documentation from the hazard identification workshop performed as part of the QRA will provide a sufficient basis for understanding hazards with a major accident potential. The list should, however, be verified from a critical point of view by the emergency response planning workgroup in case essential major accident hazards have not been included.

20.1.2.3 Establish Defined Situations of Hazards and Accidents (DSHA) and Analyse Course of Events

As highlighted in Sect. 9.4.2, in the NORSOK Z-013 standard [1], the selection of DSHA should include:

- major accidents including the dimensioning of accidental events identified in the QRA,
- accidental events that appear in the QRA without being identified as major accidents, as long as they represent separate challenges to the emergency preparedness, including accidental events with an annual probability lower than 10^{-4},
- events that have been experienced in comparable activities,
- acute pollution,
- events for which emergency preparedness exists according to normal practice,
- a temporary risk increase e.g. drifting objects, man overboard, unstable well in connection with well intervention, and environmental conditions, etc.

The rationale for the above list is that, in addition to major accident critical hazards, there may be additional hazards, not necessarily analysed in the QRA, that need to be included in the emergency preparedness analysis. Examples are toxic release, fire in an auxiliary system/area, fire in the living quarters as well as loss of position for floating offshore installations. Also, discharges to the environment such as oil spills and spills of chemical compounds need to be considered in the emergency preparedness analysis. There may also be hazards related to health care such as man overboard and acute medical incidents, and hazards related to intentional acts such as terrorism, threats and criminal actions. Finally, there may be additional hazards in particular phases or operational conditions. An example of this is a radiation accident relevant when tools with radioactive material are present at the installation (for example, for measuring or well logging). In general, it is expected that all hazards that may result in a challenge for the emergency preparedness organisation are included in the emergency response planning. This includes hazards with a low risk contribution due to low probabilities, such as severe hazardous events with a probability lower than 10^{-4} as well as hazards associated with large uncertainties.

20.1.2.4 Identify and Evaluate Governing and Specific Performance Requirements, Response Strategies, Measures and Solutions

This step includes a systematic evaluation of each of the representative situations, one at a time, for each of the emergency response phases: detection and alert, danger limitation, rescue, escape and evacuation, and normalisation. For each of the combinations of the above, performance requirements are established based on authority regulations/guidelines, applicable standards, the operating company's emergency preparedness principles and internal governing documents. Examples of performance requirements are that we should be able to pick up one person falling into the sea within a specific number of minutes or that we will be able to evacuate all personnel before impact with a vessel on a collision course. Performance requirements should be formulated in terms of capacity/functionality (e.g. two persons having a specific competence/course), availability (e.g. 24/7/365) and response time (within x minutes).

For each of the scenarios, response strategies should be discussed and decided upon, and the resources/measures/solutions required to meet the performance measure should be allocated. Resources can be operational, organisational and technical, or in most cases, a combination of the three categories. To ensure that all three categories are considered, it can be useful to ask the question 'who does what with which equipment?'. Then 'who' will be the organisational barrier element, 'what' will be the operational element and 'with which equipment' will be the technical barrier element.

Potential split-scenarios should be addressed. These are scenarios where personnel are not able to reach the primary means of evacuation. In such case, additional means of evacuation or operational restrictions may be required.

20.1.2.5   Establish Emergency Preparedness

It is the operating company's responsibility to dimension the emergency response to ensure that representative situations are handled according to the performance requirements defined by the emergency preparedness analysis workgroup. As part of this work, it should be evaluated whether there may be incompatible functions, e.g. that personnel are allocated to several tasks that may be needed at the same point of time in case of an emergency. In such situations, back-up personnel may be needed.

With hindsight, it is often seen that, in situations where an emergency organisation is not sufficiently dimensioned, this was because (a) the scenario that occurred was more severe than the representative scenario considered in the establishment of the emergency response, or (b) the performance requirements were not sufficiently robust to deal with the event that occurred. Examples of root causes are that the workgroup did not have sufficient knowledge about scenarios that could be expected and how they might develop, or that the members of the workgroup did not have the authority to approve the agreed performance requirements. Sometimes the management did not recognize the emergency analysis as a sole basis for emergency response planning, and therefore did not approve implementation of the suggestions given by the workgroup.

20.1.3   Emergency Response Plan

The emergency response plan describes how the emergency response should be carried out in practice. The document is intended to be used as a tool for the personnel involved in the emergency response, and therefore it should be short, simple and to the point. The target person(s) for the emergency response plan should be personnel with the necessary competence and training, who know their duties and who understand the roles and responsibilities of the other parties involved. This means that text explaining why measures are implemented should not be included, as this will result in a too comprehensive document. In many cases, the emergency response plan can consist of dedicated check lists for each of the emergency response functions. Such checklists should cover all the emergency preparedness phases; detection and alert, danger limitation, rescue, escape and evacuation, and normalisation.

In emergency response planning, a distinction is commonly made between tactical, operational and strategic levels. In oil and gas activities, these are often referred to as 1st line, 2nd line and 3rd line, correspondingly. The tactical level involves the emergency preparedness organisation at the offshore facility, such as officers in the emergency response room and local emergency response resources on the offshore installation. The operational level involves personnel in an emergency response room onshore, supporting the offshore organisation and providing external resources such as helicopters, vessels, evacuation centres etc. The strategic level includes the management of the responsible company. They take care of next of kin, information to the media, information and involvement of external companies and other stakeholders, etc.

20.1.4   Emergency Training and Exercises

Training and exercises are needed to prepare emergency response personnel for the tasks they are supposed to carry out in an emergency. Training for being involved in an emergency response situation can be compared to learning how to drive a car: in the same way as an unskilled driver needs to allocate cognitive capacity to driving the car, most unskilled personnel involved in emergency response training will spend their cognitive capacity on software systems and other parts of the 'structure', and will struggle to get the big overview and be able to think through what may be around the next corner. To be able to achieve a sufficient overview of the situation, and to prepare for surprises, the knowledge needs to be internalised. Training and exercise is a key to obtain this effect.

For coordinators and other staff in an emergency response team 'outside the battlefield', effective training can be achieved without involving real resources such as helicopters and vessels. Instead, the response team can communicate with actors pretending to represent the real resources. Such training can be surprisingly effective relative to the limited amount of resources spent on the exercise.

Scenarios to be used for training should be chosen among the scenarios and hazards addressed in the emergency preparedness analysis. This ensures that representative scenarios are being exercised. If emergency response personnel are trained to handle representative events, it is likely that they will be able to use the experience gained to handle other accidental events as well. The real situation will most probably not be the same as the chosen scenarios, but when the personnel involved in the emergency response organisation understand their own roles and responsibilities, and are trained to handle representative scenarios, it is likely that they will manage to improvise and expand if the real situation requires.

20.1.5 Emergency Response Planning in a Resilience Perspective

The establishment of an emergency preparedness response described in Sects. 20.1.1–20.1.4 is based on an idea that relevant hazards can be identified and planned for beforehand. This idea can of course be questioned. Even on an offshore installation, with a limited area far away from shore, there may be surprises and accidental events that have not been considered beforehand. This can be used as an argument that emergency response planning should not be 'specific', focusing on hazards and scenarios, but instead be 'generic', ensuring that the people involved are able to cope with any kind of situation, including surprises that have not been seen in the past. We will not elaborate on this perspective in the book. However, the reader should be aware that there is more than one perspective on how to obtain high quality emergency response planning.

20.2 Philosophy of Emergency Response

The philosophy of emergency response planning is that emergency actions shall be planned for a variety of scenarios. The Norwegian regulations require that emergency actions are planned for:

- Major accident scenarios
- Less extensive accidental scenarios
- Scenarios that represent temporary increase in risk.

Even if the risk level is very low, there is still an obligation to provide emergency response plans for actions to be taken in the case of emergencies.

This is essential, emergency plans and scenarios can not be omitted on the basis of a low probability. It is not the purpose of this chapter to propose solutions or ideas that change this essential premise.

What the remaining parts of this chapter sets out to establish, are some essential parameters in the emergency response planning; these parameters may be decided on in a risk based context, which implies that values are chosen in accordance with the most severe conditions that may reasonably occur, based on input from modelling of accidental scenarios in a risk assessment approach. In some of the cases this will challenge the sophistication of the modelling of accident scenarios in risk assessment, modelling of accident scenarios is sometimes too coarse in order for the risk models to be used for scenario definition in emergency response planning. This is further discussed in Sect. 20.4 below.

The parameters that are proposed for a risk based approach are the response times for emergency response actions and the required capacity of these actions. The philosophy of deciding on these parameters in a risk based context has been discussed with experts in PSA, they have agreed that the distinction made here is logical, the need for emergency response actions can not be decided upon based on risk assessment, but some parameters may be decided upon in this manner. This topic is discussed in the following.

20.3   Risk Based Emergency Response Times and Capacity

After 2000, the Norwegian petroleum industry established four areas with cooperation between operators with respect to external marine and airborne resources, known as the 'area-based emergency response' [2]. The following scenarios form the basis of this resource planning:

1. Person over board in connection with work over sea
2. Persons needing rescue after helicopter ditching
3. Persons needing rescue after emergency evacuation (to sea)
4. Collision hazard
5. Acute oil spill
6. Installation fire with external firefighting needs
7. Acute injury or illness with the need for external medical response
8. Helicopter crash on installation with severely injured personnel.

The essential parts of this emergency response planning are risk-based. This is applicable to Scenarios 2, 3, 6, 7 and 8, for which the connection to risk assessment is somewhat indirect; see further discussion in Sects. 20.5, 20.6 and 20.7. For the assessment of the rescue capacity of personnel after emergency evacuation to sea, this directly depends on the risk analysis, as discussed in Sect. 20.4. However, some simplified rules are also available, as shown in Sect. 20.4.5.

20.4   Risk-Based Rescue Capacity After Evacuation

Several accidents worldwide have demonstrated clearly that it is essential to have rescue capacity available in the proximity of installations in remote areas, in order to take care of survivors, whether they are injured or uninjured. Some may argue that evacuation is in itself an emergency resource and that it should not be required to have emergency response solutions in the case of the failure of emergency resources. However, this attitude is too laid back. Experience from several accidents demonstrates the importance of the capacity to rescue personnel from the sea, from rafts and from lifeboats after emergency evacuation. This principle is also already adopted on offshore installations. There are secondary (and possibly tertiary) evacuation means installed for the case of the failure of the primary evacuation means.

Therefore, an approach to risk-based rescue capacity following emergency evacuation to sea from Norwegian installations has been developed. It should be emphasised that the requirement to provide rescue capacity is not subject to risk assessment, but rather to risk-based dimensioning because of the differences that exist between installations with respect to the possibilities for a safe escape to a shelter area on the installation and provisions for safe evacuation from an installation.

The following steps are completed in order to determine a risk-based rescue capacity for offshore installations:

1. Assumptions, premises and results from QRA are reviewed to identify any requirements that have been implicitly assumed in order to calculate risk levels. If assumptions relating to the rescue of personnel have been made, these must be adhered to in order for the risk analysis results to be valid. Indirectly, such assumptions and premises may therefore be some of the conditions for the risk levels to be tolerable.
2. The highest number of personnel that have been assumed according to Step 1 to be rescued in any accident scenario is the minimum rescue capacity requirement.
3. A separate analysis of escape and evacuation is performed in order to reflect the suitability of the arrangements for escape and evacuation on the installation. The results are then used to determine the required capacity based on the defined probability of exceedance.
4. If the value from Step 3 exceeds the value from Step 2, then this value is the risk-based rescue capacity, otherwise the value from Step 2 is retained.

These four steps represent two alternative approaches to risk-based rescue capacity following emergency evacuation. It is not required to use both approaches, but at least one of them has to be used if the approach is to be risk-based. The analysis in Step 3 may be a separate escape and evacuation study or be part of a detailed QRA with sufficiently detailed analysis of the escape and evacuation scenarios. The Steps 1 through 4 are discussed in more depth in the following.

20.4.1   Assumptions in QRA Studies

The risk analysis will often have assumptions, premises and/or intermediate results that show details of the assumed performance of the evacuation and rescue of personnel. These assumptions are used in the calculation of personnel risk (FAR, IR or equivalent), and they form the basis for how risk tolerability criteria are satisfied. When these preconditions are met, the risk will be tolerable in accordance with these criteria. The assumptions involved are usually the following scenarios:

- Groups of people who do not reach primary evacuation means
- Number of people in such groups
- Percentage of people who end up in the sea or rafts
- Groups of people who can use primary evacuation means, but who end up in the sea or in rafts or damaged evacuation means
- Number of people in such groups
- Percentage assumed to be rescued in accordance with the risk analysis.

These assumptions should not be assessed in terms of probability. They are the assumptions of the QRA that must be met for the risk to be in accordance with the results of the risk analysis. This is often a precondition of a tolerable risk level. No such scenarios require rescue subsequent to the use of secondary (or tertiary) evacuation means.

In some cases, the QRA may have made explicit assumptions about people who die during the lifeboat evacuation. This may be personnel injured for various reasons that need to be picked-up from the sea, liferafts or lifeboats rapidly, in order to avoid fatalities. Such assumptions may provide explicit basis for dimensioning of the emergency response capacity. Risk-based dimensioning may not be necessary in such cases.

For those who do not reach the primary means of evacuation, it is assumed that the rescue capacity is equal to this number, irrespective of what fate they have been given in QRA studies. It would be impossible to know who would be in need of rapid rescue and who would be lost in any case; therefore the capacity would need to allow for all of these to be rescued rapidly. Such calculations may be carried out using conservative assumptions. It may therefore be relevant to decide on somewhat lower values than all of those who in the QRA were assumed not to reach the primary evacuation means. Such deviations must then be thoroughly justified. This approach is based on the following assumptions:

- All persons in lifeboats, liferafts or the sea have survival suits on.
- Survival suits must be deployed on the installation so that any group that according to the risk analysis can be prevented from reaching the muster stations can access them.

20.4.2   Escape and Evacuation Robustness

This section discusses the technical differences between different types of production facilities that are important for dimensioning the rescue capacity:

- People who do not reach the primary evacuation means
- People can be injured during evacuation by lifeboat or other means.

The former aspect relates primarily to escape routes on the facility, while the second aspect is related to the actual evacuation. Each aspect is discussed separately below.

20.4.2.1 Escape on the Installation

If escape routes are well protected and logically arranged, is it realistic to expect that personnel are not "caught" without the ability to reach the muster areas/stations. In such cases, the QRA study may imply that no groups that need to use the secondary or tertiary evacuation means. In practice, this may still be possible, even though the analysis may not be sufficiently detailed to identify such scenarios. The installations are likely to be quite different with respect to these conditions. The evaluations of these aspects should as far as possible be based on detailed risk analysis studies focusing on the EER.

20.4.2.2 Lifeboat Evacuation

Conventional lifeboats are generally considered to be prone to being damaged against the installation structure or from the waves during deployment. This is the main difference between conventional lifeboats and freefall lifeboats/lifeboats of the skid launch type.

Lifeboats of the freefall type are most common on permanent production facilities in the Norwegian sector, while the skid type is more common on floating production facilities. It is usually considered that the likelihood of damage to the sea is very small for such lifeboats. Moreover, owing to their shape, they can speed away from the facility when hitting the sea level, so the chance of drifting back and be damaged against the structure is considered to be significantly less. It should be noted that no faults or severe injuries have occurred during thousands of training drops during almost 30 years.

Conventional lifeboats have been improved in order to prevent damage by sea impact during deployment, including hooks that can be released even with significant loads. The main problem for conventional lifeboats is often considered to be the risk of damage to the lifeboat by collision against the structure, if the engine fails to start. This applies to both fixed and floating structures. Allowance for such events must be given in the case of conventional lifeboats, persons in the sea or in damaged lifeboats would be the inevitable result if this occurred.

Bridge-connected installations usually imply that lifeboat evacuation has a very low probability, personnel will cross over the bridge as a first action and lifeboat evacuation will only be required if the conditions on all bridge-linked platforms become unbearable. This possibility can usually be disregarded. There may, however, still be scenarios where personnel cannot find protected escape routes to the shelter area, so that an escape to the sea or by liferaft may be required.

20.4.3   Principles for Probabilistic Pick-Up Calculations

The risk-based capacity for the rescue of personnel from the sea in an emergency evacuation is in principle based on each installation's probability of getting groups of people who need fast rescue. The possible causes that can lead to such needs include:

- People who end up in the sea because the evacuation means are damaged during deployment
- People who are injured in an intact lifeboat for example because it is damaged due to collision with the facility or because some people are not sufficiently strapped in during the freefall launch of the lifeboats.

Conventional lifeboats have been considered to be considerably more susceptible to being damaged against the installation or waves during launching as described above.

There are therefore considerable differences between individual installations with respect to the probability that there will be personnel in the sea, and if so, how many this is likely to be. In most areas offshore, there will be differences between facilities.

The relationship between the number of persons in the water during an emergency evacuation and the associated probability will have to be determined for each installation. A principal illustration of such a relationship is shown in Fig. 20.2, for three assumed installations, with the following evacuation possibilities:

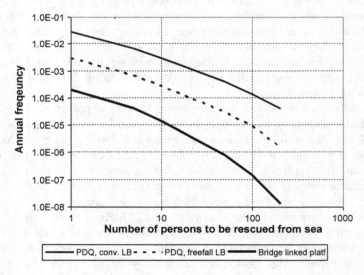

Fig. 20.2 Relationship between the number of persons to be rescued from sea and exceedance probability (idealised example)

- Standalone platform with conventional lifeboats
- Standalone platform with freefall lifeboats
- Bridge linked platforms irrespective of the types of lifeboats.

Please note that this diagram only illustrates the fundamental relationships, so there are three curves in the same diagram. For an installation only one curve will apply, and there is no reason to expect that the curves for an installation in practice will be as "smooth" as those shown in Fig. 20.2.

Figure 20.2 shows the cumulative frequency of exceedance for three theoretical installations, two of which are stand-alone installations for production, drilling and quarters (PDQ), and the last is for bridge linked installations, irrespective of type of lifeboats. When the frequency of at least 10 people in the sea is 1.4×10^{-5} per year (curve for bridge-linked platforms), this implies that the frequency of all accident scenarios involving 10 or more people in the sea in an emergency evacuation is 1.4×10^{-5} per year. Similarly, there is a frequency of 2×10^{-4} per year for bridge-linked platforms for accidents with one or more persons in the water. This means that the frequency of accidents from 1 to 9 persons in the water is 1.86×10^{-4} per year.

The following data must be provided for the calculation of the accident probabilities for each of the installations involved:

- Frequency of the various evacuation scenarios
- Extent of the personnel consequences of evacuation scenarios.

20.4.4   Probability Limit for Determining Dimensioning Scenarios

There are no tolerance limits for the probability used as the basis for dimensioning the capacity of the emergency response system. In some cases limits around 10^{-4} per year have been used in emergency analysis, without a clear rationale for the selection of such limits.

It is emphasised that rescue operations that have been assumed to be successful by the QRA must be met in full, without any probability assessment (see Sect. 20.4.1). The assessments discussed here are therefore not required to meet the risk tolerance criteria, but are additional. Since this approach is an extension of the QRA, it is considered to be acceptable to determine the required capacity based on a probabilistic analysis.

Section 17.3 discussed the frequencies for the various emergency evacuation scenarios on the NCS as a whole. These values show that the scenarios involving the rescue of personnel generally have a frequency of the order of 0.1 per year. For some scenarios, it should be noted that the events occurred a long time ago (e.g., emergency evacuation), so that the actual frequency is probably lower for facilities designed to current needs.

On this basis it can be argued that the implicit limit for the events that form the basis for the design of emergency response plans is equivalent to 0.1 per year for the Norwegian sector.

It must be recognised that risk analysis and statistics can never be exhaustive; analyses will always have significant limitations with respect to the events analysed, while statistics for rare events are usually quite uncertain. The limit for when you do not choose to dimension a particular scenario should therefore be somewhat lower. Section 17.3 discussed evacuation frequencies for the Norwegian sector, indicating an annual value in the order of $1-5 \times 10^{-4}$.

It is unclear whether the limit of 10^{-3} per platform per year for the scenarios that are ignored is reasonable or whether 10^{-4} per platform per year is a more appropriate limit. It is well known that the limit of 10^{-4} per platform per year often used in other contexts for the selection of such dimensioning scenarios. When this is to be considered, it must be assumed that the scenarios considered represent risks to personnel that in principle are tolerable. The events that form the basis must be prevented sufficiently by reducing the probability and/or consequence reducing actions, so that the residual risk is acceptable. Moreover, these events represent the failure of the emergency systems.

It is therefore unreasonable to apply the same strict requirements to the establishment of the emergency response to these incidents, since we already have a tolerable risk level. It is also relevant to recall that for the assumptions made in the analysis made for the level of risk to be acceptable, they must apply directly to emergency requirements without any form of probability assessment. Still, it should be an assessment of the limits to be used on the following areas:

• Frequency = 10^{-3} per year	Highest frequency that may be applied in order to establish the dimensioning capacity
• Frequency in interval 10^{-4}–10^{-3} per year	Separate assessment to be made reflecting the specific conditions in order to determine the highest frequency that may be applied in order to establish the dimensioning capacity
• Frequency = 10^{-4} per year	Lowest frequency that may be applied in order to establish the dimensioning capacity

20.4.5   Simplified Rules for Dimensioning Pick-Up Capacity

The guidelines for the area-based emergency response include simplified rules for the determination of the required pick-up capacity from the start. Interestingly, only simplified rules were used during the first 10 years or so of application. These simplified rules omit the use of risk assessments, and stipulate conservative rules for rescue capacity. For most installations, these rules imply that helicopter ditching will be the determining scenario; this is most likely the reason why only simplified rules have been used. These rules are presented in Table 20.1.

Table 20.1 Simplified rules for determining the rescue capacity of personnel in sea after an emergency evacuation

Installation characteristics	Functional rescue capacity requirements for survivors in the sea after an emergency evacuation (wearing survival suits)
Production installation with freefall lifeboats, irrespective of manning level	5% of POB as injured persons in lifeboat, liferaft or the sea
FPSO with well protected escape tunnel and shelter area forward of all hydrocarbon containing areas, with freefall lifeboats	5% of POB as injured persons in lifeboat, liferaft or the sea
Production installation with bridge connection to other installation, irrespective of evacuation means	Rescue capacity not to be provided
Installation with conventional lifeboats, no bridge connection, irrespective of manning level	25% of POB as injured persons (not exceeding capacity of one full lifeboat) in lifeboat, liferaft or the sea

20.5 Risk Based External Fire Fighting

If external firefighting capacity is needed, this will define requirements for standby vessels or corresponding functionality with fire water monitors (usually firefighting Class 2 or 3 according to classification requirements) within a certain response time.

In order to determine if external firefighting is needed, this has to be determined from the installations' QRA studies. As far as is known, there is only one installation in the Norwegian sector which is dependent on external fire fighting in order to protect its structural integrity for a sufficiently long period during a large fire. This is a steel hull tension-leg platform.

20.6 Rescue of Personnel in Helicopter Accidents

The requirements to provide rescue capacity to personnel who end up in the sea or in liferafts after helicopter accidents are also based on risk assessment, but not installation-specific risk assessment. Chapter 13 demonstrated that the risk for offshore employees during helicopter transportation is the highest that they are exposed to, when counted on a per hour exposure basis.

Therefore, there must be capacity to rescue all onboard a helicopter within 120 min. With the helicopters used currently, the maximum number of personnel onboard is 21 persons. The rescue of 21 persons within 120 min is therefore the requirement.

The 120-min response time is dependent on the capabilities of the survival suits worn by passengers. The approach adopted is that there should be a 1.5 safety factor. This implies that if the response time is 120 min, then the suits have to be verified for at least 180 min. This is not a problem in the North Sea and Norwegian Sea, where the water temperature is never lower than 2 °C. The suits worn in Norway have to be tested for 6 h protection against hypothermia, according to Norwegian Oil and Gas requirements. For Barents Sea and Arctic Ocean, this may be a problem.

The rescue capability can be provided by SAR-helicopters located offshore or by FRC from the installation. The use of FRC will have significant limitations by wind and waves. The responsibility for the operators of offshore installations in Norway is limited to the safety zone around installations. When a SAR-helicopter is available, this implies that personnel outside the safety zone may also be assisted in time.

20.7 External Medical Assistance to Injured and Ill Persons

One of the most important aspects of the area based-emergency response cooperation is the possibility of providing external assistance to seriously ill or injured personnel on installations. There is a requirement to be able to bring in advanced thrombolytic equipment within 60 min and to deliver seriously injured or ill persons to an onshore hospital within 3 h [2].

This is not based on a specific risk assessment, either general or installation specific. However, it is known that there are three to four serious injuries in the Norwegian sector each month as well as several cases of serious illness. A number of statistics are presented in Ref. [3], which also documents that so-called 'yellow or red' missions are flown once per week per area with emergency cooperation.

20.8 Area Based Emergency Response Planning—Results

Vinnem [3], also presents a summary of the frequencies for the scenarios that are included in the area-based emergency response cooperation. These values are to some extent based on relatively coarse calculations. It is also essential to emphasise that some of these scenarios use the same emergency resources. Scenario 7 involves SAR helicopters (as well as normal transport helicopters), while Scenario 6 is only related to standby vessels. The overview is presented in Fig. 20.3. Scenario 1 has a typical frequency of 1–3 per year for the entire Norwegian sector.

The last time there was an emergency helicopter landing during transit over sea in the Norwegian sector was in 2002, but the pilots this time found a tanker with a helideck to land on. The last time there was an emergency landing on water in the Norwegian sector was in 1996, while there was an emergency landing on water in UK in 2009 and two such landings in 2012 (see Sect. 13.2).

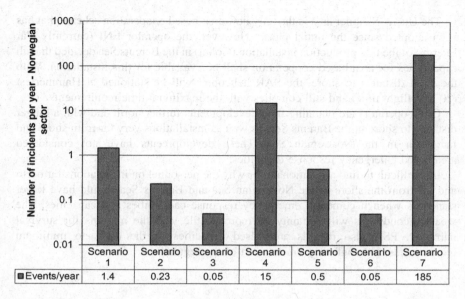

Fig. 20.3 Typical frequencies for the Norwegian sector for scenarios with area-based requirements

The last emergency evacuation to sea from an installation in the Norwegian sector was in October 1985 on the Halten bank, with a shallow gas blowout on semi-submersible West Vanguard (later renamed Ocean Vanguard).

The value for Scenario 7 is by far the highest, and that scenario has the best statistical basis for offshore-based SAR helicopters. Because of this high frequency, the benefit in terms of persons whose live were saved by rapid transport to onshore hospitals is formidable. If we assume that only 1% of those transported to onshore hospital are saved, this is still about two lives per year. If we assume 5%, then the number becomes nine lives each year.

20.9   External Emergency Response Planning in Arctic Conditions

The area-based emergency response cooperation was established about 20 years ago. One area is in the Norwegian Sea (Halten area) and three areas are in the North Sea, with two in the northern Norwegian North Sea and one in the southern Norwegian North Sea. Some installations in the central Norwegian North Sea were for some years covered by the UK-based JIGSAW SAR helicopter cooperation; this has now been replaced by a new area called Utsira–Sleipner, which will have an offshore based helicopter on the Johan Sverdrup field when in operation.

The Utsira–Sleipner area is the only new area-based cooperation scheme that has been adopted since the initial phase. However, the operator ENI (currently Vår Energy) of the first production installation (Goliat) in the Barents Sea decided that all capabilities for area based cooperation shall be available for this installation. With the short distance to shore, the SAR helicopter will be stationed at Hammerfest (20-min flight time) and still complies with the maximum time requirements.

Other operators are planning field developments further north and with a longer distance to shore in the Barents Sea, as well as installations very far from shore and far north in the Norwegian Sea. These developments have not considered area-based emergency response capacities.

It is difficult to find any rationale for why the personnel on installations far north and far from the shore in the Norwegian Sea and Barents Sea should have lower standards when it comes to emergency response capabilities and capacities. The weather conditions will certainly be more hostile, and the margins for survival narrower. PSA also regards area-based capacities as the de facto minimum requirements, although they are not part of the regulations.

20.10   Evaluation of the Area-Based Emergency Response

In 2018 the status of the area-based emergency response systems was evaluated by Safetec [4]. The purpose of the survey was to highlight the status, development and possible differences between the different area-based emergency response systems on the NCS. The study was based on interviews with representatives from operating companies and shipowners.

The survey concludes that, in general, the area-based emergency response systems on the NCS work as intended and that the systems are managed by the operating companies in a proper way. The survey also gives some suggestions on how the current status can be improved. For further details, the reader is referred to the survey report [4].

References

1. Standard Norway (2010) Risk and emergency preparedness analysis, NORSOK Standard Z-013
2. Norwegian Oil and Gas (2012) Recommended guidelines for establishing area based emergency response, 10 Sept 2012
3. Vinnem JE (2012) Area based emergency response planning on the Norwegian continental shelf, supporting documentation of premises and scientific assessments in Norwegian oil and gas, Guideline No. 64, 7 Dec 2012
4. Ranum SA, Wold MS, Johansen TS, Vinnem JE (2018) Evaluering av Samarbeid om Beredskap (områdeberedskap), (in Norwegian only), Report no. ST-13755-2, Rev. 2. 07 Dec 2018

Part IV
Risk Assessment and Monitoring in Operations Phase

Chapter 21
Use of Risk Analysis During the Operations Phase

21.1 Study Updating

21.1.1 Overview

Most analysts will usually associate study updating with the design and construction phases. This is just touched upon. During new developments, study updating will be carried out in several stages during the engineering phases. The final updating of risk [and emergency preparedness] analysis should be carried out towards the end of these phases. Emergency preparedness analysis is not the subject of this book but is included for the sake of completeness. Study updating during engineering usually involves:

- Update quantified risk analysis to reflect the chosen solutions and systems.
- Document the assumptions from the QRA in a suitable way, as input to the subsequent emergency preparedness analysis.
- [Carry out the final emergency preparedness analysis.]
- [Document the results from the emergency preparedness analysis in a suitable way, for all design accidental events and DFU, possible causes, and effects of accidents for use in the operational phase.]

The need for updating the QRA study during operations is determined by the extent of modifications (see Sect. 21.9 concerning the actual modification work itself.). Included in this phase are normal operation, inspection, maintenance, and limited modifications. The objectives of the studies are:

- To update risk [and emergency preparedness] analysis in order to ensure that they reflect relevant technical and operational aspects.
- To ensure that the risk level is kept under control.
- To ensure that operational personnel are familiar with the most important risk factors and their significance in achieving a tolerable risk level.

© Springer-Verlag London Ltd., part of Springer Nature 2020
J.-E. Vinnem and W. Røed, *Offshore Risk Assessment Vol. 2*, Springer Series
in Reliability Engineering, https://doi.org/10.1007/978-1-4471-7448-6_21

- To ensure that risk aspects in connection with ongoing operations and work tasks are being assessed and that necessary risk reducing measures are implemented.
- To ensure that the risk level is monitored according to updated risk analysis databases, tools, methods and experience.

Qualitative studies should be carried out when planning and preparing for work tasks that have vital importance for operational safety. There are a few formal requirements relating to the use of qualitative studies, such as Safe Job Analysis. It is important that the workforce and other operational personnel are actively involved in the work.

21.1.2   Scope of Updating

Studies during engineering phases should not be limited to the production installation, but should also cover nearby vessels and installations if they are close enough to be affected by accidental effects or if they may pose a threat for the production installation(s).

Updating of risk [and emergency preparedness] analyses should identify the needs for further risk reducing measures such as emergency preparedness measures, and identify areas for particular attention in the HES management work. QRA studies should also be updated on the basis of experience from near-misses and accidents that have occurred, organisational changes, and changes to regulations. The updating of analyses includes, inter alia, an updating of:

1. The installation and operations in accordance with the development of the activity.
2. Assumptions and premises that the earlier analysis has used. Further development of these as assumptions may also be needed.
3. Whether the risk associated with planned operations, or new equipment, has been assessed at an earlier stage.
4. The data basis used as influenced by new experience, new knowledge or changes in the databases that have been used, including revision of experience data from own operations.
5. The methodology used.
6. The analysis results in the light of possible changes to the operators'/owner's risk tolerance criteria for the installation or operations.

21.1.3   *Frequency of Updating*

The operator/owner should formulate minimum requirements for the frequency of updating of the quantified risk analyses and emergency preparedness analysis unless technical or operational circumstances require more frequent updating.

The UK Safety Case Regulations [1] have quite rigid requirements for updating safety cases, and thus also the QRA studies. This has been shown to ensure that updating is given the necessary attention. It appears that operators otherwise often fail to see the need for updating QRA studies. Statoil (now Equinor) has a recommendation for evaluation of the need for updating every three years.

The Guidance to the Safety Case Regulations [1] refers to the 'salami' principle, by which many minor modifications over time may result in quite substantial overall effects on the risk level. One of the objectives of the strict requirements of HSE is to ensure that such overall effects are not overlooked.

One of the important aspects of updating during the operations phase is to reflect specific changes in data relating to relevant equipment, facilities, and operations. This may result in the updated risk level being reduced if operation specific data can be utilised. There is often a tendency to use conservative data and assumptions when studies are performed prior to the operations phase. Updating in the light of operational experience (see Sect. 2.4.3) may therefore give a more representative risk level.

It should be noted, however, that when such an update is made, the risk of accidents is not changed as such, but a less conservative prediction of the risk level is produced.

The Safety Case Regulations required previously that no longer than three years pass between each time the Safety Cases (and QRA if relevant) were updated and resubmitted. The 3-yearly resubmissions apparently were a significant administrative burden on the authorities. The requirement to resubmit every three years has now been replaced by a more thorough review each five years, but without resubmission. The need to always keep the safety case up to date remains.

21.2   Risk Analysis of Operational Improvement

Methodologies for operational risk analysis were outlined in Sect. 15.3. One of the objectives of the approach is related to how analysis of operational improvements may be carried out. Some of the case studies conducted in the project were also aimed at this.

21.2.1 Overview of BORA Case Studies

Case studies were planned and conducted as part of the BORA methodology development, in order to explore how effectively the approach could be applied, the availability of sources for assessment of the status of the RIFs, and the type of results that may be achieved through these studies. The case studies were conducted for stand-alone production installations in the Norwegian sector, Case study 1 for an old installation with steel jacket support structure, and Case study 2 for a relatively new floating production installation. The following scenarios were analysed in detail:

- Case study 1:

  - Scenario A: Leak due to valve(s) in wrong position after maintenance
  - Scenario B: Leak due to incorrect fitting of flanges or bolts during maintenance
  - Scenario C: Leak due to internal corrosion
  - Scenario D: Leak due to external corrosion (qualitatively).

- Case study 2:

  - Scenario A: Leak due to valve(s) in wrong position after maintenance
  - Scenario B: Leak due failure prior to or during disassembling of HC-system.

Case study 1 is presented in detail in Ref. [2]. Case study 1 was limited to containment barriers, whereas Case study 2 also considered a selection of consequence barriers i.e., barrier functions intended to limit the consequences of a hydrocarbon leak, if it occurs.

One of the advantages of the BORA approach is that sensitivity studies are easy to carry out to evaluate the effect of potential actions in order to reduce risk. As an illustration of how this is performed, consider Table 21.1, which summarises some results from sensitivity studies.

The value for Scenario A, 0.20, implies that if the probability that a checker will fail to detect a valve in the wrong position when the isolation plan is not used can be reduced by a factor of 10, then the probability of a leak due to Scenario A will be

Table 21.1 Illustration of results from sensitivity study—effect of reducing probabilities with factor of 10

Parameter	Scenario A	Scenario B
Probability that a checker fails to detect valve in wrong position when isolation plan is not used	0.20	
Probability that a checker will fail to detect valve in wrong position after maintenance if control of work is performed	0.11	
Probability of failure to use plan correctly		0.12
Probability of failure to detect pressurised system		0.10

reduced by a factor of 5 (or down to 20% of original value), when no other changes to probabilities are made. The values shown in Table 21.1 are those that have the most extensive effects, for the two scenarios considered in the case study.

21.2.2 Risk_OMT Case Studies

Case studies were also carried out in the Risk_OMT project [3]. The following were the risk reducing measures considered:

1a. Work process training: Work on normally pressurised equipment
1b. Change the procedures for safe job analysis and pre-work dialogue with a greater emphasis on hydrocarbon leaks (major accidents)
2. Increase emphasis on leaks (major accident potential) in the training of managers and executives
3. Compliance program: Conduct training in action compliance according to the Statoil (now Equinor) "A-standard" action pattern
4a. Increase focus on the psychosocial work environment through greater degree of involvement across levels and disciplines
4b. A subset of 4a is to improve the involvement of contractors
5. Improve the availability and faster updating of technical documentation
6. Improve the labelling of process equipment, i.e. more uniform labelling in accordance with technical documentation, in combination with improvements in established practice of radio communication
7. Improve the management of change, especially routines for quality control and the handover from modification projects to operation
8. Formalise requirements to the work process 'Work on normally pressurised equipment' in the form of new procedures
9. Develop procedures for the preparation and use of specific checklists for drainage/sampling.

The quantitative effect of these improvements was lower than expected. This is probably because of the misrepresentation of the number of barriers made in the modelling, as discussed in Sect. 15.3.3 in relation to Fig. 15.6. This misrepresentation implies that two independent barriers have been assumed in virtually all the scenarios, whereas later information suggests that two barriers are only available in one third of the scenarios. As a result each barrier element becomes more critical, which should imply that improvement in a barrier element should have a higher effect on the overall result.

21.2.3   HRA in QRA

The possible use of HRA in QRA studies was described by Gould et al. [4]. SPAR-H [5] was suggested as a promising approach, but it was realised that the conditions of applications in the offshore petroleum industry are quite different from those in the nuclear power industry, where control room post initiator actions are the main focus areas. The main application areas in the offshore petroleum industry are associated with the execution of work permits (WPs), isolation plans in process plants according to plans and procedures as well as various control tasks.

An HRA method especially adapted for the oil and gas industry is Petro-HRA [6]. Petro-HRA is based on the well-known SPAR-H method, and although the list of performance shaping factors (PSFs) that is utilized by the method is not exhaustive, it is considered a representative selection of factors for the oil and gas industry. The seven steps in the Petro-HRA method are briefly described as follows [7]:

1. Scenario definition:
   Define the scenario to be analysed, and the scope and boundaries of the analysis. This is an essential step that significantly shapes the subsequent qualitative and quantitative analyses.
2. Qualitative data collection:
   Collect specific and focused data to enable a detailed task description, including information about factors that may affect human performance and the outcome of the scenario, usually done via scenario walk/talk-through, observation, interview and documentation review.
3. Task analysis:
   Describe the actions that are performed for the tasks in the scenario. A systematic way of organizing information about the tasks to help the analyst understand how the scenario is likely to unfold.
4. Human error identification:
   Identify and describe potential errors related to the tasks, consequences of each error, recovery opportunities and performance shaping factors (PSFs) that may have an impact on error probability.
5. Human error modelling:
   Model of the tasks to logically describe how the HFE can occur in the scenario, and to clarify the links between the errors. The model logic can then be used to calculate the Human Error Probability (HEP) for the HFE (in Step 6).
6. Human error quantification:
   Quantify each chosen task or event based on a nominal value and an evaluated set of PSFs.
7. Human error reduction:
   Optional step, aiming at the development of risk-informed recommendations for improvements to prevent the occurrence of human error or mitigate the consequences.

A key goal for the Petro-HRA project according to Ref. [8], was to establish an industry standard for HRA in the petroleum industry. A variety of other HRA methods have been used by analysts prior to the development of Petro-HRA. This resulted in analyses with varying degrees of qualitative analysis, with different quantitative estimates. So far, users of the Petro-HRA method for commercial analyses have indicated that the provision of a complete method description has improved the general quality of the analyses, and that this will lead to improved consistency between analysts in the long term.

21.3 Risk Analysis in Operational Decision-Making

PSA has repeatedly claimed that the risk assessment tools used by the Norwegian petroleum industry are unsuitable for operational decision-making. This was emphasised through a survey of operational personnel on installations, conducted by PSA in 2009–10, which pointed to a need for the further development of risk analysis tools. Such tools need to be usable as inputs into the day-to-day decisions on installations, such as those regarding minor modifications, maintenance and interventions. The same observation would also be applicable for drilling operations. Until recently, nobody has taken the complaints made by authorities seriously.

This is an area where there are large differences between nuclear power plants and offshore installations with respect to the development of on-line risk monitoring and so-called living Probabilistic Safety Assessment (see further discussion in Sect. 21.4).

The following illustrates the typical decision-making contexts relating to activities and manual intervention in the process area on offshore installations:

- Approval of single WPs or a program of WPs, within defined categories (work on process equipment, hot work, lifting, etc).
- Decisions to perform or disregard the leak testing of isolation valves, where the system is not suitable for such tests, or the test method is not optimal.
- Decisions relating to execution of or interval for the recertification of PSV, when allowance is made for the fact that several leaks have been caused by the interventions required to recertify PSVs.
- Decisions relating to the execution of or interval for the testing of PSV, when experience shows that testing often causes trips, which may result in leaks.
- Use of personnel lift for work at a height (short term with increased ignition probability), or alternatively building scaffolding that takes longer and increases the occupational risk for personnel.
- Decisions relating to how to install long process lines (i.e. modification work), either with hot work (increased ignition risk during installation) or with 'cold' installation methods which may increase leak probability over the remaining life cycle.

Risk analysis has been used in the petroleum industry for more than 30 years, so the outset there should be very good opportunities for developing risk analysis studies that are suitable for these needs. When the methodology for petroleum installations was first developed, it was largely based òn the use of PSA studies in the nuclear industry.

However, there are very significant differences between how QRA (TRA) studies are carried out in the petroleum industry and how corresponding studies (Probabilistic Safety Assessments) are carried out in the nuclear industry. This may be regarded as surprising, because of the common approaches taken around 1980.

The explanation is however, that QRA studies in the petroleum sector and Probabilistic Safety Assessments studies for nuclear power plants were developed in opposite directions in the period 1980–2000, where the latter placed increased focus on modelling the causes of initiating events (FTA studies), while the former focused on modelling the consequences of hydrocarbon leaks, based extensively on computational fluid dynamics (CFD) for modelling gas dispersion, fire and explosion.

It is therefore not obvious that extension of present analytical methods to living risk analysis will be easy and possible to implement with limited resources. It should be emphasised that the current needs cover qualitative as well as quantitative analysis.

21.4 Living Risk Analysis (Risk Monitors) in Operational Phase

Risk Monitors for nuclear power plants have been used since 1988. The first Risk Monitor was used in the UK. The following is based on the papers by Puglia and Atefi [9], Majdara and Nematollahi [10] and NEA [11].

Risk Monitors are not available for offshore installations in general, although some minor attempts have been referenced in the literature [12]. It is not expected that Risk Monitors for offshore installations will be developed in the near future because of the extensive differences between how QRA studies are performed for offshore installations and nuclear power plants. However, several risk management experts think that such monitors should be attempted in the offshore industry. The discussion below nevertheless refers to their use for nuclear power plants.

The term Risk Monitor according to the IAEA safety glossary is defined as: 'A plant-specific real-time analysis tool used to determine the instantaneous risk based on the actual status of the systems and components. At any given time, the risk monitor reflects the current plant configuration in terms of the known status of the various systems and/or components, e.g., whether there are any components out of service for maintenance or tests. The model used by the risk monitor is based on, and is consistent with, the Living Probabilistic Safety Assessment for the facility'.

RELCON (now part of Lloyds Risk Consulting) developed a Risk Monitor called RiskSpectrum's RiskWatcher®. This software works using a probabilistic safety analysis model. The program inputs include the operational state of the plant, equipment outages, system configuration, periodic tests and environmental conditions. Another risk monitor called RiskVu was developed by Isograph.

Modifications are a common aspect of the operation of offshore and onshore petroleum facilities in contrast to the nuclear power plants, where stable operations are one of the main safety features. Maintenance and testing are, by contrast, regular activities, as emphasised by the Risk Monitor definition above.

The types of activities for which risk monitoring should provide aid to decision-making [6] include:

(1) outage planning for the testing and maintenance of components;
(2) setting and complying with technical specification requirements;
(3) risk-focused inspections;
(4) identification of areas of plant vulnerability because of hardware, human, or procedural deficiencies;
(5) assessment of the importance of various safety-related issues to the plant operational risk; and
(6) prioritisation of the plant procedural or design modifications because of various safety-related shortcomings identified.

The following are mentioned as crucial aspects during the development of risk monitoring:

• Issues relating to tolerance criteria
• Modelling aspects relating to fault trees and event trees. This is specific to the probabilistic safety analysis studies of nuclear power plants, and QRA studies for petroleum installations do not have the same features at present.

21.4.1 MIRMAP Project

The experience with risk monitoring in the nuclear sector was the starting point for a research project (MIRMAP—Modelling Instantaneous Risk for Major Accident Prevention, [13]. The starting point for the project was the different types of decisions taken in the daily operation of a petroleum facility (onshore or offshore), which are significantly different from the types of decisions in a design project. Risk models have been developed with the intention of giving up-to-date risk information with limited effort and sufficiently quickly to be available when the decisions are being made. Risk in operation will vary from day to day and even from hour to hour, depending on the type of ongoing activity during operation and maintenance.

The QRA models used during design perform an averaging of risk over a long period of time, usually for a 'typical' year in the operation of the plant, which is useless in an operational context. But this is adequate as a basis for developing a design that can be safe 'on average' over the lifetime of the installation.

The work in the MIRMAP project is based on the common barrier function structure outline for flammable substances used throughout the petroleum industry (see Fig. 21.1), where 'BF' denotes 'Barrier Function'. BF5, prevent fatalities, is not addressed in the MIRMAP project.

In daily operations, there is a significant amount of ongoing operational and maintenance work. In addition, there are barrier degradations that either exist in parallel with this work or are caused by the work task itself. Any work at a process facility can be broken down into subtasks or 'activities'. For example, consider the recertification of a process safety valve (PSV), which involves activities such as isolation of the hydrocarbon segment, construction of scaffolding, reinstatement of the hydrocarbon segment etc. All activities are categorized into two categories:

- A1: introduces a hazard that may affect the integrity of a barrier
- A2: represents a condition that directly impairs/weakens a barrier element

'Disconnection of a gas detector' is an example of a category A2 activity that directly weakens gas detection. On the other hand, 'critical lifting' is an example of a category A1 activity that introduces a hazard that can lead to a process leak due to dropped or swinging loads, an impairment of BF1. More examples are provided in Ref. [10].

Through a thorough review of each barrier function and its associated systems and elements, all possible activities or conditions that may directly or indirectly represent or lead to an impairment or deviation in the barrier are identified. This exhaustive activity list may also correspond to different types of operational deviations for a process facility, for instance functional tests, ongoing maintenance work, alarms from condition monitoring services, possible notifications from the control system, etc.

Risk influencing factors describe the nature and degree to which the RIF s affect the safe execution of the task with respect to major accident risk and thereby the risk level. The following are examples of risk influencing factors that are considered:

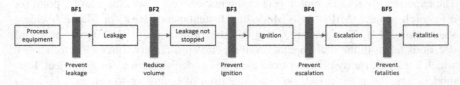

Fig. 21.1 Barrier functions for process leak accident scenario

- Type A1 activity

  - Competence, process work
  - Competence, mechanical work
  - Time pressure
  - Design/human machine interface (HMI)
  - Documentation
  - Isolation plans

- Type A2 activity

  - Redundancy
  - Compensating measures

The complete MIRMAP model consists of event trees, fault trees and influence diagrams, and has extensive links to activity data, status of barrier elements, use of scaffolding and tarpaulins, etc. An example is shown in Ref. [10]. The final model will need to be tailored to the actual facility, as well as the details of the barrier elements and the administrative systems that are used to provide data and will need to be calibrated for the facility in question.

The model is considered to have its primary applicability in the planning process, in order to rank individual activities, identify high risk activities that need additional attention and suggest possible risk reducing measures. This can be useful early and all through the planning process until a final decision is made to approve the execution of an activity. This should contribute to reducing the need to postpone or even stop activities late in the planning process. The approach may also be used to evaluate the effect of combining activities and highlight possible interaction effects that may contribute to increased risk (e.g. leaks and hot work in the same area and at the same time).

Planning of operations and maintenance is usually an iterative process, with annual plans, quarterly plans, 2 weeks plans and daily plans. The MIRMAP approach is probably most useful in the later planning stages, when most of the detailed input required is available. At this stage the approach may contribute to risk reduction, but also possibly to the improvement of the planning process, reducing the need for late rescheduling and thus potential inefficiencies in the use of resources.

21.4.2 Risk Monitoring for DP Systems

Risk monitoring for DP systems is discussed in Sect. 10.4.16.

21.5 Use of Sensitivity Studies for Safety Systems Improvement

Prior to one of the updates of the Frigg Central Complex[1] (FCC) Safety Case it was found that the fire water supply needed to be reassessed [14]. This arose due to the need to install more equipment on one of the platforms. In this work, it was determined that one of the installations, the so-called 'TCP2' processing platform, needed to have further segregation of fire areas in order to limit the fire water supply needs. Even so, a new fire water supply was also required. An increase in the fire water capacity of 46,000 L/min, almost double the previous capacity, was found to be necessary for the total Frigg Central Complex, consisting of the three bridge linked installations, the TCP2 process platform, the TP1 riser platform, and the QP quarters platform.

Experience from similar upgrading tasks of old installations in the offshore industry has shown that such improvements often have knock-on effects for other safety systems and functions (area segregation by means of fire walls may sometimes introduce a ventilation problem). It was therefore decided to evaluate all safety systems on TCP2 in order to decide if any further upgrading was necessary, and to ensure that unwanted negative effects were not generated.

Several other safety systems and functions were later found to need upgrading, resulting in a need for comparative studies to assess the possibilities for risk reduction. Comparative studies were conducted, supported by specialist calculations including explosion simulations and structural fire response. Some of the work carried out during these studies is outlined in Ref. [10].

21.5.1 Risk Management Objectives

The TCP2 platform, which at the time was the main processing platform of the Frigg Central Complex, was designed in the mid-1970s, when offshore safety standards were quite different from today practice. Under Norwegian legislation, it has not generally been required to upgrade technical equipment on old installations to comply with present standards.

However, in the UK, there is a need for all platforms, irrespective of their age, to comply with the risk based approach to safety by demonstrating through the Safety Case that risks to personnel are ALARP. In Norway, there is a similar requirement to demonstrate that the risk level is acceptable in relation to the company's risk tolerance criteria. Risk quantification was used in relation to the upgrading of the equipment on the Frigg installations for the following purposes, which are outlined further in the following paragraphs:

---

[1]Frigg field production stopped in October 2004, and most of the installations have been removed, with exception of three concrete structures.

- To decide on the required extent of risk reduction
- To choose the right composition of technical and procedural measures
- To demonstrate that risks are as low as reasonably practicable.

21.5.1.1 Decision on Necessary Risk Reduction

As the Frigg Field was approaching the planned end of the field life, there was a difficult balance to be maintained between the need to improve the safety level and what was rational based on field economy. The decision as to what is necessary and what is a sufficient risk reduction was therefore very demanding. Relatively advanced approaches were therefore required to determine the merit of any risk reducing actions. The results from sensitivity studies, as shown in Fig. 19.1, demonstrate the difficulty in decision-making.

Figure 19.1 showed what the effect on the risk levels would be, if a certain parameter value in the event tree is changed to 50% of the original value. Reduction of the hot work activity by 50% implies about 18% reduction of the immediate fatality risk to personnel, and about 32% reduction of the frequency of significant material damage cases. Only two of the aspects considered have anything above an insignificant effect on the risk levels for personnel, namely reduction in hot work activities and reduction of explosion overpressure.

In the case of the Frigg installations, the decision on which safety systems to improve became even more difficult, because it was demonstrated in the original Safety Case that the risk level for personnel was quite low [10], especially when considering the age of the installations. However, as could be expected for an ageing installation, the risk of damage to the installation resulting in loss of assets was quite a bit higher.

21.5.1.2 Technical Versus Procedural Measures

There is often a limit to what can be done in terms of technical upgrading of an existing installation, without running into costs that are difficult to justify.

Procedural risk reduction measures can, however, often be implemented at a low initial cost, although there is often limited scope regarding what these procedures can achieve. Further, a problem with such improvements is that the reliability of the procedure in achieving the planned risk reduction may be questionable. This is particularly the case if the procedure is intended to control actions during an emergency. One example of such a measure is the required response time to allow initiation of manual blowdown.

In the case discussed here there was strong emphasis from the very beginning that suboptimisation of this kind could not be allowed. The possible negative effects of area segregation on ventilation were avoided by the use of arrangements which had only a minimum effect on ventilation.

21.5.2   Case Study: Effect of Improved Blowdown

A new flare stack was installed on the TCP2 process platform in 1995 in connection with the new oil processing module. This enabled a reduction in the blowdown times on the old part of TCP2. In some cases these were quite long, and not in line with current practice.

Process studies had shown, however, that the blowdown flow restrictions were associated with valves and piping systems, which would require extensive replacement if blowdown times were to be substantially reduced. This is very expensive on a live platform, and also requires periods of full production shutdown.

Another significant effect of such a blowdown modification would be extensive hot works, in terms of welding and cutting. This is in itself one of the main sources of risk, and was an aspect that would cause considerable concern.

Thus blowdown improvements needed in depth-consideration, in which the overall objective was to compare the benefit gained against the cost and increased risk during offshore construction. Quantification of the risk reduction potential was therefore required.

21.5.2.1   Improvement Potential

The blowdown times for the process equipment receiving gas and oil from the new satellite fields were, with one minor exception, noted to be in line with current practice i.e., capable of reducing pressure to below 7 bar (or half of initial value) in no more than 15 min.

On a qualitative basis it was noted that blowdown would reduce the duration of a fire and thereby also reduce the likelihood of the fire escalating from the initial process segment. The development of the fire in the first few minutes is therefore virtually unaffected by initiation of blowdown.

Since the TCP2 platform is bridge-connected to the other FCC platforms, it had been demonstrated in the QRA that almost all (99%) the risk to personnel occurs in the first five minutes. Thus it was unlikely that improvements to the blowdown system would affect risk to personnel. The main effort was therefore spent on evaluating the material damage risk.

21.5.2.2   Cases Considered

Two cases of possible improvement to the blowdown system were considered in relation to the original condition of the system;

- Reduction of the blowdown time to 15 min (7 bar) for all segments where applicable.

- Reduction of the blowdown time to 7.5 min for four segments that, from the QRA, had been determined to be most critical i.e., with the highest fire frequency. In this case, other segments retain their original blowdown time.

The behaviour and duration of the fire resulting from these options was modelled directly in the flow calculations, and was in need of no further attention. It was, however, realised that fairly detailed modelling of escalation from one segment to another would be required in order to enable a realistic reflection of the differences between the cases studied. This required further development of the fire escalation model is described below.

It should be noted that no further development of escalation due to explosion was considered necessary. This model assumes, somewhat simplistically and conservatively, that at an overpressure at 0.3 barg or above there will be an immediate rupture of piping which results in escalation.

21.5.2.3 Modelling of Fire Escalation

In order to determine when escalation occurs due to failure of piping engulfed in a fire, a non–linear structural analysis of typical pipe sizes was undertaken. This is described in Sect. 15.10.

21.5.2.4 Results and Evaluation

It was found that the risk to assets was only marginally affected, by reduced blowdown times, with less than 2% reduction in the frequency of total loss of the platform. These results were somewhat surprising, and some comments are appropriate regarding special aspects of the TCP2 platform which have influenced these results.

It should first of all be noted that the platform mainly has an open layout, without clear separation of segments and systems. This implies that it is quite likely that a fire in one segment will impinge upon small sized piping from another system resulting in a likely failure in a short time (about 2–5 min, see Fig. 15.25), irrespective of whether the blowdown is within 7.5 or 15 min.

For blowdown to be effective in controlling escalation of fire, the blowdown times have to be less than 5 min. Alternatively, longer blowdown times may be combined with passive fire protection of small bore piping. In the present case, none of these actions were practical or economical to implement. It should also be observed that the escalation of accidents is strongly influenced by explosion overpressure, which actually was most important for escalation on the TCP2 platform.

The conclusion of this evaluation was therefore that the costs of the measures to improve the blowdown system were in gross disproportion to the benefits that could be expected. This was obvious as far as the risk to personnel was concerned, but when the economic benefits were compared to the costs, it was also shown to be the case for risk to assets. The economic benefits in this regard are the values of reduced accident costs (resulting from shorter fire duration) and likewise reduced reconstruction times. For TCP2 it was therefore decided to employ compensating actions and essentially leave the blowdown system in its initial state, apart from a minor replacement of one blowdown valve.

This clear conclusion was made without consideration of the increased risk from hot work if the work was undertaken. Later it was calculated that the changes of risk for every 1000 h of extra or reduced hot work are as follows:

- 7.4% of AIR
- 13% of frequency for severe/significant damage
- 7.0% of frequency of total loss.

This implies that if changes to the blowdown system required more than 1000 h of open flame hot work, then several years would pass before the total accumulated effect was a positive risk reduction.

The compensating measure chosen was to increase the explosion venting areas by removing a significant amount of the cladded outer walls. Through CFD explosion simulation and ventilation studies this was shown to have good effects for explosion overpressure as well as ventilation rates.

21.5.3 Acceptable Internal Leak Rates of Isolation Valves

Section 47 of the Activities regulations [15] calls for a maintenance programme to be used. The Guidelines require the following with respect to isolation valves (ESV) in the emergency shut-down system:

b) the emergency shut-down system is verified in accordance with the safety integrity levels set on the basis of the IEC 61508 standard and Norwegian Oil and Gas (*sic*) Guideline 070. For plants that are not covered by this standard and this guideline, the operability should be verified through a full-scale function test at least once each year. The test should cover all parts of the safety function, including closing of valves. The test should also include measurement of interior leakage through closed valves......

It is not uncommon that there are minor internal leaks in the isolation valves. The next question is then what are acceptable leak rates. Such an evaluation normally has to distinguish between isolation valves on risers and isolation valves between different process segments in the process system. Often there is also an evaluation to be made of what are the isolation valves in the process systems that have high importance from a safety point of view, not necessarily all isolation valves in the process part have the same importance. Usually all isolation valves on risers (i.e. pipelines and flowlines) are important from a safety point of view. The valves with high importance shall be tested and criteria for acceptable leak rate have to be formulated.

Some principles have been formulated in order to develop the criteria for acceptable internal leak rate in isolation valves. The scenario considered is a fire scenario. If there is an internal leak, there is an associated increase in risk, ΔR, which should be insignificant, interpreted as less than 1% increase. It is further assumed that pressure relief is activated within 10 min. The difference between a case with tight and leaking isolation valve is indicated in Fig. 21.2.

It is evident from Fig. 21.2 that there is no difference in the initial phase, but the difference starts when the depressurisation is started. The significance of the difference increases with the time passed after initiation of depressurisation. This implies that the largest fire loads in the initial phase are not affected by this difference. The immediate fatalities in the initial phase are also not affected.

The main difference will be a long lasting limiting fire, which may have effects in relation to further weakening of the structure and possible further escalation through new leaks. The duration of this phase is strongly dependent on the volume upstream of the leaking valve.

When this approach has been used in practice, it has resulted in different values for isolation valves against risers/pipelines (large reservoirs) and the isolation valves within the process systems. Typical values are indicated in Table 21.2.

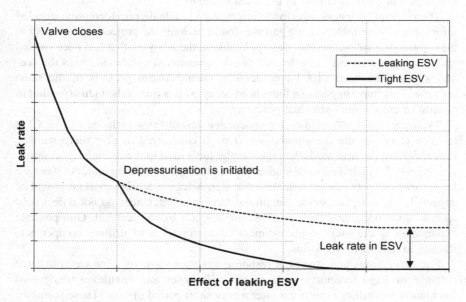

Fig. 21.2 Idealised effect of tight and leaking isolation valve

Table 21.2 Typical acceptable and unacceptable leak rates in isolation valves

Typical valve	Unacceptable leak rate (kg/s)	Leak rate calling for actions to be considered (kg/s)	Acceptable leak rate (kg/s)
Riser ESV	>0.5	0.05–0.5	<0.05
Process ESVs			
Least critical	>11.8	2.4–11.8	<2.4
Most critical	>3.0	0.6–3.0	<0.6

21.6   Drilling with Jack-Up in Cantilever Over Steel Jacket

A study was performed for PSA in 2014 [16] in order to investigate the fire loads on structures when a jack-up mobile drilling unit is drilling with the derrick in cantilever modus over the wellhead area of a fixed steel jacket installation. The work consisted of a review of selected risk, fire and structural response analyses from eight field developments on the NCS, none of which had their own drilling facilities. The need for well drilling and completion is ensured by using a mobile drilling unit which is positioned next to the fixed installation.

The relevant operators who made information available provided quite detailed information. Nevertheless, a big gap was found between the projects with respect to the information submitted that was applicable to the analyses that had been carried out in the projects and the extent of the documentation submitted. Also the scenarios that were used varied considerably, even between projects of the same operator. This may suggest that there is no accepted practice in the industry either in general or among the individual operators.

Typically, the QRA studies for production installations included use of CFD tools to carry out the fire simulations. This is considered to give more realistic illustrations of the heat loads than the simpler tools used in the drilling rig analyses. All projects did include special studies of fire loads on the jack-up unit, but there are large variations between the scenarios that were selected. Some variations would be expected, due to field specific conditions, but other differences do not reflect field specific conditions, and the variations are difficult to comprehend. One possible explanation is the lack of involvement of personnel with drilling competence, leading to unrealistic scenarios.

Fire simulations and structural response analyses show in some cases that the collapse of large structures such as the drill tower and cantilever (uncovered foundation for drilling) can occur after a very short period of time. These scenarios are not necessarily catastrophic to the drilling rig. But it has to be taken into account that the scenarios may lead to the rapid loss of escape routes from the drilling and wellhead areas at a sufficiently high frequency to exceed the acceptance criterion.

For the production installations, these events are considered to be catastrophic in the sense that they imply a loss of global load structural capacity (structural collapse) and/or loss of accommodation with a safe haven and control rooms. But such scenarios are not considered dimensional due to the frequency falling below the limit by which such loss of integrity become dimensional (such as 10^{-4} per year). However, several of the reviewed analyses show frequencies for these events that are so close to the limit that they are in practice either on or above the limit when uncertainties are considered. A general feature is that robustness and uncertainty in data bases are inadequately discussed in the reviewed studies, as are ALARP assessments and associated measures.

The common challenge appears to be that the combined operations were not accounted for during the design of any of the installations, and thus when analysed it is far too late to provide good protective measures without incurring very extensive costs. This leads the projects to try to show that these scenarios are negligible, in order to avoid ending up with costly design modifications.

21.7 Case Study—Cost Benefit Analysis

Some of the principles of cost benefit analysis are presented through a case study, which is focused on whether additional passive fire protection should be provided. It should be emphasised, however, that demonstration of ALARP (As Low As Reasonably Practicable) is considerably more extensive than just performing a cost benefit analysis; see Ref. [17].

It should be noted that the example was prepared in 1998, where the period after year 2000 was considered 'future'. The periods have now been adjusted and assumes 2015 and onwards as 'future'. In the current version 2015 is still regarded as 'future', implying that the analysis is seen from year 2014.

21.7.1 Field Data

The overall field characteristics first need to be defined. The oil production profile (gas production disregarded), and assumed oil price curve are shown in Fig. 21.3. A plateau production period of four years is followed by a period of steeply declining production, perhaps due to increased water production.

Figure 21.4 shows a comparison of the base case production curve with the deferred production curve following a 1 year delay in 2018. The plateau production level is also extended into 2019 and the production curve in the following years is shifted one year to the right i.e., the production volume in year 2020 with the delay corresponds to the production in year 2019 without the delay.

The important economical value in this regard is the production delay and the resulting loss of interest from the later production of the same volume. The interest

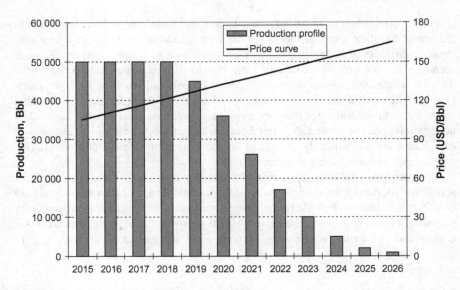

Fig. 21.3 Production profile and assumed oil price curve

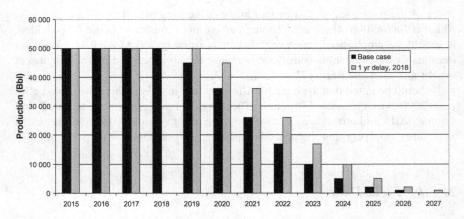

Fig. 21.4 Base case production curve and deferred production curve with 1 year delay in year 2018

rate used in the calculations is 10%. It should be noted that cost inflation is not included in this illustration.

The economic effects of material damage are shown in Table 21.3, for the failure categories affected. Four material damage categories are used (only the middle two shown in Table 21.3):

Table 21.3 Material damage characteristics

Damage	Reconstruction cost (MUSD)	Full production delay (months)
Significant	300	3
Severe	1,500	12

Minor: Local equipment
Significant: One module
Severe: Two modules or more
Total: Entire platform.

21.7.2 Definition of Risk Reducing Measure

The RRM considered is the application of passive fire protection on a selected part of the platform structure considered to be the most critical. The following costs apply to the RRM:

Investment cost: 6 MUSD
Annual cost of inspection: 0.06 MUSD

21.7.3 Risk Reducing Potentials

The RRM has the effect of reducing the damage in some accident scenarios from 'Severe' damage (to several modules) to 'Significant' damage (to one module). There is also a small reduction in the annual PLL. The changes to the damage frequencies are as follows:

- Frequency of Severe damage reduces by: 1×10^{-3} per year
- Frequency of Significant damage increases by: 3.5×10^{-4} per year
- Annual PLL value reduces by: 3.5×10^{-3} per year.

All the changes are rather limited (less than 10% of total platform values). There is no effect on the level of risk to environment.

21.7.4   Overall Approach to Comparison of Costs and Benefits

If the risk level s are in the 'Intolerable' or 'Tolerable' regions of the ALARP diagram, then no further consideration is needed. If we assume that the risk levels are not in either of these regions, then the overall approach can be summarised as follows:

- The second evaluation is to determine if the RRM is according to good design or operational practice. If it is in accordance with good practice, and the cost is limited (as in the present case, 6 MUSD), then it may be concluded that the measure should be implemented.
- If it is not obvious that the measure is required to comply with good practice, or if it is judged that the cost is too extensive, then the cost benefit approach will be required.
- Cost benefit analysis is undertaken for material damage and production delay first.
- If economical benefits exceed costs, then this RRM should be implemented without further consideration of the risk to personnel.
- If costs exceed economical benefits, the net present cost (NPV) is calculated and compared with the total change in PLL over the field life.
- The costs per averted statistical fatality have to be considered. If the costs are in gross disproportion to the benefits, then this RRM should not be implemented.

If there had been an environmental aspect of this RRM, the economical value of the spill should be integrated into the calculation after consideration of the material damage, but before the risk to personnel. Neither insurance, tax savings, nor partner share aspects of cost are included, this may be regarded as a societal consideration. It should be noted that loss of prestige, contract obligations etc., are not taken into consideration. The following should, however, be noted in relation to intangible aspects where the relation between costs and benefits may be quite special. An example may illustrate this.

A field may be owned by a partner group of typically four to six companies, where the field operator may have a 30% ownership (or less) in the field. It is not uncommon that the tax on profit in the petroleum sector is quite high, say 60–80%. If we use 30% ownership and 70% tax, the operator implicitly carries only 9% of any expenditure (assuming that he has an annual profit). But if the decision is made **not** to install a particular risk reduction measure and an accident occurs, which the measure could have eliminated (or reduced), then it is quite likely that the field operator will carry all of the 'blame' as assessed by the media and public opinion, if the accident occurs.

The interest of the field operator will therefore often be best served by the implementation of risk reducing measures.

Another aspect is that the government often is the major loser in the event of accidents due to reduced tax income. This is not insured, and is another reason why the consideration of risk reducing measures should be carried out in a societal context without consideration of tax and insurance issues.

21.7.5   Modelling of Benefits

Table 21.4 presents the relative changes in production costs for the entire field.life, according to the year in which the accident occurs.

For significant damage it is seen that there is a considerable loss the year the accident occurs. The value of the lost production (the year the accident occurs) increases with the oil price for the duration of the plateau period, and decreases when the production volume in the year is lower. In most of the cases, the nominal value of the delayed production is higher, due to the assumed increase in oil price.

When the loss occurs during the plateau production period, the lost production cannot be reclaimed until the plateau period has ended. For severe damage it is seen that the extraordinary costs are the material damage, plus the value of lost production. This will lead to increased production after the plateau period and thereby provide extra income in later years for the duration of the production period. The value of the lost production (the year the accident occurs) increases with the oil price for the duration of the plateau period and decreases when the production volume in the year is lower. As an illustration, the values for 2015 are calculated as follows:

- Significant damage:

  - 300 MUSD in direct cost
  - 12,500 bbls lost production per day for 90 days with a value of 110 USD/bbl equals a total of 123.75 MUSD.

- Severe damage:

  - 1,500 MUSD in direct cost
  - 50,000 bbls lost production each for 365 days with a value of 110 USD/bbl of total 2007.5 MUSD.

In the case of significant damage, the deferred production will be produced after the production plateau period is finished, by prolonging the plateau level with one extra year, and increased production the following year. The values (negative loss i.e., extra income) in the illustration are seen in years 2018 and 2019, which have an extra income of 144 MUSD, in contrast to the production loss of 123.75 MUSD in year 2016 (both 2015 year values, since no inflation is considered).

It is implied by the tables that the accident occurs in the beginning of the year considered. A refinement to the calculation would be to assume that the accident occurs in the middle of the year.

The values for both the damage categories are now multiplied by the annual changes in frequencies, in order to get the expected values, as shown in Table 21.5.

Table 21.4 Changes in production cost according to year the accident occurs for 'significant' and 'severe' damage during the field life

Significant damage in year	2015	2016	2017	2018	2019	2020	2021	2022	2023	2024	2025	2026
2015	423.8	0	0	-56.3	-87.8	0	0	0	0	0	0	
2016		429.4	0	-56.3	-87.8	0	0	0	0	0	0	
2017			435	-56.3	-87.8	0	0	0	0	0	0	
2018				426.6	-131.6	0	0	0	0	0	0	
2019					405.3	-109.4	0	0	0	0	0	
2020						379	-81.9	0	0	0	0	
2021							353.6	-55.5	0	0	0	
2022								332.6	-33.8	0	0	
2023									316.9	-17.44	0	
2024										306.98	14.4	
2025											303.6	
Severe damage in year												
2015	3507	0	0	-228.1	-427.1	-492.8	-459.9	-370.5	-273.8	-169.7	-58.4	-60.2
2016		3598.8	0	-228.1	-427.1	-492.8	-459.9	-370.5	-273.8	-169.7	-58.4	-60.2
2017			3690	-228.1	-427.1	-492.8	-459.9	-370.5	-273.8	-169.7	-58.4	-60.2
2018				3553	-664.3	-1183	-357.7	0	0	0	0	0
2019					3208	-1183	-613.2	0	0	0	0	0
2020						2781.2	-1329	0	0	0	0	0
2021							2369	-899.7	0	0	0	0
2022								2029	-547.5	0	0	0
2023									1773	-282.9	0	0
2024										1613	58.4	0
2025											1558	0

Finally, the Net Present Values of the changes are calculated for the profiles shown in Table 21.5. The resulting NPV values are shown in Table 21.6. These expected values are calculated for occurrence of the accident in each year. Thus, these values are additive.

The accumulated benefit from Table 21.6 is 4.67 MUSD, which is the economic benefit of the risk reducing measure.

21.7.6   Modelling of Costs

The initial investment cost is 6 MUSD and the annual operational cost is 0.06 MUSD per year. The Net Present Value is calculated for the operational costs 0.41 MUSD. The total cost of the RRM is thus 6.4 MUSD.

21.7.7   Results

The following comparison may now be made on an economical basis (all NPV based):

Benefit: 5.99 MUSD
Cost: 6.41 MUSD

The net investment is therefore 0.42 MUSD (NPV). The corresponding PLL reduction is 0.0035 fatalities per year, or 0.042 fatalities averted over 12 years. This corresponds to an investment of 10.1 MUSD per averted statistical fatality.

21.7.8   Sensitivity Study

Figure 21.5 presents a sensitivity analysis for the variations in net investment and the ratio of this investment per averted statistical fatality over the field life, depending on the interest rate used in the calculations.

It is seen that with 7% interest there is a negative cost per averted life, implying that there is an income. When interest increases to 13%, the value of investment per life averted increases to 20.6 million USD per averted life. This shows that the calculations are very sensitive to the interest rate.

It is often important to carry out extensive sensitivity studies for variations in cost per averted life as a function of variations in parameter values. An example of the results from sensitivity studies is presented in Fig. 21.6.

The variations are calculated by multiplying one parameter at a time by factors of 2.0 and 0.5 and recalculating the cost per averted life over 12 years. Figure 21.5 shows variations around the expected value, 10.8 MUSD, or around 60 MNOK

Table 21.5 Expected changes of loss (income) for each year, given the year of occurrence

Expected value, cost changes (MUSD)	2015	2016	2017	2018	2019	2020	2021	2022	2023	2024	2025	2026
2015	-1.65	0	0	0.14	0.247	0.17	0.16	0.13	0.10	0.06	0.02	0.02
2016	0	-1.69	0	0.14	0.247	0.17	0.16	0.13	0.10	0.06	0.02	0.02
2017	0	0	-1.73	0.14	0.247	0.17	0.16	0.13	0.10	0.06	0.02	0.02
2018	0	0	0	-1.67	0.36	0.41	0.13	0	0	0	0	0
2019	0	0	0	0	-1.53	0.52	0.21	0	0	0	0	0
2010	0	0	0	0	0	-1.35	0.55	0	0	0	0	0
2021	0	0	0	0	0	0	-1.18	0.37	0	0	0	0
2022	0	0	0	0	0	0	0	-1.04	0.23	0	0	0
2023	0	0	0	0	0	0	0	0	-0.94	0.12	0	0
2024	0	0	0	0	0	0	0	0	0	-0.87	0.03	0
2025	0	0	0	0	0	0	0	0	0	0	-0.85	0

Table 21.6 NPV for each year

Year	Relative NPVs (MUSD)	Year	Relative NPVs (MUSD)
2015	−0.94	2021	−0.43
2016	−0.84	2022	−0.39
2017	−0.74	2023	−0.35
2018	−0.62	2024	−0.35
2019	−0.54	2025	−0.30
2020	−0.48		

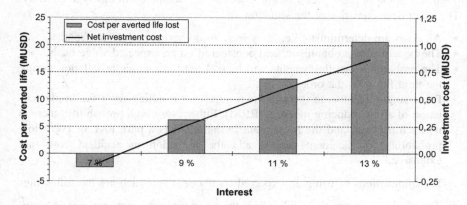

Fig. 21.5 Sensitivity study for different interest levels

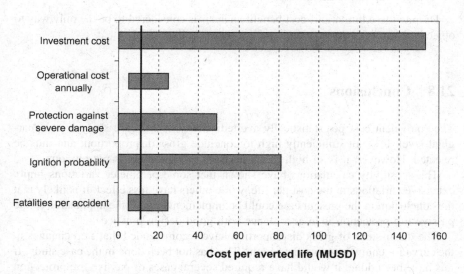

Fig. 21.6 Sensitivity study for different parameter variations

with the current exchange rate. When the values in Fig. 21.6 drop to zero, this suggests that the actual cost per averted life is negative, which implies that there are higher reduced asset costs than the actual investment and annual operating costs.

21.7.9   Discussion and Evaluation

It should be noted that this cost benefit evaluation is somewhat special in its premises. Some of these premises are however, very common in this type of study. The main characteristics of the situation are:

- All costs are deterministic i.e., they have to be borne, mainly up front.
- The benefits are probabilistic, and considered on an expected value basis.
- The probability that the benefits will be realised for a single installation during the field life is in the order of 1%.
- For a single installation there are two outcomes resulting from the implementation of a risk reducing measure (RRM). Either (very high probability) there is no such accident and the investment in the RRM is a pure loss, or (very low probability) an accident occurs, and the reduction in accident costs are extensive.

The implications of using an expected cost/benefit approach are the following:

- The approach is strictly valid only if the operator has a high number of corresponding installations.
- It is presupposed that the operator is financially capable of surviving the year with the highest nominal loss.

Despite these limitations cost benefit analysis is considered to be the only way to objectively analyse the economic values of a possible RRM.

21.8   Conclusions

The equivalent cost per statistically averted fatality in this case study is at a marginal level. It is not sufficiently high to conclude gross disproportion, and thus be rejected. However, it is so high that a decision to implement is not obvious.

The sensitivity calculations have shown that some parameter variations imply extensive variations in the cost per life, while others have less effect. It is likely that the conclusion in the present case could be implementation, if the application of the passive fire protection is in accordance with good design practice.

The discussions of gross disproportion have recommended that step changes in the curve for equivalent costs be sought. This has not been done in the case study. If this have been done, it would have required several cases of passive fire protection

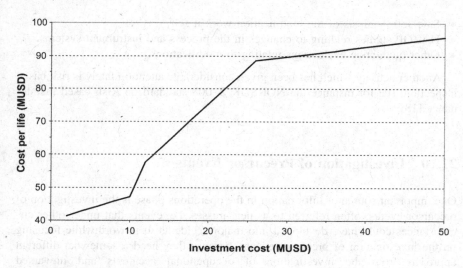

Fig. 21.7 Equivalent cost per life saved for the case study

coverage to be analysed for the costs and benefits, and the equivalent cost per life saved calculated in the same way as above.

A diagram similar to Fig. 21.7 would have resulted. It shows the resulting equivalent cost per statistical life saved (in MUSD) as a function of the extent of passive fire protection applied to the structure, measured as an investment cost (MUSD).

The step change in the curve occurs from 10 MUSD to about 25 MUSD, implying that an investment up to 10 MUSD is certainly not in gross disproportion as it corresponds to a value of 47 MUSD per life saved.

The upper step change for the curve, corresponding to 26 MUSD invested and 89 MUSD per life saved, certainly indicates that costs per life saved exceeding 89 MUSD are in gross disproportion. Whereas, in the interval from 10 to 26 MUSD, the limit for gross disproportion would be open to evaluations according to the specifics of the case, and no general rules can be stated.

21.9   Analysis of Maintenance Activities

Minor modifications to facilities are such that in isolation they do not represent any marked influence on the risk level. As such, quantified risk assessment is not very significant when considering the risk associated with such activities. The main tools for minor modification risk control are usually qualitative evaluations (not covered in this book), including:

- Safe Job Analysis of the modification work it self.
- HAZOP studies relating to changes in the process and instrument systems.
- Other qualitative evaluations relating to modifications.

Another activity which has been given considerable attention lately is risk based inspection, see for instance the NORSOK Z–008 standard on Risk based maintenance [18].

21.10   Investigation of Precursor Events

One important source of information in the operations phase is the investigation of precursor events, often referred to as near-misses, i.e. events that under other circumstances could have developed into major accidents. It is worthwhile focusing on the investigation of precursor events, because they need a somewhat different approach from the investigation of occupational accidents and unwanted occurrences.

21.10.1   Authority Requirements

Management and information duty regulations [19] have investigation requirements in §20, which underline that the actual course of events as well as 'other potential courses of events and consequences' shall be investigated, in addition to the 'human, technical and organisational causes of the hazard and accident situation, as well as in which processes and at what level the causes exist' and several other aspects. The purpose of this recording and analysis is according to §20 to prevent recurrence. This implies that learning from accidents and incidents is the main purpose.

The PSA initiated a process around 2000 [20], whereby the so-called 'MTO approach' was introduced into the Norwegian offshore industry as a means of moving accident and incident investigations forward and increasing the scope beyond just looking at the technical causes, which was the most common approach at the time. This resulted in the clear development of investigations in the Norwegian petroleum industry, whereby HOFs are focused upon in addition to technical causes, and the understanding of the 'root causes' has improved significantly since then.

21.10.2 *Authority Investigations in the Norwegian Petroleum Sector*

Authority investigations in the petroleum sector in Norway fall into two categories:

(a) Investigations of accidents that are considered to be so severe from a national perspective that an investigation commission appointed by the Ministry of Work is warranted in order to enable the performance of the supervisory authority to also be investigated as a possible causal factor and potential for change and improvement. The last time such an investigation was conducted was in 1985, when the semi-submersible unit West Vanguard had an ignited shallow gas blowout [21].

(b) Selected accidents and precursor events of lower importance are investigated by an investigation committee appointed by and from within the PSA on an ad hoc basis, selected based on severity, accident potential and importance for industry. A large group of PSA employees are competent in systematic investigation, and committee members are selected on an ad hoc basis from this group, according to what type of event has occurred and the availability of personnel. PSA performs around five to 10 investigations per year, each usually requiring three people. With up to 100 employees to choose from, it will be several years before the average PSA employee participates in an investigation.

The investigation arrangements in the petroleum sector in Norway are distinct from the corresponding arrangements in the transportation sector in Norway, and it has thus been debated if they are preferable or not. For the transportation sector, the independent official Accident Investigation Board Norway (AIBN) performs all official investigations of serious accidents in the marine, air, road and rail transportation in Norway [22]. This is a professional investigation body, and it has been suggested that investigations in the petroleum sector should be transferred to this board.

PSA publishes its reports within three to four months, whereas the AIBN often takes up to one year. This is one of the arguments against transferring this responsibility for the petroleum sector to the AIBN. PSA also argues that accident investigation is part of its regular supervisory activities and that there are mutual benefits from the integration of investigations and other supervisory activities.

PSA has been criticised for being too focused on the identification of deviations from the regulations associated with the operations leading up to the accident or incident. Further, PSA investigations often have insufficient focus on the identification of root causes, especially in the organisational sphere. AIBN reports are often claimed to be better in this regard.

21.10.3 Company Investigation Practices

Company investigations are usually conducted according to internal guidelines at two or three levels, according to the severity of the potential risk of the event. In a three-level structure, the following would be typical levels of investigations:

(i)   Installation/plant internal
(ii) Field or area unit
(iii) Corporate level

The most serious accidents or precursor events are investigated at the corporate level by specialists, including investigation method specialists. The lowest level is limited to personnel with those qualifications that are available on the installation and usually do not include personnel with investigation competencies; this is rarely termed an 'investigation', but rather an 'examination' or similar. Sometimes there is a fourth level, level zero, which is not an investigation, but rather an informal 'in-depth study' performed by one or two people who are familiar with the installation/plant.

However, even though the approach is the same, there are significant differences in how investigations are conducted using the MTO approach. PSA investigations focus on deviations from regulatory requirements and on where improvements would be beneficial. By contrast, company investigations often focus on root causes, although the quality of such discussions is strongly dependent on the depth of the understanding of the term 'root causes' by members of the committee. Ten years ago, it was common to find 'failure to follow procedures' as a typical company-stated root cause. The competency of committee members has improved over the years, and a typical root cause in current investigations is 'weak safety culture'. In this regard, investigation practice has improved. 'Root causes' are not a focal topic of this paper even though they are essential in an investigation. However, the consideration of differences between investigations at different levels (company, authority, professional entity) is included in order to illustrate differences in the quality of the investigation process.

One interesting development over the past couple of years is that company reports are published much more frequently. This applies to the investigation of the hydrocarbon leak on Gullfaks B 4.12.2010 [23] and investigations by Statoil (now Equinor) of two cases of falling objects, although no injuries occurred in these two cases. The report on the former was followed by the publishing of the PSA report [24] on the same incident a couple of weeks later. The PSA emphasised that it disagreed strongly with the findings of the internal investigation, implying that its investigation had found several additional failures, especially in the organisational field. It was further claimed that the two investigations also disagreed on the potential consequences of the incident.

21.10.4 Improvements in Investigation Practices Relating to HC Leaks

Two improvements may improve investigation practices in the industry with respect to learning from the investigation of unignited hydrocarbon leaks as precursor events. This was to some extent discussed in Ref. [25]

(a) Many of the hydrocarbon leaks (>0.1 kg/s) have causes in the manual practices for intervention on pressurised equipment. This would improve the learning from such events if the failures with respect to WP and isolation plan implementation were classified according to the phases shown in Fig. 6.12.
(b) It has been demonstrated in Ref. [20] that the majority of the investigations under-communicate the major accident potential in hydrocarbon leaks, if they had ignited. The recommendation has been made to communicate the major accident potential in a more transparent manner.

21.11 Overall Analysis of Modifications

21.11.1 Overview

The modifications discussed in this section are medium sized modifications, which range from modifications undertaken as part of regular maintenance work, to projects involving the addition of new facilities and structures (see following section Tie-in of New Facilities). A modification project will normally include study phase, engineering, fabrication, installation, completion and operation. Risk assessments, risk acceptance, and emergency preparedness shall address all phases involved.

The studies should include all relevant installations engaged in the activities, including mobile units and vessels that may be involved in operations and possibly also nearby vessels and installations if they are close enough to be affected by accidental effects.

During the study phase the feasibility of the planned modifications should be assessed with respect to safety and risk acceptance. For smaller modifications this may be a qualitative risk analysis, while for larger modifications and new structures, quantitative risk assessment may be needed. It may be sufficient to update the parts of an existing installation's QRA that are affected by the modification. For modification of process systems a HAZOP is required.

A separate risk [and emergency preparedness] analysis should be undertaken to cover the period when the modification work takes place on the installation. In these analyses the additional risks arising from the modification work itself should be added to the existing risk level on the installation. This needs to be compared to the risk tolerance criteria for the installation.

For smaller modifications, when it is obvious that the risk tolerance criteria will be met, a qualitative risk [and emergency preparedness] analysis may be sufficient. The quantitative effect of the modification on the risk level may then be calculated as part of the regular updating of the quantified risk analysis and emergency preparedness analysis for the installation.

The analyses should identify operations which require Safe Job Analysis. The following aspects which may affect the risk level should be considered if they are relevant:

- Increased number of personnel onboard during modification work.
- Increased number of personnel in hazardous areas.
- Risks associated with simultaneous operations during installation, modification and commissioning.
- Modifications to process systems which may cause hydrocarbon leaks if work permits and isolation plans are not followed closely by all involved.
- Use of hot-work particularly in respect of the use of welded or flanged connections.
- Effect of the use of habitats for hot-work.
- Dropped objects during installation work.
- Temporary unavailability of safety systems during the modification work.
- Effect of the modification work on the ESD system and process safety.
- Increase in number of leak sources and explosion loads due to more equipment.
- Increased extent of helicopter shuttling to neighbouring platforms or to the shore.
- Human errors during the work.

21.11.2   Modification Risk in a Life Cycle Perspective

Modifications are usually carried out while the installation is in normal operation and therefore result in an increase in the activity level on or around the installation. The period of the work may be relatively short or somewhat longer. Risk analysis should be used to ensure that the modification work does not in itself give rise to an unacceptable risk level.

The following parameters should be considered when evaluating the temporary increase of risk during modification work:

- The maximum increase of the risk level. Evaluations have to be made whether the peak level is tolerable even though the duration may be short. This may influence the decision whether to carry out the modification work as planned or whether compensating actions or restrictions in the normal operations need to be implemented.

- The duration of the activities causing the increased risk level. The effect of this increased risk duration on the total exposure of personnel to risk needs to be considered in relation to the risk tolerance criteria.
- Whether the increase of risk is local i.e., affects a particular area of the installation, or applies globally. A local increase of risk may be more easily accepted or even compensated for, as opposed to a global increase of risk.
- It may also be relevant to consider what effect the modification will have on the normal operation, e.g. whether a certain risk reduction is achieved. Some increase of risk during modification may be easier to accept if the resulting effect is a significant reduction of risk during normal operations.

The total risk exposure will depend on both the possible increase of risk during modification and the possible reduction of risk resulting from that modification. Figure 21.8 shows the risk level for two different situations. The risk level without risk reducing measures is shown together with the risk levels during and after the implementation of risk reducing measures. The diagram shows both the instantaneous values of risk as well as the accumulated risk values for both alternatives.

The diagram shows that there is a significant increase of the risk level during the modification work. The risk level after the measures is slightly reduced from the level before the risk reducing measures. With the numbers used [which are not unrealistic for a major modification], a long period will need to elapse before the accumulated risk level is lower for the alternative with risk reduction implemented.

If the residual life time is less than ten years, it would not be worth implementing the risk reducing measures.

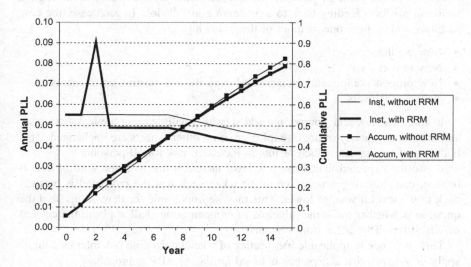

Fig. 21.8 Example of variation of risk level during implementation of risk reducing measures, and effect on accumulated risk over time, with and without RRMs

The final decision regarding acceptance will have to be based on an overall evaluation of all parameters. Life cycle considerations should be used particularly in relation to modification works. There will often be trade-offs between risk increasing and decreasing aspects, such as welded piping connections as opposed to flanged piping connections. A welded connection will imply lower risk in operation due to reduced leak frequency, but may imply a higher risk during modification due to hot work.

Establishment of a procedure for integrating risk analysis competence into the early phases of modifications planning, is a core element in HES management. This should ensure that risk analysis aspects are included as an integral part of the planning and execution of modification work.

21.12   Tie-in of New Facilities

From a HES management point of view new facilities will usually be considered as a new project i.e., risk assessments for engineering, fabrication and installation should be carried out in accordance with the practice for new projects.

One aspect which is often discussed in relation to the tie-in of new facilities is how to consider the possible increase in risk to personnel on the installation, resulting from the new facilities. This is the opposite of what is usually expected, that risk should be decreased over time, as new improvements are implemented, either as technical or operational barriers.

The tie-in of new facilities will often involve production from new fields or new reservoirs, either through platform completed wells, subsea wells, or through a wellhead platform feeding back to a processing installation. In each case, the new facilities will involve one or more of the following:

- New wellheads
- New import risers
- New process components or increased capacities of existing components
- New export risers.

In all cases, new leak sources are added, which will increase the leak frequency and thus the frequency of fire or explosion, if all other aspects are unchanged. This is the usual mechanism which results in increased risk for all personnel.

It is often expected that if the risk level increases due to some kind of deterioration, compensating actions should be taken such that the risk level is brought back to the level it was, or lower. This may be impossible for new tie-ins, and the question is whether the same principle of compensation shall apply in the case of new facilities. This is far from obvious.

This challenge is applicable irrespective of whether absolute risk tolerance limits apply or whether risk acceptance is based on the ALARP approach.

References

1. HSE (2015) The offshore installations (Offshore Safety Directive) (Safety Case etc.) regulations 2015. Health and Safety Executive, HMSO, London
2. Sklet T, Vinnem JE, Aven T (2006) Barrier and operational risk analysis of hydrocarbon releases (BORA–release). Part II, Results from a case study. J Hazard Mater A 137:692–708
3. Gran BA, et al (2012) Evaluation of the risk model of maintenance work on major process equipment on offshore petroleum installations. Loss Prev Process Ind 25(3):582–593
4. Gould KS, Ringstad AJ, van de Merwe K (2012) Human reliability analysis in major accident risk analyses in the Norwegian petroleum industry. Paper presented at HFC forum, Halden, Norway, 17–18 Oct 2012
5. Boring RL, Blackman HS (2007) The origins of the SPAR-H method's performance shaping factor multipliers. Paper presented at the 8th IEEE conference on human factors and power plants
6. Bye A, Laumann K, Taylor C, Rasmussen M, Øie S, Van De Merwe K, Øien K, Boring R, Paltrinieri N, Wærø I, Massaiu S, Gould K (2017) The petro-HRA guideline. Institute for Energy Technology, Halden
7. Taylor C, Øie S, Paltrinieri N (2016) Human reliability in the petroleum industry: a case study of the petro-HRA method. Presented at ESREL 2016, Glasgow, Scotland, 25–29 Sept 2016
8. Taylor C, Øie S, Gould K (2018) Lessons learned from applying a new HRA method for the petroleum industry. Reliab Eng Syst Safety 49 (article in press). https://doi.org/10.1016/j.ress.2018.10.001
9. Puglia WJ, Atefi B (1995) Examination of issues related to the development and implementation of real-time operational safety monitoring tools in the nuclear power industry. Reliab Eng Syst Safety 49, 189–199
10. Majdara A, Nematollahi MR (2008) Development and application of a risk assessment tool. Reliab Eng Syst Safety 93, 1130–1137
11. NEA (2005) CSNI technical opinion papers: #7—living PSA and its use in the nuclear safety decision-making process; #8—development and use of risk monitors at nuclear power plants, NEA No. 4411. Nuclear Energy Agency, OECD
12. Alme IA, He X, Fylking TB, Sörman J (2012) BOP tisk and reliability model to give critical decision support for offshore drilling operations. Presented at PSAM/11ESREL2012, Helsinki, Finland
13. Haugen S, Edwin NJ, Vinnem JE, Brautaset O, Nyheim OM, Zhu T, Tuft V (2016) Activity-based risk analysis for process plant operations. Institution of Chemical Engineers Symposium Series, vol 2016, January (161)
14. Vinnem JE, Pedersen JI, Rosenthal P (1996) Efficient risk management: use of computerized QRA model for safety improvements to an existing installation. In: 3rd international conference on health, safety and environment in oil and gas exploration and production, New Orleans, USA. SPE paper 35775
15. PSA (2011) Regulations relating to conducting petroleum activities and at certain onshore facilities (the activities regulations)
16. Viddal S, Holmen HK, Vembe BE, Vinnem JE, Amdahl J, Wiencke HS (2014) Fire loads with drilling over installation in cantilever mode. Report no: PS-1071613-RE-01, Proactima
17. Aven T, Vinnem JE (2007) Risk management, with applications from the offshore petroleum industry. Springer, London
18. Standard Norway (2011) Risk based maintenance and consequence classification, NORSOK Standard Z-008, Rev. 3, June 2011
19. PSA (2011) Regulations relating to management and the duty to provide information in the petroleum activities and at certain onshore facilities (the Management Regulations)
20. Bento J-P (2000) Human-Technology-Organisation; MTO-analysis of event reports. OD-00–2 (In Swedish). Restricted

21. NOU (1986) Uncontrolled blowout on mobile drilling unit West Vanguard 6th October 1985 (in Norwegian only). Norwegian Ministry of Justice
22. AIBN (2011) AIBN mandate. http://www.aibn.no/about-us/mandate. Accessed 26 Mar 2011
23. Equinor (2011) Investigation report, COA INV, gas leak on Gullfaks B (in Norwegian with English summary), Statoil. www.statoil.com
24. PSA (2011) Investigation—gas leak on Gullfaks B. www.psa.no
25. Vinnem JE (2013) Use of accident precursor event investigations in the understanding of major hazard risk potential in the Norwegian offshore industry. J Risk Reliab 227(1):66–79

Chapter 22
Management of Risk and Emergency Preparedness for Future NUI Production Concepts

22.1 Framework Conditions for Future Production

Safetec Nordic issued a report to the Norwegian Ministry of Labour and Social Affairs early in 2018, focused on future development concepts and the use of construction and maintenance vessels in the period until 2030 [1]. The purpose of the study is to describe the current modes of operation and evaluate future developments, given changes in environmental, climate, technology, economic, competence and regulatory framework conditions.

The report gives a relatively detailed overview of the framework conditions for the Norwegian petroleum industry and how they are expected to affect the development of future fields in the period up to 2030. Oil price development is the factor that has the greatest impact on the future economic framework conditions on the NCS. Historically, oil prices have been very volatile and thus difficult to predict. The report suggests that petroleum supply factors indicate that the price of oil and gas will not increase dramatically over the longer term.

The future global environmental policy will affect how early fossil energy sources meet strong and real competition from other energy sources. With a more binding Paris agreement, this may mean that CO_2 emissions are priced as considerably more expensive than is done through today's quota market. In that case, oil and gas will have a clear competitive impact from wind and solar power, which will accelerate the phasing-in of new energy sources.

There are strong interest groups (NGOs) that wants to phase out Norwegian petroleum production in the next decade, but no responsible politician has accepted this view so far. International society still needs petroleum products for many decades, and the Norwegian production of petroleum is one of the most environmentally friendly worldwide, with virtually all new installations powered from the national grid, which is mainly hydro power. For politicians it is also necessary to consider the very high importance of petroleum production for the Norwegian national economy, such that phasing out will need to be very gradual. The taxation of profits in the Norwegian petroleum sector is very high (78%), which implies also that a large portion of the investments in new facilities is financed by the State. This also calls for gradual phasing out. We have therefore considered that new petroleum production facilities will still be developed for many years yet, in future with possibly new technology to reduce further the climate effects of production operations.

The Norwegian petroleum sector has high labour costs for the individual employee compared with other industries in Norway [1]. Companies on the NCS also have high indirect personnel costs, primarily related to social expenses, training costs and employer's fees. The high labour costs are essential parts of the operating costs (OPEX).

Also the capital investment costs (CAPEX) associated with personnel are high in the Norwegian sector, mainly due to the working environment requirements, which apply particularly to the standards of the accommodation facilities. This means that when personnel staying permanently, both CAPEX and OPEX are significantly reduced. The required maintenance volume has to be significantly reduced in order for the removal of personnel to be feasible (see also Sects. 22.7.5 and 22.7.6). Thus the CAPEX and OPEX are also reduced in this way.

Many of the safety systems installed on permanently manned offshore installations have their primary function to protect personnel until they can be safely evacuated from the installation. Thus many of the safety systems may be removed or significantly simplified on normally unmanned installations, which reduces the maintenance volume, CAPEX and OPEX.

NUI facilities will with reduced CAPEX and OPEX be favoured over subsea facilities in many cases, because subsea facilities often have higher costs, CAPEX and OPEX.

The fall in the oil price has contributed to a clearer focus on cutting the costs associated with social benefits. This is expressed in the case of indirect personnel costs that were experienced between 2015 and 2016. Labour costs in the industry are expected to largely follow developments for the rest of the economy in the future. Major changes in oil prices and technological change are the two factors that can largely change this.

The role of the major international companies has been significantly replaced by medium and small companies during the last ten years, including pension funds and investment companies. So far the new companies have been more focused on new investment than the majors have been lately. An increasing, high rate of development is expected up to and including 2022/23, mainly with conventional manned fixed and floating production facilities. After 2023 there is greater uncertainty, and so far there are only three discoveries in the Barents Sea that can provide stand-alone developments [1], all with floating production facilities and subsea wells.

The report projects that the use of simpler, normally unmanned facilities (NUI) with minimal equipment is expected to increase in the future. With simpler facilities, walk-to-work vessels follow the accommodation facilities for the crew and workshop facilities. If the most technologically optimistic experts' ambitions are fulfilled, it will only be possible to build new fields with manned facilities after 2030 if mega discoveries like the Norwegian field Johan Sverdrup are made.

The average size of new reservoirs found in Norwegian offshore has fallen dramatically over several decades and there are now quite small reservoirs [1]. Since 2013, no new reservoirs that can be developed as a stand-alone development with manned production facilities have been found, but there is an almost three digit number of small deposits that have been found recently. Some of them may be developed as satellites to existing facilities, if the distance is not too great. Others have no chance of development, unless the unit cost of production can be reduced substantially. This is where Normally Unmanned Production facilities may come in with a substantial cost reduction.

There is some experience with Normally Unmanned Installations (NUI) for wellhead purposes in the Norwegian sector. Earlier, these were sometimes developed with the same safety functions and barriers as a manned installation, because manning was also required quite frequently. Recently, much simpler installations are being built for the provision of dry wellheads (still considerably cheaper than subsea wellheads) with easy access. With simpler installations there is limited need for maintenance, and manning can be limited to short periods some few times per year, with personnel staying on Work-To-Work (W2W) vessels, which are bridge connected to the installation during manned periods.

Presently, there is limited but valuable experience with these simple wellhead installations (see also Sect. 22.6), but not with production installations without permanent manning, unless they are part of a production complex, where all the accommodation is on one installation, and all other installations in the complex are only manned for operation and maintenance purposes.

But there are other regions which have experience with unmanned production facilities, notably in the Gulf of Mexico and in the southern North Sea UK sector. These are normally in more benign waters, relatively simple installations, and mainly from an era where the regulatory and internal requirements were much less demanding. It is therefore considered that there is very limited applicability of the experience from these operations for future NUI production installations in harsh environment and deep water locations.

22.2 Future Offshore Production Concepts

22.2.1 Overview

Future production concepts will extend the present concepts with simple wellhead installations to more complex installations, where separation of oil, gas and water may take place, as well as drying, cooling and compression of gas for export or injection.

It is expected that the development will be stepwise to some extent, allowing experience to be gained from one step until the next is designed. The first few steps have already been taken:

- Some simple installations that initially had an 'on-installation process control room', have had their control rooms relocated to an office onshore ('onshore control room'), which is operated as if it was still on the installation (including work rotations), but with significantly reduced operational (manning) costs. Equinor's Valemon platform was initially operated as a manned installation, but has since the end of 2017 been operated as a normally unmanned installation.
- One NUI wellhead installation has started operations (not as NUI the first year, due to jack-up rig drilling production wells in cantilever modus, see also Sect. 21.7), while another similar installation is being constructed.
- When an installation is operated initially as a manned installation, it implies that all safety systems (barrier elements) will be installed according to regulations and normal practice for the initial period. These safety systems are obviously kept functional also after demanning, making it an abnormal unmanned installation.
- Some installations are planned with only one shift of operational crew, implying that it will be manned for two weeks, then unmanned for four weeks, before being manned again for another two weeks, etc. Such installations obviously have all safety systems installed for the manned periods, also resulting in an abnormal, partially unmanned installation.

The wellhead NUI installations are minimal installations without safety systems except for well control. This implies that safety systems to protect personnel are non-existent. The protection of personnel in an emergency is dependent on removing personnel from the installation immediately, over the bridge to the W2W vessel. This has been accepted by PSA for these simple installations, if the bridge to the W2W vessel is connected continuously when the installation is manned, in order to provide the shortest possible time to move personnel away to a sheltered place in the case of emergency.

It is not obvious that the same will apply for a NUI production installation, which even without many of the safety systems will be significantly more complex, at least in the sense that the number of possible leak points will increase considerably.

A presentation from Statoil (now Equinor [2]) shows a stepwise development of normally unmanned production installations:

- Early phase projects: Unmanned production platform, supported by host
- Future projects: Stand-alone unmanned production platform

  - Initially: fixed installations
  - Long term: floating installations

With respect to floating unmanned production platforms, it is indicated that the water depths in the Barents Sea area would be the initial step (depths in the range 300–500 m) and concepts for deeper waters would be the second step.

Liquid production requires either an export pipeline or storage and off-take by shuttle tankers. Pipelines will rarely be available, such that storage and off-take by shuttle tankers will probably be the rule. Storage may be provided by storage tankers or may be subsea tanks, although the latter is so far unproven technology.

All these steps are indicated to occur before 2030. This may appear to be somewhat optimistic, if the intention is to take it stepwise and gain experience from one step until taking the next step.

The benefits of unmanned production installations are:

- About 30% lower capital investment
- About 50% lower operating cost
- Low carbon

  – 30–50% reduction in CO_2 intensity

It is argued that unmanned production installations are always beneficial with respect to safety, because fewer personnel are exposed to the risks involved in the transportation to/from and the stay offshore. This may be true in a societal context, but for an individual who works on an unmanned installation it might not be true, if for instance the risk per hour of exposure on the unmanned installation is several times higher than for the manned installations. The safety arguments used are therefore too simplistic to be convincing. It would be expected that risk would be higher on an unmanned installation, perhaps not ten times higher, but that would depend entirely on the installation and its operations when manned.

When autonomous ships are considered, it is considered likely that a small crew will be onboard, at least for the foreseeable future. This is not a likely option for normally unmanned installations, because if even a small permanent crew were present, all the safety functions for the protection of personnel would become necessary, in addition to all the requirements for the working environment inside the accommodation. The cost increase would be phenomenal.

Another case is for some recent installations where the control room has been moved onshore after some time of conventional operation with an on-installation control centre. Then all safety systems and accommodation in accordance with applicable requirements have been implemented. Some of these installations have been made partly unmanned with permanent on-shore control centre, in the sense that there is only one maintenance crew employed, who will perform 2 weeks maintenance work on the installation, then go off shift for 4 weeks (2–4 rotation, common on production installations in Norway), and come back for 2 weeks maintenance work.

22.2.2 Legislation and Overall Principles

22.2.2.1 Norwegian Regulations

The Facilities regulations [3] have for some time had a clause for 'simpler facilities' without accommodation with the following requirements:

- They shall be designed to fulfil the requirements for acceptable risk
- A prudent working environment and satisfactory hygienic conditions shall also be facilitated during stays on the facility
- Where the facilities regulations describe specific solutions, simpler solutions can be chosen than those prescribed, provided these solutions can be proven satisfactory through special assessments.

The guidelines to this clause give some examples of where simpler solutions may be considered, the majority of which are associated with working environment conditions, as well as fire and gas detection, active fire protection, rescue of personnel and evacuation of personnel in the case of an emergency. But it is important to note that the regulatory reference is to simplified solutions, not removal of these systems.

Recent amendments (from 1.4.2019?) of the Framework regulations [4] allow W2W vessels to be used for the provision of accommodation facilities for personnel working on simpler facilities. Both petroleum and working environment regulations will apply for these employees 24/7 during these periods, including the time they spend on the W2W vessel. There are also requirements for the bridge connection (motion-compensated walkway) and support systems between the vessel and the installation. The required DP class of W2W vessels is DP Class 2, implying redundancy of all active components.

The main challenge with this regulatory requirement for W2W vessels is noise reduction for sleeping cabins, where the offshore regime is stricter than the maritime regime. PSA has apparently accepted that existing vessels will not be required to improve noise protection.

22.2.2.2 UK Regulations

There are no specific UK regulations pertaining to normally unmanned installations. There are many normally unmanned installations in the southern North Sea, but these are from the time before early 1990s, when the first edition of the Safety Case regulations was introduced.

22.2.2.3   Safety Protection Philosophy

The main safety protection philosophy of normally unmanned installations which has been adopted by the industry so far, is the following:

- Manned mode:

  - Get people off the installation as quickly as possible using the gangway to the W2W vessel, with protection of the routes to the gangway if needed.
  - After removal of people the installation may be treated as unmanned.

- Unmanned mode:

  - Protect against environmental spills and escalation of asset damage.

Such philosophy is very different from the current philosophy for manned installations, which is focused on extensive protection of personnel. This is the consistent approach in many countries such as the UK, Norway, EU, Australia, Brazil and Canada, although the implementation is different, as outlined in Sect. 1.4. Some requirements have been stated by the Norwegian authorities for 'simpler facilities', but these are quite vague and leave most of the assessments to each case individually. At the moment 'simpler facilities' imply quite basic wellhead installations without any treatment of the well-stream. More advanced NUI facilities may be developed in the near future, but this is not addressed specifically in the present regulations.

Some safety systems will be needed in order to protect personnel when leaving the installation and to protect the environment and assets in unmanned mode. In this case however, safety systems may be acceptable with lower reliability, such as less redundancy. The degree of simplification of safety systems is particularly crucial, because safety systems with a high maintenance need have a strong influence on the volume of manhours needed, and are thus vital for the manning level and the frequency of manned periods.

It would also be expected that preventive maintenance of safety systems is, to a large extent, replaced by break-down maintenance, in order to limit the duration of manned periods. This may, on the other hand imply that production may be shut down for somewhat longer periods, especially during the actual repair work of safety systems. Mobilisation of a team to visit a NUI in case of shutdown due to failure may also take some time, depending on the availability of suitable W2W vessels, personnel, suitable weather conditions (especially during winter), etc.

22.2.3   Safety Systems

The facilities regulations [3] have very generalized requirements for 'simpler facilities' in the Norwegian sector (see Sect. 22.2.2.1) which in practice means normally unmanned installations. The requirements are very vague, stating that simpler solutions may be chosen, based on a specific analysis.

When talking to facilities designers one can get the impression that they want design extensive NUI facilities for production and processing with a very bare minimum of safety systems, for instance completely without any active or passive fire and blast protection systems. This is virtually the opposite of how we interpret the message from authorities; 'start with all systems, then maybe some can be simplified based on case-by-case analysis'. It would thus not be surprising if the sanction of the first project to choose such a solution is going to face some challenges in order to agree acceptable solutions with the authorities.

It has been suggested that the need for safety systems is governed by the need to protect the environment against accidental spill. This is not very explicit, because the only system strictly speaking to be required in this case would be a drain system. It is unlikely that this will be the only safety system needed.

But for the significant savings that are outlined in Sect. 22.2.1 above, there has to be an extensive reduction in the use of safety systems. The amount of maintenance work on the installation has to be reduced to only a small fraction of the volume traditionally carried out on manned installations, in order to achieve a 50% reduction of operating costs. An initial review is discussed in Sects. 22.2.4–22.2.8 below.

It is therefore considered that evaluation of which safety systems (or barrier functions and elements) should be installed on an advanced NUI facility for production and processing would be a difficult task to carry out. This chapter will attempt to provide further input to such evaluations through the following:

- Verbal discussion of the potential to simplify current practice for manned installations (see Sects. 22.2.4–22.2.8).
- Performance of a case study for a fictitious NUI facility with HC processing and with three different cases for the extent of safety systems and consequent manning level (see Sect. 22.3).

22.2.4 Well Control Barrier Elements

The well control barrier elements are all directed at the prevention of blowouts or well leaks. The most typical barrier elements are the Downhole Safety Valve (DHSV) and the isolation valves in the Christmas tree, master Valve (MV) and wing Valve (WV).

These barrier elements are considered to be required in any case in order to protect against blowouts and significant environmental spill. They are therefore omitted from further discussion, because there is no uncertainty about the need for them.

The implication of this is that the remaining barrier functions discussed in this section are limited to the process area of the installation. For obvious reasons, the applicable barrier elements are all technical, as they have to function most of the time when the installation is unmanned.

22.2.5 Prevent Loss of Containment Barrier Function

The barrier function to prevent loss of containment is the first barrier function, shown on the left side of the bow-tie diagram (see Figs. 14.1 and 15.1). Several of the barrier elements in this barrier function are usually operational (see Chap. 6). These barrier elements are not discussed in the present section, which is all focused on technical systems and the discussion of whether they should be installed on a normally unmanned installation.

22.2.5.1 Process Control Systems

These barrier elements are those that are required according to ISO 10418 [5], including pressure and liquid level control, as well as the Pressure Safety Valve (PSV).

These barrier functions are required for process control, also when unmanned, in order to protect the environment and assets. It would therefore be expected that these systems are installed in any case on a normally unmanned production installation.

22.2.5.2 Blowdown and Flare Systems

These barrier elements are also according to ISO 10418 [5]. The flare system is needed to handle pressure upsets, also when unmanned. The blowdown system is also essential in order to limit escalation in the case of a fire in the process systems. It may on the other hand e argued that it may be a challenge to determine which sections to depressurize in a remote control situation, such that the advantage is somewhat more uncertain.

It is therefore considered that a flare system would always be installed, also on a NUI production installation. The blowdown system could possibly be considered optional.

22.2.5.3 Drain System

This system is also according to ISO 10418 [5] in order to handle liquid spills in a controlled manner and prevent small spills from reaching the sea and causing pollution to the environment.

Protection of the environment is also important for an unmanned installation. It is therefore considered that the drain system is required on a NUI production installation.

22.2.5.4   ESD Valves

Provision of an ESD system is also according to ISO 10418 [5]. The limitation of each segment by 'internal' ESVs is more open to consideration. There are traditionally quite small segments on installations in the Norwegian sector, often with a significant number of ESVs in the process systems. This is certainly an advantage to limit escalation, but such limitation will be less important when the installation is normally unmanned.

It is therefore considered that internal ESVs in the process area could be omitted for an unmanned production installation.

Even if 'internal' process ESVs are omitted, barriers against the large reservoirs will be needed in any case, such as against the petroleum reservoir in the underground (see Sect. 22.2.4 for well barriers). Also ESVs on gas risers are needed to protect against large reservoir, but may be omitted for an unmanned installation.

22.2.5.5   Gas Detection

The gas detection system is also according to ISO 10418 [5]. Even in an unmanned mode, it will be important to detect gas leaks, in order to avoid further consequences to assets.

The recommended Safety Integrity Level (SIL) is SIL2 according to Norwegian Oil and Gas Guideline 070 [6] for manned production installations (see also Sect. 19.7). It may be considered that SIL1 is sufficient for a normally unmanned installation. If so, this would be a cheaper solution with respect to installation as well as operations costs.

22.2.6   Prevent Ignition Barrier Function

The barrier function to prevent ignition is the second most important barrier function, following the LoC barrier function (see Sect. 22.2.5). The importance of this barrier function is underlined by the fact that if ignition of an HC leak does not occur, then the consequences of the HC leak or release are usually quite small.

22.2.6.1   Ex-protected Equipment

Ex-protected electrical equipment (according to the ATEX directive [7]) is a very critical aspect in order to prevent ignition in the case of HC leaks and releases. There are several protection principles used, depending on the equipment type and circumstances. As an illustration, Ex-p implies overpressure protection, in order to prevent inflammable materials entering the enclosure through the use of overpressure.

It is considered that Ex-protected equipment will be required also in unmanned mode, due to the need to prevent fire and explosion with significant asset damage potential.

22.2.6.2 Hot Work Activities

Hot work activities are the other main potential source of ignition of HC leaks, typically involving welding arc or steel cutting with an open flame. Also the use of some equipment for heating, grinding, etc. may be relevant, but these are often categorized as 'hot work Class B', because of lower temperatures and less ignition potential. These operations are obviously manual activities that only are relevant when the installations are manned.

It is important to stop all hot work activities immediately if a HC leak occurs. It is therefore common that the electrical supply to outlets where hot work equipment may be connected is stopped once gas detection is confirmed. This shows the importance of gas detection.

It is therefore considered that simultaneous production and maintenance in manned periods could not be allowed if no gas detection was installed.

22.2.7 Prevent Escalation Barrier Function

The barrier function to prevent escalation is important, especially for the risk to personnel and the main safety functions. It is also important to prevent extensive asset damage. The risk to personnel and main safety functions is not critical during unmanned phases, but asset protection of the investment is important in all phases.

22.2.7.1 Fire Detection

The fire detection system is also according to ISO 10418 [5]. Even in an unmanned mode, it will be important to detect fires, in order to isolate and extinguish them. This assumes however, that there is the possibility to extinguish fires, or at least to use fire water for cooling purposes. The implication here is that no fire detection is required if no active fire-fighting system is installed. This is probably a very special case.

The recommended level is SIL2 for fire detection according to Norwegian Oil and Gas Guideline 070 [6] for manned production installations. SIL1 may be considered sufficient for a normally unmanned installation. If so, this would be a cheaper solution with respect to installation as well as operations costs.

22.2.7.2 Fire Water Systems

Fire water systems on a manned installation implies usually at least one fire water ring main, for the supply to active fire protection systems, with the purpose of cooling equipment and structures and if possible, also to extinguish the fire. The main purpose of cooling is to prevent escalation. Such prevention is primarily focused on asset protection, for a normally unmanned installation. Even when the installation is manned, it will be expected that personnel have been evacuated to the support vessel (W2W) before any escalation occurs. This implies that active fire protection is mainly an asset protection issue, possibly with one exception.

On some installations it may be required to install water curtains in order to protect escape ways for personnel who want to reach a bridge landing and evacuate to the W2W vessel. If such water curtain is required, then the fire water system becomes more than just asset protection.

Some companies have suggested that active fire-fighting from the W2W vessel could also play a role. This does not appear to be a viable solution. This is first of all because the W2W vessel will have important tasks to carry out to protect personnel in the case of a fire on the NUI when persons are present. Secondly, if active fire protection is required to protect personnel, the fire water supply from the W2W vessel is not sufficiently reliable to be accepted as a replacement for on-installation fire water supply.

22.2.7.3 Fire Water Supply

If a fire water system is considered to be needed (see Sect. 22.2.7.2), then a fire water supply is also needed. The requirements for fire water supply are quite extensive in the Norwegian regulations and practice, usually implying that separate fire water pumps are installed, often with a configuration like $5 \times 50\%$ capacity pumps, due to the need to have 200% capacity available even during maintenance of one pump.

It would be reasonable to reduce the minimum requirements for capacity and redundancy for normally NUI installations. The recommended level is SIL2 for fire water supply according to Norwegian Oil and Gas Guideline 070 [6] for manned production installations. SIL1 may be considered sufficient for a normally unmanned installation. It might also be considered to reduce the capacity requirement from 200% capacity in relation to the two largest fire areas. Both these two aspects would result in significant savings with respect to installation as well as operations costs.

The fire water supply is usually a separate system on Norwegian installations. Combination with other systems has been attempted in a few cases, but with very extensive challenges. Such combination should be easier to implement in practice with reduced requirements on an unmanned installation. It would therefore be realistic to try to combine with, for instance, cooling water pumps.

22.2.7.4 Deluge System

Deluge protection is normally installed for area protection and object protection in process areas and takes the form of extensive systems with high maintenance needs relating to piping, nozzles and deluge activation valves.

The need to reduce maintenance needs on a normally unmanned installation may trigger the need to replace deluge systems by simpler solutions in order to reduce costs as well as the maintenance volume. The use of remotely controlled monitors may be one possible solution to replace deluge systems with simpler solutions.

22.2.7.5 Foam Systems

Foam systems are important to extinguish liquid fires on installations, where the foam system may be used to provide a layer to prevent air supply to the fire in the liquid and thus extinguish the fire. This will require extensive systems which would not be wanted on an unmanned installation.

If the deluge system is replaced by the use of remotely controlled monitors (see Sect. 22.2.7.4), then these may be combined with a foam capability in the areas where liquid fire may occur.

22.2.7.6 Monitors and Hoses

Fire monitors may offer flexible solutions as outlined for deluge and foam systems in Sects. 22.2.7.4 and 22.2.7.5 above. Normal practice on manned installations in the Norwegian sector is also to install hose stations in the process and utility areas, in order to allow manual fire-fighting in areas where fixed systems are unavailable or insufficient. This is not needed in the normally unmanned phases, and they do not have any significant use during manned periods, as the prevailing philosophy will be to evacuate the installation as quickly as possible. Hose stations would therefore not be needed.

22.2.7.7 Water Curtains

Water curtains have not been commonly installed on new manned installations in the Norwegian sector during the last few years, but were more common in the early years of activity in the North Sea. Fire and blast walls have probably replaced the use of water curtains on modern installations.

Water curtains may be more interesting to use in the case of normally unmanned installations, as an easy solution to protect personnel from heat loads on their way to the bridge landing to evacuate the installation. Water curtains will imply the need for a fire water system and a reliable supply of fire water (see Sect. 22.2.7.3).

22.2.7.8   Fire Protection in Electric Rooms

The modern approach to fire protection in electric rooms is that of early detection in order to initiate early action. Damage due to fires in electric equipment may require long repair times even if they do not have any effect on personnel, and rapid response is therefore essential, also in an asset management context. This may imply that fire detection and extinction will be important also for an unmanned installation.

22.2.7.9   Fire and Blast Walls

Fire and blast walls are installed on Norwegian manned installations in order to protect personnel before escape and evacuation has been completed as well as to limit escalation to other areas of the installation. Another objective may be to reduce the extent of the fire areas, and thus reduce the size of the required fire water supply pumps.

It is not required to protect personnel from escalation on a normally unmanned installation, when we assume that rapid evacuation to the W2W vessel will be the main principle to protect personnel. It is further assumed to be less important to limit the size of the fire areas.

Thus it may be considered that fire and blast walls are not so important for an unmanned installation. It would nevertheless be an advantage to limit escalation from an asset protection point of view, also for an unmanned installation. Therefore, fire and blast walls may be installed in some special cases, but as a general rule, it is considered not to be installed on an unmanned installation.

22.2.7.10   Structural Fire Protection

Structural [passive] fire protection may be installed on primary and secondary structural members, as well as on piping and equipment supports in the process areas, in order to prevent rapid escalation and to limit the fire potential and/or duration.
Not required to protect personnel from escalation.

It is not required to protect personnel from escalation on a normally unmanned installation, as argued with respect to fire and blast walls above in Sect. 22.2.7.9. It would still be an advantage to limit escalation, also when unmanned, but it will also increase the need for maintenance and inspection. It is therefore considered that the norm would be that no structural fire protection is applied.

22.2.8   Prevent Fatalities Barrier Function

The barrier function to prevent fatalities is the 'final' function in the sequence to protect personnel. There are several barrier elements that need to be provided for a manned installation. An unmanned installation has considerably more limited needs in this regard, as outlined in the following subsections.

22.2.8.1   Lifeboats

The main approach to rapid evacuation of personnel is to provide a bridge to the W2W vessel. If the bridge is connected continuously, then there is no potential waiting time before the bridge may be usable. There may be reasons why it is preferred not to have the bridge connected all the time, in which case there may be a short waiting time before the bridge is reconnected. A dedicated analysis of the site specific aspects would be needed to investigate if the bridge may take 1–2 min (typically) for reconnection, without severe effects for personnel.

The main approach to the protection of personnel on unmanned installations is in any case that lifeboats are not needed on a normally unmanned installation.

22.2.8.2   Liferafts

Liferafts are as secondary means of evacuation as a backup for the primary means, usually lifeboats. Since the primary means of evacuation is over the bridge to the W2W vessel, secondary means should not be required. Selantic escape chute (see Sect. 22.2.8.3) should be a better solution if some reserve systems is required.

22.2.8.3   Selantic

If an emergency analysis shows that there is a need for an alternative solution if the bridge to the W2W vessel is not reachable, then a Selantic system, with escape chute and liferafts is probably the most suitable solution.

22.2.8.4   Escape Routes

The escape routes needed on a normally unmanned installation are those that lead to the bridge landing, in order to get to the W2W vessel. The escape routes should be shielded from severe accidental effects during the period it would take for the maximum number of personnel to escape from their workplaces to the bridge landing. The shielding may for instance be provided by using water curtains.

22.3 Case Study—Risk Levels on NUI Production Installations

22.3.1 Study Cases

The case study is performed for a fictitious unmanned processing installation. The process equipment has been based on a manned production installation, and adjusted to reflect a simplified, unmanned installation. The intention is that the risk level should be typical in an overall sense, but the design would not be representative.

When it comes to safety systems, there are three cases which have been quantified in this case study for comparison, based on the evaluations made in Sects. 22.2.3 through 22.2.8 above. The following cases are defined for comparison:

- Case max: all safety systems as if it was manned
- Case min: an absolute minimum set of safety systems
- Case limited: limited extent of safety systems, but more than minimum.

The detailed definition of the safety systems that are assumed to be installed in the unmanned processing installation in the three cases is shown in Table 22.1.

It is shown that well control barrier elements are included in all three cases, but there are significant differences when it comes to barrier elements in the process area. 'Case min' has very few safety systems (barrier elements); flare, drain, ESV on gas riser, fire monitors, water curtain and fire protection in electric rooms. 'Case limited' has several additional safety systems; gas and fire detection, fire water system and a Selantic chute. One might think that 'Case max' is unrealistic, because nobody would install all these safety systems on a normally unmanned installation. But this is precisely the case for installations that have been operated as a manned installation for an initial period, before being demanned and made normally unmanned or partially manned. There are already some few installations in the Norwegian sector that are operated in that manner.

It should be noted that the present case study is limited to hazards related to the process area, i.e. fire and explosion risk due to HC leaks from process components, which may have effects for personnel if the leaks occur during manned periods. Accidental effects are limited to assets if fire or explosion occurs in the unmanned periods.

22.3.2 Manning Levels and Patterns

Manning levels will depend on the need for interventions and maintenance tasks. The frequency of manning is therefore closely correlated with the number of systems needing maintenance. Occasional visits may also be required to start up production after a shut-in or break-down. It is assumed that preventive maintenance

Table 22.1 Overview of study cases for risk quantification with barrier elements included

Safety system/element	Case max	Case min	Case limited
Well control (MV, WV, DHSV)	Yes	Yes	Yes
Prevent LoC			
RP14C (PSH, PSL, PSV, etc.)	Yes	Yes	Yes
Blowdown and flare systems	Yes	Flare	Flare
Drain system	Yes	Yes	Yes
Process ESD valves	Yes		
Gas detection	Yes		Yes
Gas riser ESV	Yes	Yes	Yes
Prevent ignition			
Ex-protected equipment	Yes	Yes	Yes
Hot work activities	Yes		
Fire detection	Yes		Yes
Prevent escalation			
Fire water systems	Yes		Yes
Fire water supply	Yes	Yes	Yes
Deluge system	Yes		Yes
Foam system	Yes		
Monitors and hoses	Yes	Yes	Yes
Water curtains	Yes	Yes	Yes
Fire protection in electric rooms	Yes	Yes	Yes
Fire and blast walls	Yes		
Structural fire protection	Yes		
Prevent fatalities			
Lifeboats	Yes		
Liferafts	Yes		
Selantic	Yes		Yes
Escape routes	Yes	Yes	Yes

will to a large extent be replaced by break-down maintenance. Case max is assumed to have one regular crew that visits the installation in each working period, corresponding to how few installations are operated, except that personnel stay on the installation in these cases. Process control is assumed to be external in all cases, i.e. from shore or from another installation. When a few process personnel work a night shift, this is assumed to be personnel preparing for interventions, corresponding to standard practice on manned installations.

The unmanned production platform is assumed to be manned as follows:

- Case max
  - Manned one third of the time, two weeks each period
  - Average manning level 20 persons, dayshift only except process personnel

- Case min

  - Manned four periods per year, two weeks each period
  - Average manning level 10 persons, dayshift only

- Case limited

  - Manned five periods per year, two weeks each period
  - Average manning level 20 persons, dayshift only except process personnel

It is assumed that each person on average spends 11 h per day on the installation when manned, and lunch is taken on the W2W vessel they are assumed to live on. This implies that the three alternatives involve the following manhours per year:

- Case max: 26,400 h/year
- Case min: 6,160 h/year
- Case limited: 15,400 h/year

Case max corresponds to one full shift having full time employment on the installation, two weeks on, four weeks off, etc. Case limited corresponds to about 60% of a shift's time being spent on the installation, with some additional work periods spent on other installations. Another alternative for Case limited is if a reduced crew (11.7 persons on average) spends all their offshore periods on the installation. Table 22.2 presents the compositions of the 10 men and 20 men crews respectively.

22.3.3 Process Leak Frequencies

Chapter 6 has demonstrated that a significant part of the process system leaks occurred as a result of manual intervention, either for maintenance or modification purposes. It is therefore expected that HC leak frequencies will be lower with less maintenance work. This has not been analysed in detail, but some assumptions have been made with respect to the effect of less maintenance and less manning, and the extent of safety systems (barrier elements) as shown in Table 22.1.

Table 22.2 Manning levels and manning group distributions

Personnel group	10 men crew		20 men crew	
	Day	Night	Day	Night
Supervisor	1	0	1	0
Process	6	2	4	0
Maintenance	8	0	4	0
Crane and deck operator	1	0	0	0
Catering	0	0	0	0
Scaffolding/ISO	0	0	0	0
Other; supplier, contractor, engineer	2	0	1	0

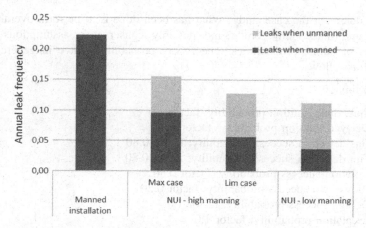

Fig. 22.1 Process area leak frequencies assumed for NUI production installation, with different manning assumptions

Figure 22.1 shows the assumed leak frequencies for the different manning cases, with contributions from manned and unmanned phases. The leftmost bar is the leak frequency when continuously manned, and the remaining cases are for NUI with different assumptions regarding the manning and extent of safety systems.

It is clearly shown that even if the installation is manned for only a small part of the year, the leak frequencies are the highest when manned, for Case max. For the lim and min cases, the leak frequencies during manned periods are less than 50%, but the relative contributions are significantly higher than the percentage of time that the installation is manned.

22.3.4   Node Probabilities

The node probabilities in the event trees reflect the safety systems that are installed on normally unmanned installations. Three cases have been considered, as discussed in Sect. 22.3.1.

Case max has all the safety systems that are commonly used on manned installations, which corresponds to the case when an installation is operated as a manned installation possibly for some years, before the control room is moved onshore and the manning is reduced to certain periods for campaign maintenance.

The lim and min cases are cases with a reduced extent of safety systems and reduced manning level and frequency.

The node probabilities could have been calculated in detail based on the reliability analysis and assumptions discussed in Sects. 22.2.5–22.2.8. This has not been

done in detail for the case study. What has been done is to perform overall evaluations and make some crude and possibly conservative assumptions about
increases of failure probabilities. The failure probabilities have in general been
changed as follows:

- Case limited:

  - Immediate ignition probability: factor 1.0
  - Delayed ignition probability: factor 2.0
  - Gas detection success probability: factor 0.80
  - Fire detection success probability: factor 0.80
  - Isolation success probability: factor 0.50
  - Blowdown success probability: factor 0.20
  - Deluge success probability: factor 1.0
  - Escalation probability: factor 1.0

- Case minimum:

  - Immediate ignition probability: factor 1.0
  - Delayed ignition probability: factor 5.0
  - Gas detection success probability: factor 0.40
  - Fire detection success probability: factor 0.60
  - Isolation success probability: factor 0.50
  - Blowdown success probability: factor 0.20
  - Deluge success probability: factor 0.40
  - Escalation probability: factor 1.0

These adjustment factors are considered to reflect changes in reaction times,
technical reliability, functionality, vulnerability to accident loads as well as manual
actions, to the extent relevant according to the assumptions discussed in
Sects. 22.2.5–22.2.8. It may also be noted that there are several of the safety
systems listed in Table 22.1 that are not reflected in the event tree, at least not
explicitly. This is typical of QRA studies.

The QRA study is made with Safetec Nordic's QRATool® which has quite
detailed event trees and consequence analysis. The nodes in the event tree for major
gas leaks are the following:

- Immediate ignition
- Gas detection successful
- Fire detection successful
- Isolation successful
- Blowdown successful
- Delayed ignition
- Escalation to other areas
- Deluge released
- Escalation to other equipment

The consequence calculations are also taken from the QRATool® for the assumed installation with relevant pressures and other operating conditions, as well as dimensions and capabilities, but these are not presented here.

22.3.5 Personnel Risk Results

The personnel risk results may be calculated on the basis of the adjusted leak frequencies as discussed above in Sect. 22.3.3, as well as the revised node probabilities as outlined in Sect. 22.3.4. The risk to personnel is obviously only relevant in the periods when the installation is manned. The risk results are in practice first calculated as if manning is constant all year around, and thereafter multiplied with the fraction of time persons are present.

The risk results presented here only reflect the fire and explosion hazards due to process systems on the topsides of the installation. This implies that the other hazards have to be added to the results in order to get the overall risk results. This would include other major hazards as well as occupational accident hazards. This has not been the purpose of the present assessment, which is mainly focused on comparison of differences in risk due to process system hazards, when the extent of safety systems may be reduced in order to reduce the scope of maintenance (see further discussion in Sect. 22.4.2). Figure 22.2 presents the different cases as outlined in Sect. 22.3.2.

The first bar in the diagram is for reference only, indicating what the FAR value would be if the installation was constantly manned. This may be used for comparison with results for the different cases.

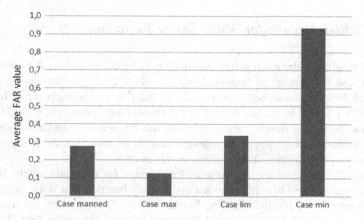

Fig. 22.2 Fatality risk results for NUI production installation (process risk only), with different manning and safety system assumptions

Table 22.3 PLL values for NUI production installation (process risk only), with different manning and safety system assumptions

Case	PLL
Case manned	$2{,}18 \times 10^{-4}$
Case max	$3{,}33 \times 10^{-5}$
Case lim	$5{,}17 \times 10^{-5}$
Case min	$5{,}76 \times 10^{-5}$

Case max is a NUI with all the safety systems of a manned installation. It is therefore expected that this will have the lowest risk level, in spite of the fact that the leak frequency is the highest during manned periods, according to Fig. 22.1.

Case lim is a NUI with reduced extent of safety systems, but no to the lowest level. Figure 22.2 shows that the FAR level is more than doubled that of Case max, but only marginally increased from the level for a manned installation.

Case min is a NUI with a very limited extent of safety systems, and thus quite limited extent of manning for maintenance purposes. Figure 22.2 shows that the FAR level for this case is almost tripled that of Case lim, and almost ten times higher than for Case max lowest level. This is a very significant increase, but it is worth noting that the value is still lower than FAR = 1.0.

PLL results are also important for risk evaluation in addition to FAR values. PLL values are shown in Table 22.3.

It is worth noting that Case max has the highest number of offshore hours with offshore risk exposure, but the lowest FAR and PLL values. FAR values for Case lim and Case min are higher than for the manned case, but the values in Table 22.3 show that all these cases have PLL values considerably below the PLL value for Case manned. Please note (as noted above) that these values reflect only the risk contributions from process systems risk, not other hazards.

22.3.6   Risk Results for Main Safety Functions

The risk results for the main safety functions may be calculated on the basis of the adjusted leak frequencies as discussed above in Sect. 22.3.3, as well as the revised node probabilities as outlined in Sect. 22.3.4. The risk for the main safety function is primarily relevant in the periods when the installation is manned. It could on the other hand be argued that it is easier to relate the frequencies to what is typical for manned installations, if they are expressed also on an annual basis, irrespective of the presence of personnel. This is done in this section.

Figure 22.3 presents the impairment frequencies for the main safety function escape ways, which is the only relevant main safety function that is applicable for a normally unmanned installation. The frequencies are factored with the fraction of the year that personnel are present in the three cases as defined in Sect. 22.3.2.

If the values were not factored with the presence of people, the values would be around 1.0×10^{-5} per year, up to 1.5×10^{-5} per year. This implies that impairment

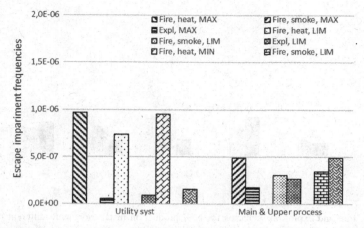

Fig. 22.3 Frequencies of impairment of escape ways from utility systems and process decks for NUI production installation, with different manning and safety system assumptions

of escape ways to the bridge landing in order to evacuate to the W2W vessel is quite a low probability, which should ensure that premises for persons to be able to leave the installation are good. This aspect has not been evaluated in detail, since the layout is adapted from a manned installation.

22.3.7   Fire and Explosion Frequencies

Fire and explosion scenarios are relevant in the unmanned and manned periods. The extent of safety systems will have a strong influence on the duration of fires and the dimensions of gas clouds. This is reflected in the fire and explosion scenarios as can be seen in Fig. 22.4.

Figure 22.4 shows that Case min implies significantly increased frequencies, especially for the gas explosion scenarios. The frequency of explosion scenarios with escalation increases by a factor of more than five.

22.3.8   Interpretation of Results

It is often claimed that QRA studies are not well suited to reflect minor changes in safety systems, as such changes will often not be visible at all when the results from a QRA study are considered. It was commented in relation to Table 22.1 above that several of the safety systems discussed for the three cases are not explicitly reflected in the event tree used in the case study.

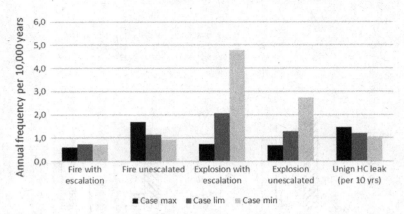

Fig. 22.4 Fire and explosion scenarios for NUI production installation, with different manning and safety system assumptions, per 10,000 installation years

The changes when comparing Case max and Case min are quite considerable in the present case, when inspecting the results shown in Figs. 22.2, 22.3 and 22.4. This implies that the risk levels for personnel in the Case min should be considered as quite high, both relatively speaking and when seen in isolation. Please see Sect. 22.4.2 below regarding the results seen in an overall context.

It may be noted that the FAR values are virtually constant if the manhours spent on the installation are changed from what was assumed for the case study (see Sect. 22.3.2). Therefore, even if somebody may think that the assumed manhours are far too high, the high values in Fig. 22.2 will still be virtually applicable. The PLL values will on the other hand be reduced correspondingly.

With results as high as those shown for Case min, it would be difficult to argue against some additional safety systems being sensible ALARP measures. It is not a quantitative consideration that will be the challenge, but a qualitative engineering consideration, which would consider that for instance fire detection or deluge systems (as in Case lim) would be sensible to install in order to reduce risk without grossly disproportionate costs, at least if installed with somewhat simplified solutions compared to those normally used on manned installations.

22.4   Overall Evaluation of Risk for NUI Production Installations

22.4.1   Background

One of the most commonly used arguments for normally unmanned installations is that risk exposure is reduced. The argument builds on working fewer manhours offshore, implying that fewer people will be exposed to hazards of offshore employment and helicopter transportation to and from offshore installations.

This is true as long as we consider personnel risk in a societal context. If we reduce the number of people working offshore by 50%, then the societal risk is also reduced by 50%, if the risk exposure per person is the same. But what if the risk exposure per person is doubled? Then the societal risk is probably unchanged.

Even if the societal risk is reduced by 50%, does that imply that each person's risk exposure is reduced? No, in this case the risk per person is unchanged, because the entire effect of societal risk reduction is due to the reduction in the number of persons exposed to the hazards of working offshore. What is good in a societal context is not necessarily good for the persons still exposed to the same (or increased?) hazards.

Another aspect should also be considered in this context. Helicopter transportation, according to Sect. 17.2.5, is the source of 44% of the total fatality risk exposure for employees on manned production installations. Personnel transport is normally carried out with vessels on normally unmanned installations. This is in principle not risk free, but the risk exposure from such personnel transportation is very low when compared to helicopter transportation. As an illustration, several hundred people have perished in helicopter accidents during more than 50 years of offshore operations in the UK and Norway, whereas none have perished from accidents on vessels in transit to offshore facilities, including the crews on supply vessels in transit. A very few accidents have occurred on supply or similar vessels in transit, but no fatalities have resulted. The replacement of helicopters by vessels for personnel transportation is therefore a significant risk exposure reduction for each person individually, not only in a societal context.

When oil companies have argued in recent years that the reduction of manning off-shore implies risk reduction for personnel, they are taking a societal consideration automatically, neglecting entirely that other considerations could be made. It is questionable whether such a limited approach is in accordance with regulations, which require the risk level for the most exposed group to be quantified.

22.4.2   Tolerance Limits Used by the Industry

Vinnem et al. [8] discussed a case where a quantitative criterion was developed for personnel attending a normally unattended installation. The installation was serviced by personnel transported by helicopter from the 'parent installation' and taken out of the workforce for that manned installation. The crew members were transported by helicopter at the start and end of each shift on the NUI (shuttling), without spending the off-shift hours there, unless helicopter landing was prevented for instance due to weather conditions. It should be noted that the installation had one free fall lifeboat. The risk tolerance criterion used was the following:

- Average FAR value not to increase by more than 25% for personnel attending the NUI, in relation to their average FAR value on the 'parent installation', including risk contribution from personnel transportation.

It could therefore be considered to use the same criterion for personnel working part time on a NUI, in addition to working on another manned installation for the remaining work periods during a full man-year. It would then also be natural to use the same tolerance limit for personnel having their full employment on the NUI in the Case max alternative.

We have also seen examples that companies use the same risk tolerance criteria for NUIs as they use for manned installations. This choice is probably made because the risk tolerance criteria are so relaxed as is discussed in Sect. 17.6. Typically a FAR value limit of 10 is used as a relaxed limit, which gives considerable flexibility for the operator. We have also seen companies use a FAR value of 25, corresponding to what is used for the most exposed group on manned installations, with the comment that time onboard the W2W vessel for transport and off-duty shall not be included. This becomes a very relaxed criterion.

The FAR values shown in Fig. 22.2 may be used to illustrate how relaxed both the FAR tolerance limits of 10 or 25 really are. Even a case with considerably fewer safety systems than the 'Case min' would satisfy these tolerance limits. This applies even if other hazards are included in the FAR value calculation, as indicated in Fig. 22.3.

22.4.3 Tolerance Limit for Case Study

The differences between the FAR and PLL values are demonstrated by referring to Fig. 22.2 and Table 22.2. Figure 22.2 shows that Case min has a significantly higher FAR value compared to Case manned. Table 22.2 shows that the PLL value of Case min is still significantly lower than for Case manned.

Figure 22.5 attempts to illustrate the differences between manned and normally unmanned installations. It is assumed for this illustration that the installation has subsea wells and thus no blowout risk affecting the installation, then all other hazards are covered by structural hazards and occupational [accident] hazards, which have been assumed to be equal in the two cases. Case manned is assumed to have helicopter transportation of personnel, which for comparative purposes is added as a constant, equivalent addition to each manhour on the installation, whereas the W2W vessel transportation for Case min is assumed to have a very low value.

It is demonstrated that the numbers in the present case imply that Case manned has a 5% higher equivalent FAR value compared to the Case min. It is on the other hand clear from Fig. 22.5 that only minor changes to some of the contributions may result in the reverse comparison between the two cases. It is therefore demonstrated that even though Case min has a significantly lower PLL value when compared to

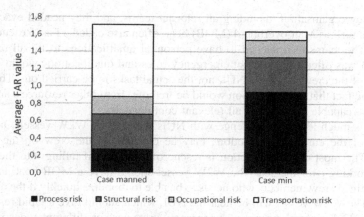

Fig. 22.5 Fatality risk results for NUI production installation, comparison of cases with assumed contributions from structural, occupational and transportation hazards

Case manned (see Table 22.3), we may easily end up with average total FAR values for Case min being lightly higher (or in fact lower as here) than Case manned. The obvious observation from this brief discussion is that only considering PLL values for comparison is too limited a scope of comparison.

The following illustration may be used to discuss aspects relating to risk tolerance for personnel manning NUI installations. The manning assumptions in Sect. 22.3.2 are based on a two-week period for each visit, i.e. the normal duration of an offshore work period in the Norwegian sector. This applies to all the different cases for maintenance and manning. In Case max the number of visits corresponds to a full offshore position, implying that those persons manning the NUI installations will not work on other installations in addition. For the two other cases, the number of visits corresponds roughly to about half a position off-shore, and it would be expected that the personnel have work on other offshore installations (most likely a manned installations) in order to fill up to a full position.

In the case with employees with work on other installations, it may be natural to consider the principle as referred to in Sect. 22.4.2 with a 25% increase due to work on a NUI facility, in addition to working on another manned installation for the remaining work periods during a full man-year. It would then also be natural to use the same tolerance limit for personnel having their full employment on the NUI in the Case max alternative.

22.5   Emergency Preparedness for NUI Installations

Emergency preparedness on normally unmanned installations in the past when personnel were transported by helicopter implied that the crew had to be 'self-contained', in the sense that they would need to fulfil all roles in the emergency organisation, including all relevant teams, and they needed evacuation means as well as Man-overboard (MOB) boat, often also called a Fast Rescue Craft (FRC). The teams also needed to have personnel qualifications to fill all relevant teams. It was often the case that emergency roles and qualifications would dictate the size of the crew visiting a NUI, not the actual tasks to be carried out. This was due to the fact that the installation would be 'remote' from other resources, and thus had to be capable of handling all relevant contingencies.

This is much less of a challenge with NUIs serviced by W2W vessels, because several of the emergency functions may be covered by the crew on the W2W vessel. The most obvious function is rescue of personnel if falling into the water when working over the sea. The W2W vessel should have a MOB-boat and specially trained crew members who needs to be able to mobilize quickly if the alarm is raised. The W2W vessel will be near the NUI, which should give a rapid response.

Table 22.4 presents proposed emergency strategies for different types of hazards. The DFUs are generated from the RNNP list [9] and the guidelines for the area based emergency preparedness collaboration [10]; see also Sect. 20.2.

It is demonstrated that the following integrity requirements are essential in order to protect personnel working on the installation:

Table 22.4 Proposed emergency strategies for NUI installation when manned

Type of hazard	DFU reference	Strategy for survivors of initial accident	Integrity requirements

Major accident initiating event	RNNP:
	1: Non-ignited HC leaks
	2: Ignited HC leaks
	   DFU 3 Well kicks/loss of
	   well control
	4: Fire/explosion in other
	   areas, flammable liquids
	7: Collision with
	   field-related vessel/-
	   installation/shuttle tanker
		Evacuate to W2W over walkway	Shielding of escape ways to bridge landing
			Walkway available for access

Possible major accident threat	5: Vessel on collision course 6: Drifting object	Evacuate to W2W over walkway	Walkway available for access
Major accident	3: Personnel in the sea during emer-gency evacuation	Launch MOB from W2W vessel	MOB boat, davit, crew
Less extensive emergency	Guideline 064: 1: Man-overboard when working over sea	Launch MOB from W2W vessel	MOB boat, davit, crew
	Loss of W2W vessel station-keeping capability	Ladder to sea to be picked up by MOB-boat	Survival suits to be available on NUI

- Escape routes to bridge landing
- Shielding of escape routes
- Gangway to W2W

The following emergency functions should be covered by personnel on the installation itself:

- Search and rescue on installation
- First aid to sick and injured

The following emergency functions should be covered by personnel on the W2W vessel:

- MOB boat with crew
- Nurse/first aid crew

It is anticipated that the NUI will have at least two walkway landing points. Which landing point should be used? If the leeward landing point is used, drift-off (or push-off from weather) of the vessel will move it away from the installation (with disengagement of the walkway), but also push gas or smoke in the direction of the vessel (very unlikely that this would be simultaneous). Location on the wind side will imply that gas and smoke should avoid blocking the landing point, but drift-off may cause damage to the walkway or its landing.

22.6 Experience Data

It is always good to find relevant experience data, in order to confirm what are applicable hazards, and illustrate conditions and essential factors. There are few incidents known from NUI installations, which are relevant for the type of NUI installations addressed in this chapter.

The UK NUI platform Erskine, operated by Chevron, experienced a fire in 2010 (23.1.2010), and the installation was unmanned at the time of the ignited leak [11]. Condensate rain-out from the leak caused a gas cloud that was not sufficiently large to result in detection by fixed gas detectors. Shutdown and pipeline blowdown were performed by the parent installation, after detection by remotely controlled cameras.

The fire was confirmed visually with a helicopter fly-by, and another helicopter fly-by 30 min later confirmed that the fire was out. The total duration of the fire was two hours. The ignition was believed to have been caused by static electricity. The platform was shut down for 10 months for repair work.

Another incident occurred on 28.7.2018 during a test programme ahead of final mobilisation of a W2W vessel planned to be used during a maintenance campaign on the NUI installation Tambar, operated by Aker BP in the Norwegian sector. The investigation report (see below) refers to experience with W2W vessels in the UK and Dutch sectors but does not present any actual experience data.

The Tambar installation is not a modern type NUI, and has a helideck, lifeboat and other safety barriers, but no accommodation. The W2W vessel was planned to be used during a modification campaign. It was preferred in the present case to use W2W vessel, as this would complete the required modification work in much shorter time and avoid extensive helicopter shuttling (see also Sect. 22.7.4). When the W2W test failed as described below, the operator reverted to the original plan, to shuttle workers back and forth to the installation for each shift, as had been done several times before.

The walkway connection was tested once, which was successful. The end of the motion-compensated walkway was connected as planned with pre-installed steel profile to act as a recipient for the walkway. The walkway end failed in the second and third tests to connect with the recipient profile as intended but hit the railings on the Tambar installation. In the third test the walkway caused damage to a cable tray, before the test program was aborted. No people were injured, but there was minor damage to the installation (railings and cable tray). Aker BP has investigated the incident, but the report is not in the public domain; PSA's report is in the public domain [12].

The tests had been planned and prepared for about 5–6 weeks during the summer of 2018. The preparations on the Tambar installation had been limited to preparing three possible connection points for the walkway, on the east, south and west sides. The steel profile to act as a recipient for the walkway end had been installed in these three locations. The intention was that the end of the walkway would push continuously during the connection with a force in the range of 4–16 kN (according to Ref. [12]), thus locking the walkway into the recipient profile, i.e. the installation.

The planning phase, according to the investigation report [12], had been less thorough than it should have been, due to time pressure and probably because it took place during the holiday period. One may get the impression that too little emphasis was put on ensuring that personnel with extensive experience with W2W-connections were participating in the hazard identification meetings, in spite of this being the first time a W2W connection would be used in the Norwegian sector during normal operation. Apparently there was insufficient focus on the hazards associated with the possibility that the walkway during landing could hit outside the intended profile and thus cause unplanned damages to nearby structures or equipment. It appears that the east landing on Tambar was the only one with no process systems in the immediate vicinity. Some wooden protective structure had in fact been installed on the south landing to protect nearby equipment.

The investigation report [12] has no focus on the operation of the motion-compensated walkway from the vessel, except noting that the movements of the vessel would influence where the walkway end would hit the installation. At the time of the test on Tambar the wind was easterly with speed 9–11 m/s, with 1.8 m Hs (significant wave height). It would therefore not appear that maloperation of the walkway took place, nor that the sea state was a challenge. On the other hand, there is evidence that connection incidents are quite frequent (see Sect. 22.7.3), implying that failure to establish the intended connection is a 'normal' occurrence during connection, even in normal sea states.

The investigation report [12] observes that the possibility that the walkway might not connect with the recipient profile and thus cause damage to nearby equipment and structures had been more or less neglected during the planning, except for the protection structures that had been arranged at the south landing point, as noted above. The investigation concludes that this is a serious planning failure. Planning failures are the main findings of the investigation [12] in the present case. It is observed that the investigation has a limited focus on the vessel and its operation.

The failed test with the walkway is important, because it reveals an important aspect, the vulnerability of the system to operational errors and possibly also environmental loads in spite of the motion-compensation and weather restrictions. PSA has formulated requirements for the DP system and walkway by referring to international standards, as already noted in Sect. 22.2.2.1. These requirements are essential, because this is the most vulnerable aspect of the W2W solution for NUI facilities.

There will usually be at least two landing points on the installation, implying that the other landing point may be available if one is damaged. This may be more exposed due to the wind direction (see also discussion in Sect. 22.5) but should be usable as a reserve option. If the walkway itself is damaged, then climbing down to the sea level or lifting off by helicopter with a winch will be the options left for personnel on board.

22.7 Challenges for Risk Management of NUIs

22.7.1 · Use of Safety Systems

A major part of this chapter has focused on the use of safety systems on normally unmanned installations. We have discussed the rationale for installing the most typical safety systems, and we have shown the results of a case study with three cases with varying extents of implementation of safety systems. These three cases demonstrated extensive variation in the risk levels for personnel in the periods when manned. But the risk levels were at the same time at relatively low levels, to the extent that the following options are possible:

- The case with a very limited extent of safety systems would also be possible if a relaxed personnel risk tolerance limit is used, such as FAR = 10.
- If ALARP evaluations are performed properly, using qualitative as well as quantitative considerations as in Ref. [13], the case with a very limited extent of safety systems could not be defended.

The Norwegian regulations for NUIs are very general and leaves a lot to interpretation, with no explicit standards available to guide the interpretation. Some design contractors have proposed very simple NUI concept proposals, with a much lesser extent of safety systems than Case min in Sect. 22.3.1. This is considered completely unrealistic in relation to Norwegian legislative requirements. This is not a really a significant problem, but at least it serves to indicate that the industry has some way to go before there are realistic expectations of what risk management for NUIs will have to include.

The way the industry is limiting the discussion of risk levels for personnel on NUI facilities only to societal consideration of risk levels, and completely neglecting the individual risk aspect is another illustration of the failure of the companies to meet authority expectations.

22.7.2 Use of W2W Vessels

One of the most recent W2W vessels is the Island Diligence with a total accommodation capacity of 100 persons, delivered in mid 2018. The vessel is 86 m long, 4.200 dwt, with a gangway with a pedestal delivered in sections in order to be flexible. This was the vessel involved in the incident referred to in Sect. 22.6.

The new PSA regulations in their guidelines to Facilities regulations [3] refer to the following standards with respect to the gangway:

(a) DNVGL-ST-0358 for design of gangway and bridge landing [14]
(b) DNVGL-RU-SHIP Part 6 Chap. 5, Sect. 5.16 [15]
(c) NORSOK R-002 Chap. 4 [16].

With respect to the stability of the W2W vessel, the following are interim guidelines for transportation and accommodation of industrial personnel onboard cargo ships, also applicable for existing vessels:

(d) NMD RSV 17–2016 Requirements regarding transport and accommodation of industrial personnel [17].

According to the what is assumed to be the latest regulation amendments (expected to be published in Q2/2019) PSA has ensured that the use of W2W vessels shall be subject to application for consent to operate such facilities and has therefore secured the mandate to ensure that the requirements are adhered to. It is believed that the intention is to ensure that vessels to be used for such operations shall follow the most recent requirements.

22.7.3 Use of Gangway from W2W Vessel

One of the contentious aspects of the use of gangway between a W2W vessel and a NUI is whether or not to have the gangway permanently connected as long as personnel are present on the NUI facility. One project has official sanction for regular use of a W2W vessel for personnel transport and accommodation, and this project has committed to having the gangway permanently connected. This has the following advantages:

- There is no delay if the gangway is needed for evacuation in the case of an emergency
- If there is a problem with the gangway or its connection, it will be detected immediately, otherwise it would not be revealed until reconnection is attempted
- It avoids evacuation to be prevented by gangway or reconnection problems.

Another oil company planned to use a W2W vessel, including the plan to disconnect the gangway after shift commencement, as well as for the lunch break and end-of-shift. Otherwise the vessel would be outside the safety zone. The application for consent to PSA was in this case withdrawn, due to the incident referred to in Sect. 22.6. The installation in question has one freefall lifeboat, which allows the vessel to be disconnected, when the bridge is not the primary means of escape in an emergency scenario. At the same time the operator was involved in a dispute with PSA about the need to apply for consent or not (see further discussion in Sect. 22.7.4). This has also given the opportunity to obtain insight into some interesting documentation.

One W2W vessel had 14 incidents with its gangway during a period of 4.5 years. In nine of these incidents there was damage to equipment and/or structures, while in the remaining five incidents the connection was lost without any other consequence. Nine of the incidents were disruption of the connection due to weather, with no or very minor damage. The remaining scenarios are 'landing failure', all with minor damage to connection point or vicinity.

The number of W2W-assignments for this vessel in the period is not known and is impossible to guestimate. The only frequency that is obvious is about one 'landing failure' in average per year during this period for this specific vessel, irrespective of the number of landings per year. When the fleet of such vessels is expected to increase in the coming years, such a frequency may be far too high.

Gangways are used in several cases in the North Sea and worldwide, for instance as a connection between flotels and production installations, either floating or fixed. These user configurations appear to have a good reputation with high reliability. But as the data above indicate, the experience from the W2W vessel is not necessarily correspondingly good. The shipowner of the W2W vessel actually in a presentation described the experience with the gangway as 'good' and well known. But the number of systems in operation is low at the moment, and the problem may not seem large. Maybe this is yet another confirmation that the offshore standards are somewhat higher than the marine standards. It is on the other hand claimed that the HES standards in the wind industry are as high as in the petroleum sector. A relatively high failure rate would not be acceptable for offshore configurations where emergency evacuation is one of the services to be performed by the gangway. This is actually common with the use of gangway on a flotel, but this does not appear to be a challenge with respect to robustness. It may seem that the combination of substantially smaller vessel (W2W) and frequent reconnections may be the challenging combination.

The gangway is usually manually manoeuvred into its landing/recipient profile, but simultaneously motion-compensated during the operation. After successful landing, a constant force is applied to hold the gangway in position in its profile, but the position may be lost in case of large vessel motions.

Typically, the maximum wave height is required to be less than 3 m H_s, during connection as well as before disconnection, although it is understood that the practical limit for disconnection of the gangway is 2.0–2.5 m H_s. Instrumentation and sensors are sometimes also used to give higher assure of successful connection. It would appear that this should be the norm rather than anything else, if a failure rate less than 1 per 1,000 attempts is the target. Apparently, some protective structures are also sometimes used to avoid the possibility to cause damage to the installation in case of connection failure [18]. None of these measures had been taken in the test at the Tambar installation, which may tie-in with the comment in the investigation report [12] which refers to inadequate planning.

It appears that, at the moment, it would be more appropriate to leave the gangway connected for as long as employees are onboard the NUI, in order to avoid the connection problems. The disadvantage in this case is that the vessel has to remain on station engaging the DP system for the duration of each shift. In case of DP instability there is the hazard that the gangway, or in the worst case the installation, could be damaged due to W2W vessel excursions.

Priority must be given to the gangway availability for personnel in the case of an emergency on the NUI facility, when rapid evacuation over the gangway may be crucial.

22.7.4 Use of W2W Vessel as Substitute for Flotel or Helicopter Shuttling

Aker BP planned during spring of 2018 to use a W2W vessel as a substitute for extensive helicopter shuttling of personnel to complete modification work on the Tambar NUI facility (see coarse description of the Tambar facility in Sect. 22.6). This was considered an efficient way to reduce modification costs, and also at the same time reduced the number of personnel exposed to significant volumes of helicopter shuttling. PSA has in the past had a considerable focus on the reduction of helicopter shuttling, which is considered to have an increased risk level, due to frequent landing and take-off (see Sect. 13.8.2). Some time ago, another operator attempted to use a W2W vessel as a substitute for a flotel, but this was also unsuccessful.

PSA informed Aker BP at the end of June 2018 that an application for consent to PSA would be required before the W2W vessel could start operating as an accommodation facility for personnel carrying out modification work on the

Tambar installation. Aker BP appealed this ruling by PSA to the Ministry of Labour and Social Affairs in a letter of 10.7.2018. Aker BP applied for consent as ordered in parallel with the appeal. The incident referred to in Sect. 22.6 occurred during tests of the W2W vessel at Tambar, and Aker BP withdrew their application for consent, and reverted to using helicopter shuttling for the outstanding modification work. The appeal was upheld, and the Ministry ruled in November 2018 in favour of PSA [19], implying that PSA's requirement to apply for consent would have applied if the proposal had not been withdrawn.

The important aspect of this conflict is associated with the application for consent, which would also imply that the petroleum and the working environment acts and their sets of regulations would apply to the W2W vessel. These acts and regulations apply to petroleum installations for production and exploration drilling and are considered to have substantial cost effects when compared to maritime regulations, Norwegian maritime regulations and (even more) international maritime regulations. Flotels used in the Norwegian sector are also considered installations and follow the set of petroleum and working environment acts and regulations.

Construction and maintenance vessels, including W2W vessels, fall outside the set of petroleum and working environment acts and regulations, but this has been an area of conflict between employees and employers (shipowners) for the past couple of years where authorities have tried, but so far failed, to find a compromise that all parties will accept (see further discussion in Sect. 22.7.9).

The expected amendments to the regulations (see Sect. 22.2.2.1) will clarify some of this uncertainty for the use of W2W vessels as accommodation facilities with respect to NUI facilities without helideck and accommodation facilities. No clarification is yet available for other construction and maintenance vessels.

The use of W2W vessels to avoid significant volume of helicopter shuttling would appear to be a good solution, but is probably not feasible until the parties can reach some form of compromise on the regulatory principles relating to the use of construction and maintenance vessels in the petroleum activities. PSA will require application for consent for vessels used for transport and accommodation during modification and/or maintenance campaigns to installations with no or insufficient quarters capacity, according to the relevant laws and regulations.

22.7.5 Maintenance Regime

The overriding maintenance objective on an unmanned production installation will be to minimise the manhours involved in the connection and disconnection, including the internal transport. This may be achieved with emphasis on the replacement rather than repair, and choice of equipment with long Mean Time Between Failure (MTBF). Application of subsea principles with replacement modules is often mentioned.

It is expected that significant tasks relating to the preventive maintenance and inspection will be concentrated on one or a few periods per year. There may be additional visits due to breakdowns or similar (see also Sect. 22.7.6). The essential aspect to consider in this context is if maintenance periods shall be carried out on a 'cold' or 'hot' installation, i.e. with or without full process shutdown and depressurisation. It the installation is 'cold', then probably more people can work in parallel and there would be no need for extra safeguards or monitoring actions. A compromise could be to have two trains with the possibility to shut-down and depressurise one train whilst the other is kept in operation.

It would also be expected that the power generation units (if shore supply by cable or similar is not feasible), critical rotating machinery, control units as well as critical valves have redundancy, such as $2 \times 100\%$, $3 \times 50\%$ or similar. The switch between redundant units could be performed automatically or manually from the shore base. It would be expected that the rotating equipment units have extensive use of condition monitoring.

Campaigns with extensive ('heavy') maintenance are expected to occur every few years (typically four to five years interval) similarly as for a manned installation, expected to be performed on a 'cold' installation.

22.7.6 Production Regularity

It is expected that the strategy in case of abnormal events or an unclear status on an unmanned installation will be to shut down the production as a precautionary action. The same is expected to occur in the case of gas or fire detection, possibly also followed by depressurization, depending on the systems available to perform such actions.

It will be expected that manning of the installation is thereafter required in order to check out the circumstances and possibly restart, unless repairs are needed. It would not be expected that manning of the facilities is decided until there is some way to verify that manning would be safe to perform. This would probably be determined based on input from sensors and cameras.

Longer response times would be expected when manning is dependent on the use of a W2W vessel. The vessel must be mobilized with relevant equipment and personnel, it takes time to go from shore to the NUI facilities, and connecting up the walkway from the W2W vessel to the NUI facilities requires a sea state with relatively good conditions, which may call for some waiting on weather. Especially in the winter months there could be substantial waiting on weather, much less in the summer months, but then the vessels may not be readily available.

In summary, the response times may lead to lower production regularity than is normally found on offshore installations in the Norwegian sector, which is typically in the range of 95–98%.

22.7.7 Cyber Risk

The control systems used in various industries, including the offshore petroleum industry are very vulnerable to cyber attacks by hackers, according to an article in the Journal of Petroleum Technology in March 2016 [20]. The article refers to several cases mentioned in Sect. 15.11.1.

These examples illustrate the vulnerabilities not only of process control systems, but also of vital marine systems. There is also an anecdote referring to a Norwegian floating installation being only a short while from being completely evacuated due to what was believed to be an ongoing cyber attack quite recently, before the suspicious network traffic found its explanation and was revealed to be harmless.

According to the article referred to above [19], many control systems use 20 or 30 year old technology, were never intended to be available on internet or through company networks, and have little or no protection against hacking or any other illegal activities.

Some of the threats and possible protections against cyber attacks are discussed in Ref. [21] for autonomous ships, and it is considered that many of these aspects are also common for offshore installations. With the large number of contractors, subcontractors and suppliers, it would appear almost impossible to completely protect against the installation of malware and threatening devices during fabrication as well as maintenance and operation.

A normally manned installation has to some extent more protection. Local control may be used to override the control system in many instances, but probably not all potential cases. If local control is initiated, it will also require that the personnel realise that a cyber attack is imminent or ongoing. There would probably be many cases where the attack is realised too late.

In the case of an unmanned installation, there is no way to exercise local control, and the control of the installation will be lost if subjected to a cyber attack.

22.7.8 Operational Control When Manned

All normally unmanned installations built to date have two control rooms, on the installation and onshore with an identical setup and control possibilities. In the past it was common that local personnel would take over control when manning the installation. This implies that local persons with the best insight into ongoing activities would be responsible for operations with personnel present. But this may not be continued in the future. The first installation to abandon this principle has been seen.

This implies that external personnel without local insight into ongoing activities will be in control also when there are personnel working on the installation. This also opens the possibility of hacking into safety systems (see Sect. 22.7.7 above) during manned periods.

It would be expected that local control would be given priority when the installation is manned, as long as this option is available. But this would probably imply two extra positions (control room personnel) to cover control functions during the dayshift when the installation is manned. If personnel were sleeping on the installation, another two more would be required in order to continue manning the control room also during the night. If personnel were sleeping on a W2W vessel there would be no need to mann the control room during the night shift, provided it is not too demanding to transfer operation between offshore and onshore each morning and evening.

It is therefore surprising that an installation does not provide the option for personnel who will stay for two weeks to take over control of operations from the shore centre. It is assumed that the extra costs of four extra personnel offshore have dictated this solution, but it is surprising that the solution has been accepted by the authorities. This is particularly surprising in light of the security (cyber) hazards as outlined in Sect. 22.7.7 above.

22.7.9   Legislation for Construction and Maintenance Vessels

Mobile units were included under the jurisdiction of the Norwegian petroleum act in the early 1990s, thus also falling under the jurisdiction of the Working environment act. The working environment law has particularly stronger requirements than the maritime law (Norwegian and international) when it comes to technical and working environment (noise in particular) requirements to the quarters facilities onboard. Also the diving activities onboard diving vessels fall under the working environment law. It was considered at the time to include other vessels also, but it was decided against it.

The employees' unions have for a few years claimed that the use of construction and maintenance vessels (often referred to as 'service operations vessel, SOV) is increasing in the Norwegian petroleum sector, to the extent that employees on production and mobile units to some extent are replaced by employees on SOV. W2W vessels belong in this category.

It has therefore been a requirement from the unions side for some few years that employees on SOV also should fall under the same jurisdiction as for mobile units. The shipowners claim that the cost penalties would be out of proportion, if the petroleum and the working environment acts were to apply for the construction and maintenance vessels. The proposal from the unions has been that all types of vessels involved in the petroleum sector should fall under the same jurisdiction as for mobile units, irrespective of what type of activity they undertake. Although this would be a clear cut definition, it may appear unreasonable that vessels involved in plain transport of supplies to installations like any other marine transportation, should be defined as petroleum activity.

The employees claim that the HES aspects are better taken care of under the petroleum and the working environment acts when compared to maritime regulations. Statistics presented in Chap. 2 confirm this to some extent. A recent report [1] has shown unclear employment conditions and regulatory framework for industrial employees (offshore personnel) working onboard SOV, with significant implications for employee health and safety [1]. This is particularly the case for vessels registered outside the Norwegian national (NOR) or international (NIS) ship registers. The report to the Ministry of labour and social affairs [1] also pointed out that petroleum authorities traditionally had very little emphasis on the use of SOV. They had for instance never performed any audits on the use of SOV, which they would be entitled to as long as the vessels have significant influence on the petroleum activities.

The authorities (petroleum as well as maritime) have never focused on trying to survey the extent of use of SOV in the petroleum sector. When it comes to man-hours on production and mobile units, all companies have to report manhours down to 15 min, but the use of SOV has never attracted any interest, nobody appears to know how many vessels are involved, how many industrial workers (petroleum personnel) and other employee groups that participate in the activities, etc. One of the recommendations made in Ref. [1] is for the authorities at least to start surveying what is the extent of such activities.

A proposal was tabled in the Norwegian parliament in 2017 but was voted down by the right wing government. The same government tabled a white paper [22] to the Norwegian parliament in 2018 about petroleum sector HES, but the issue relating to legislation for SOV was omitted from the discussion, because the Ministry of labour and social affairs had formed a tri partite working group in 2017 with employers, employees and authorities to try to resolve the issue. This working group has worked all through 2018, and presented its final report in July 2019 (Ref. [23]), but without reaching consensus or a compromise.

A separate solution has been found for W2W vessels, as documented in changes to the framework regulations 26th April 2019 (Ref. [24], see also Sect. 22.2.2.1).

Safetec in their report the Ministry (Ref. [1], see also Sect. 22.1) recommended that Norwegian authorities should take an initial step to start collecting statistics about the activity with use of SOV, types of activities, vessel-days, manhours, etc. There is at present no overview of what the activity levels are, if they are increasing or declining or stable. PSA knows manhours on all types of installations down to every manhour but knows nothing about the use of SOV and associated manpower.

References

1. Bye J et al (2018) Study of with new operational concepts in the petroleum industry (in Norwegian only). Safetec report ST-12578-2, 8 February 2018
2. Samuelsberg A (2017) Statoil remotely operated factory (ROFTM) safe, high value and low carbon intensity concepts. Equinor, 8 August 2017
3. PSA (2017) Regulations relating to design and outfitting of facilities, etc. in the petroleum activities (the facilities regulations). PSA, Last amended 18 December 2017

4. PSA (2017) Regulations relating to health, safety and the environment in the petroleum activities and at certain onshore facilities (the framework regulations). PSA, last mended 15 December 2017
5. ISO (2003) Analysis, design, installation and testing of basic surface safety systems for offshore production platforms. International Standards Organisation, Geneva. ISO10418:2003
6. Norwegian Oil and Gas (2018) 070 application of IEC 61508 and IEC 61511 in the Norwegian petroleum industry (recommended SIL requirements), 26 June 2018
7. EU (2014) Equipment for potentially explosive atmospheres (ATEX), ATEX Directive 2014/34/EU https://ec.europa.eu/growth/sectors/mechanical-engineering/atex_en
8. Vinnem JE, Pedersen JI, Rosenthal P (1996) Efficient risk management: use of computerized QRA model for safety improvements to an existing installation. In: Presented at SPE 3rd international conference on health, safety and environment, New Orleans, June 1996
9. PSA (2017) Trends in risk levels in the petroleum activity (in Norwegian only), Main report, 26 April 2018. PSA
10. Norwegian Oil and Gas (2013) Recommended guidelines for establishing area-based emergency preparedness, guideline 064 (Rev. 3), 19 March 2013. Norwegian Oil and Gas
11. Fossan I (2016) Modelling of ignition sources on offshore oil and gas facilities—MISOF, Lloyds Register, 25 November 2016
12. PSA (2019). Report of the investigation into the walkway incident on Tambar, 16 January 2019. PSA
13. Aven T, Vinnem JE (2007) Risk management, with applications from the offshore oil and gas industry. Springer
14. DNV GL (2015) DNV GL-ST-0358 certification of offshore gangways for personnel transfer, December 2015
15. DNV GL (2016) DNV GL-RU-SHIP rules for classification: ships, October 2015, Last amended January 2016
16. Standards Norway (2017) R-002 lifting equipment (edition 2017). Standards Norway
17. Norwegian Maritime Authority (2016) Requirements regarding transport and accommodation of industrial personnel. Norwegian Maritime Authority, RSV 17–2016, 27 December 2016
18. Uptime International (2019) Private communication, 18 February 2019
19. Ministry of Labour and Social Affairs (2018) Decision regarding appeal—Tambar (in Norwegian only), letter to Aker BP, dated 21 November 2018
20. Jacobs T (2016) Industrial-sized cyber attacks threaten the upstream sector. J Pet Technol
21. Vinnem JE, Utne IB (2018) Risk from cyberattacks on autonomous ships. In: Presented at ESREL2018, Trondheim, Norway
22. Ministry Ministry of Labour and Social Affairs (2018) Health, safety and environment in the petroleum industry, report (White Paper) to the Storting, Meld. St 12 (2017–2018) https://www.regjeringen.no/contentassets/258cadcb3cca4e3c87c858fd787e0f75/en-gb/pdfs/stm201720180012000engpdfs.pdf
23. Ministry Ministry of Labour and Social Affairs (2019) The use of service vessels in the petroleum activity on the Norwegian Continental Shelf, 2 July 2019 https://www.regjeringen.no/contentassets/e5dbfd19f8be4d81b632405253ce43b0/rapport_flerbruksfartoy.pdf
24. PSA (2019) Changes to the framework regulations, 26 April 2019

Chapter 23
Use of Risk Indicators for Major Hazard Risk

23.1 Background

23.1.1 Historical Development

The BP Texas City refinery disaster in 2005 created a high awareness that the management of major hazards is not the same as the management of occupational hazards. Several papers and other publications have been issued in the last ten years, including a recent special issue of Safety Science [1] devoted to this topic. Indicators of major hazard risk are among the topics with high attention.

However, earlier efforts were made in this field; indeed, the author developed major hazard indicators for a large Norwegian offshore field operator in the mid-1990s [2] and Øien and Sklet [3] suggested an alternative approach some years later.

There was a dispute between the major stakeholders in the Norwegian petroleum sector in the latter part of the 1990s. Representatives of unions and authorities were extremely concerned that the risk levels were increasing in offshore operations, while company management and their representatives were equally claiming that 'safety had never been better'. Owing to this dispute and other factors, there was considerable mistrust between the parties and a lack of constructive communication about sensitive issues.

There was a need to have unbiased and as far as possible, objective information about actual conditions and developments. The authorities, the Norwegian Petroleum Directorate (NPD) at the time, now the PSA, defined a project (RNNP) of extended indicators, in order to fulfil these needs.

The project was initiated in 1999 and progressed throughout 2000 in order to develop the necessary methods. The first report was presented early in 2001, based on data for the period 1996–2000.

It was shown that the use of risk indicators is an effective way to follow up QRA studies during the operations phase. It is therefore recommended that the risk level

© Springer-Verlag London Ltd., part of Springer Nature 2020
J.-E. Vinnem and W. Røed, *Offshore Risk Assessment Vol. 2*, Springer Series in Reliability Engineering, https://doi.org/10.1007/978-1-4471-7448-6_23

in the operations phase be monitored in order to identify risk level changes. Risk analyses should therefore be capable of identifying the parameters or indicators that have a strong impact on risk level as well as the effect changes will have on risk level. This will enable the effective monitoring of risk level. Examples of such indicators include:

- Hydrocarbon leak frequencies
- Extent of hot work
- Availability of critical safety systems
- Essential RIFs with implications for MO-factors in accident causation.

The starting point is that these indicators have a certain value as presumed or established in the risk analysis. Changes in these values will imply change in risk level. The intention is therefore to receive an early warning of any trends which finally may result in inability to meet risk tolerance criteria. The trends for risk indicators must be monitored relatively continuously, which requires that the indicators are suitable to identify changes within a relatively short time span.

Monitoring risk indicators focuses attention on crucial risk factors and provides a tool for identifying and rectifying deviations from the assumptions made in the risk analysis.

23.1.2 The 'PFEER' Approach to Risk Monitoring

The PFEER [4] regulations require PSs to be defined and monitored in relation to the so-called PFEER assessment. Usually, this is implemented using both 'high level' and 'low level' PS. Performance indicators are described as measurements required to demonstrate that the assumptions and premises of QRA are fulfilled. The indicators also need to be suitable for monitoring during normal operations. It is a further requirement that the consequences are evaluated, if the monitoring of platform operations leads to an observation of the exceedence of an indicator.

23.1.2.1 High Level PS

High level PSs are usually the risk tolerance criteria. Performance monitoring using these criteria is usually not possible over a relatively short time span. If the fatality risk level (PLL value) on the installation is 0.062 fatalities per year, then this is sometimes claimed to correspond to just over 16 years between fatalities. There are two problems with such a characterisation:

- Such statements about risk convey very strongly the impression that the risk estimates are statistical predictions of a 'real' risk level, and not according to the Bayesian approach as proposed in Chap. 2.

- The actual expected period will in fact be longer, because of the accidents with more than one fatality per accident (i.e. average in excess of one fatality per accident).

23.1.2.2 Low Level PS

Low level PSs relate to the most important assumptions and premises in risk assessments and evaluations. Performance indicators are usually defined as the basis for comparison between observed values and PSs. The definition of PSs is often based on the relevant QRA for the installation, but UK authorities have also indicated that qualitative evaluations should be given appropriate emphasis.

The following defines the performance indicators used for the Frigg Field installations, as presented in Totals HES Annual Report for 1997 [5]:

- Frequency of gas leaks
- Extent of hot work activity
- Automatic gas detection
- Automatic fire detection
- Availability of pipeline ESD valves
- Fire water supply
- Availability of activation valves in the deluge system
- Availability of emergency lighting in accommodation
- Pick-up time for 'man-over-board' scenario
- Mobilisation time for smoke diving team
- Mustering of all personnel in muster area.

Most of these parameters are based on QRAs for installations, except the last four items, the first of which is based on a qualitative evaluation, while the remaining three are important emergency preparedness aspects.

23.1.2.3 Other Practices

No requirements similar to the UK PFEER assessments exist in the Norwegian regulatory regime. In a paper [6], however, the use of indicators by ConocoPhillips on one Ekofisk platform was described. In the paper by Øien et al. [6], the following indicators were referred to:

- Number of experienced process operators in the area
- Total number of personnel on the platform
- Number of critical faults on electrical equipment
- Total hot work time per period
- Number of times that any applicable hot work restrictions are exceeded
- Number of leaks per period

- Number of experienced control room operators
- Number of safety-critical failures of process components
- Number of aging-induced faults with potential safety-related consequences.

Key Performance Indicators (KPIs) are also used internally by operating companies for a variety of purposes. Obvious examples are production rates, injection rates, maintenance hours, maintenance back-logs and production downtime. It is therefore important to stress that the use of performance indicators is not limited to regulatory needs but is certainly a part of the day-to-day operations of any oil and gas company.

23.1.3   Objectives

The overall objectives of the use of performance indicators for HES reasons are:

- To enable the HES management of the company.
- To measure the effectiveness of its HES management activities.
- To identify priority areas for further efforts in the HES field.
- To spot as early as possible, degradation of equipment or lack of adherence to procedures that are vital for safety.
- To satisfy regulatory systems with respect to monitoring HES performance.

The management of major hazards requires a continuous high attention level to be maintained even despite many years of operation without accidents or near-misses. It is common that motivation and attention may reduce over the years, when an installation has operated without serious events for a long time.

The objectives are somewhat different, in the sense that the requirements of the company may favour one (or a few) overall indicators, whereas the use of overall indicators may not be equally important in a regulatory context. The overall objectives may be broken down into refined specifications. For all types of accidents performance indicators should comply with the following requirements:

- The indicators should address a range of incidents, from the insignificant near misses to the most severe and complex accident sequences.
- The hazard potential of an incident should be indicated, in addition to the actual outcome of the incident.
- When it comes to indicators that relate to potential major accidents, performance indicators should be capable of reflecting the importance of the incident in question. Unauthorised hot work in a safety critical area is more critical than the failure of one out of three parallel (100% capacity) fire water pumps.
- Indicators reflecting individual systems are required along with a limited set of overall indicators that present an aggregated picture. Thus two levels of indicators are natural, with the highest level needed for presentation of an overall indication of the safety level related to potential major accidents.

- It should be emphasised that indicators need to cover 'M', 'O' and 'T' aspects, not only technical aspects as is often the case.

The discussion later in this section is devoted to the needs of the operating company in order to have one or a few indicators that express the resulting level of safety with respect to avoiding major accidents. This implies that the last three specifications in the list above are those mainly addressed in the following.

Developing an approach to performance indicators for major hazard incidents is presented in two stages:

- First the different parameters that should be followed up are discussed, and a procedure for selection is proposed.
- Then a proposal on how these could be combined into overall performance indicators is discussed.

23.2 Need for Specific Indicators for Major Hazard Risk

Major accidents are very rare in high reliability industries; indeed, the last major accident in offshore operations on the NCS with fatalities occurred in 1985. The Snorre Alpha subsea gas blowout in November 2004 may or may not be counted as a major accident, it certainly had the potential to cause at least five fatalities if it had been ignited, but even the PSA investigation report concludes that it was not a major accident.

Even major accident precursor events are quite rare, typically in the order of one event per installation per year. A precursor event in this context is an event that may provide a warning about possible more severe accidents in the future. An unignited hydrocarbon leak may warn that an ignited leak could occur sometime in the future. It is therefore crucial to maintain motivation and awareness that we have indicators that change before the frequency of major accidents has changed. The next precursor event may be the next major accident if the battery of mitigation barriers on offshore installations has a complete failure.

There have been risk indicators for occupational accidents for many years, if we go about 20 years back it was assumed that indicators based on occupational injuries also provided an indication of major hazard risk. During the past 10–15 years, and especially in the wake of the Texas City Refinery disaster [7], it has been realised that major hazard risk needs separate indicators. The investigation found that this invalid assumption about a strong correlation between occupational and major hazards was one of the factors leading to conditions that would allow such an accident to occur.

Therefore, what is special about major hazard risk? The most obvious factor is that major accidents usually occur rarely. Another particularity about major hazards is that potential causes are often specific to major accidents and often have little in common with occupational accidents and other limited accidents that are more frequent than major accidents. This is usually because of different mechanisms that may cause unwanted consequences.

Consider hydrocarbon leaks as an illustration. There is usually a distinction between leaks with a flowrate above 0.1 kg/s and those with flowrate less than 0.1 kg/s. There are usually few of those above 0.1 kg/s, and many more smaller leaks below 0.1 kg/s. Leaks with a flowrate above 0.1 kg/s are considered to be major hazard precursors, because the possible gas cloud generated is sufficiently large to cause an explosion or fire which may lead to the uncontrolled development of the accidental event, following the severe failure of consequence reducing barriers.

It might be assumed that the number of leaks below 0.1 kg/s is a good indicator of more serious leaks. Generally this is not the case, however. The reasoning for this is parallel to why injury to personnel is not a good indicator of major hazard risk; the causal mechanisms of the occurrences are significantly different. Minor leaks (<0.1 kg/s) are often caused by rubber seals around valve stems becoming more brittle over time or because of high temperature, thus allowing minor seepages of gas or oil around the valve stem. Such seepages may often be fixed easily by tightening the stem seal or replacing it. More importantly, such mechanisms do not have the potential for high leak rates except in very rare circumstances, and thus they are rarely precursors of more serious leaks. Large leaks are often caused by mechanisms such as valves left open during intervention, the inadvertent dismantling of pressurised equipment, rupture of a gasket or holes due to corrosion. The causal mechanisms are thus distinctly different. Accordingly, minor leaks would not constitute a good indicator of more serious leaks.

It can be argued that there is considerable similarity between occupational and major accidents when it comes to the root causes of organisational nature. This is discussed more in depth in Sect. 23.8.2. Otherwise, the similarity would be expected to be very limited. Therefore, indicators of personal injuries have very limited applicability for monitoring major hazard risk.

It can also be argued that safety culture is the common factor between occupational accidents and major accidents. The culture is considered to be a common factor, but it is not sufficient to defend that occupational accidents and major accidents be considered together.

Because even major accident precursors are rare occurrences, the data sources are significantly different. Occupational accidents occur regularly and imply that a significant volume of data is established in a reasonable timeframe, even for a single installation. Major accident precursors require that data are collected almost at the national level in order to have a significant volume in a reasonable time.

23.2.1   Indicator Concepts and Definitions

The distinction between leading and lagging indicators has been debated for some time, such as in Ref. [1]. Kjellén [8] states: 'A leading safety performance indicator is, in this interpretation, an indicator that changes before the actual risk level has changed.' According to Kjellén [8] this definition is consistent with what is commonly used in economy, but significantly different from what other researchers in safety use. Kjellén argues further that what are now typically called 'leading indicators' were previously called 'proactive indicators'.

This definition is supported. It is particularly important for major hazards to focus on indicators that change before the actual risk level has changed. Major accidents and even precursor events are quite rare, as noted above.

Precursor event occurrences are obviously lagging indicators, according to the definition above. The risk level has changed, at least temporarily, when a precursor event occurs. What constitutes a leading indicator seems to be more controversial.

Barrier indicators are usually based on the periodic test of barrier elements as part of preventative maintenance schemes, using 'man made' activation signals or stimuli (such as test gas releases). This type of test must not be confused with barrier testing which can be used by the industry as an additional recording, when the actual performance of barrier elements is recorded during a precursor event, such as a gas leak. The second type of test is principally not achieved without an increase in risk, and it should not be considered to be a relevant basis for a leading indicator. However, the tests that are part of preventative maintenance are carried out without any coupling to increased risk, and these should be considered to be leading indicators according to the definition stated by Kjellén.

It is commonly accepted that 'leading' indicators are clearly preferred over 'lagging' indicators. This implies that there is more motivation in reporting performance of preventative measures compared with the occurrence of near misses and incidents. We believe this is particularly crucial for major hazard risk indicators in high reliability industries where even precursor events may be rare.

Ale [9] argues that no indicator for which the values are based on observations over time can be regarded as leading. According to this interpretation, none of the indicators used in RNNP (see Sect. 23.3) could be called 'leading'. The definition proposed by Kjellén [8] implies that barrier test indicators are leading indicators, which is consistent with the view adopted in RNNP. It is also argued that the performance of these barrier systems strongly influences the likelihood that an incident in the future would develop into a major disaster if it occurred. The leading indicators in RNNP are indicators of barrier systems that aim at preventing future incidents escalating into major accidents. This view is also shared by Zwetsloot [10], who refers to indicators based on testing barrier systems as 'output' indicators, as opposed to 'outcome' indicators, which are the only true lagging indicators.

Owing to the rarity of precursor events in the Norwegian sector, barrier indicators are given high priority in RNNP. They were not developed from the start of the Risk Level work, but once the collection of event based indicators had been firmly established. Emphasis has been placed on barrier elements that are associated with the prevention of fire and explosion, but structural and marine system barriers are also addressed to some extent. The discussion in the following is limited to barrier elements that are associated with the prevention of fire and explosion. A discussion of one barrier indicator for structural failure may be found in Ref. [11].

The Risk Level work has therefore also included 'leading' or proactive indicators, where this has been possible and realistic. The following are those that are considered to be 'leading' indicators:

- Indicators based on the performance of barriers that are installed in order to protect against major hazards
- Indicators based on the assessment of management aspects of the chemical work environment as well as noise exposure
- Indicators reflecting the quality of operational barrier elements based on questionnaire surveys.

The first type of indicator is discussed below, whereas the other two types are not considered any further in this paper. A brief overview of lagging indicators for major hazard risk is also presented in the following section.

23.2.2   Criteria for the Assessment of Major Hazard Risk Indicators

The list below was established partly based on work by Kjellén [12]. This list is general and may apply to any risk indicators. The work in Ref. [12] is not particularly focused on major hazards, and it is therefore applicable to all types of risk indicators:

- Observable and quantifiable
- Sensitive to change
- Transparent and easily understood
- Robust against manipulation
- Valid.

A more comprehensive set of criteria was presented by Rockwell [13], these are claimed still to be valid by Kjellén [12]. This list needs to be expanded and tailored to major hazards, which are special in some senses. Major hazards in the offshore industry are rare events at a national level. However, even major hazard precursors (hydrocarbon leaks, well kicks, ships on a collision course) are rare on a single installation. For an offshore employee the return period between the occurrences of major hazard precursors may typically be in the range of 10–15 years. The criteria for major hazards risk indicators are therefore discussed in a wider context.

It must be realistic to observe and measure performance by applying a recognised data collection method and scale of measurement. Usually, indicators are expressed on a ratio scale of measurement, such as the number of hydrocarbon leaks (>0.1 kg/s) per installation year or per million manhours. It is difficult to establish a data collection method that provides reliable data that corresponds to the quantity we would like to observe. For example, measuring the true number of leaks is in practice difficult. Recording events may be poor, and the data may be contaminated by extraneous factors. In one case in Norway about ten years ago, a hydrocarbon leak and a personnel injury occurred in the same event. For a long time, the injury was focused on, and the hydrocarbon leak was overlooked. It is therefore essential that the data we want to record are easily observable, such that disagreements about whether the event occurred or not can be avoided. Limiting data collection to those events with medium and large severity usually improves the reliability significantly.

Indicators should preferably be intuitive in the sense that what is measured is considered intuitively by the workforce to be important for the prevention of major accidents. When about 20 years ago the number of years in the same position (e.g. as process operator) was proposed as a major hazard risk indicator [14], this was not intuitively accepted as important by many employees offshore. Yet we see proposals such as this being repeated frequently, probably reflecting the fact that data availability is given higher priority than relevance.

It is further preferable that major hazard risk indicators do not require complex calculations; this is also related to intuitiveness. If the number of observations goes down, it should correspond to an improvement. If complex calculations are required, confidence may be lost.

Psychological and organisational reasons could in many cases result in too low reporting. As an example, we may think of an organisational incentive structure where the absence of incidents is rewarded (e.g. by a bonus scheme). Then we may experience that some incidents are not reported as the incentive structure is interpreted as rewarding the absence of reported incidents.

Indicators should reflect as closely as possible the hazard mechanisms. If they do, then they may contribute to maintaining awareness about risk mechanisms. The number of hydrocarbon leaks is thus a good major hazard risk indicator, because hydrocarbon leaks are among the most hazardous incidents on offshore installations.

A risk indicator must be sensitive to change. It must allow for early warning by capturing changes in a socio-technical system that have significant effects on accident risks. Clearly, the number of occupational accidents leading to fatalities would not normally be sufficiently sensitive to change. A "good" set of indicators would reflect changes in risk as well as point to aspects where improvements should be sought. Usually the number of observations (incidents, barrier faults, etc.) for the system being considered should be in the order of a dozen or so per period (such as a year).

Risk indicators should not allow the organisation to "look good" by for example, changing reporting behaviour rather than making the necessary basic changes that reduce accidents. Thus, risk indicators must also be robust to manipulation.

This leads to the requirement of validity, which is a critical point in the evaluation of the goodness of an indicator. Is the indicator valid for major accident risk? Does it actually measure what it is intended to measure? Consider for example the indicator defined by the number of lost time injuries. Clearly, this indicator says something about accident risk, but of course, accident risk is more than the number of lost time injuries, so this indicator cannot be used alone to conclude on the accident risk level as a whole. The validity of a statement concerning accident risk based on observations of the injury rate only would thus in most cases be low. By restricting attention to this specific type of injuries, there should be no validity problem in this respect. However, we would still have a problem in concluding on any development in injury risk based on the observations of the indicator. This is discussed further by Aven [15].

The following is a summary of the criteria discussed in this section. This should be addressed when reviewing the current use of indicators or new proposals:

(a) Combination of lagging and leading indicators
(b) Easily observable performance
(c) Intuitive indicators
(d) Not require complex calculations
(e) Not be influenced by campaigns that give conflicting signals
(f) Reflect hazard mechanisms
(g) Sensitive to change
(h) Show trends
(i) Robust to manipulation
(j) Validity for major hazard risk.

23.3 Major Hazard Risk Indicators at a National Level

The NPD (PSA from 2004) took in 1999 an initiative to develop risk indicators for the Norwegian petroleum industry at the national level [16], including indicators of major hazard risk. This work was unique in its focus at a national level, and it has been important for the development of major hazard indicators for the offshore petroleum industry. This section describes the Risk Level project, with the main emphasis on major hazard risk.

23.3.1 Objectives, Scope of the Work and Stakeholders

23.3.1.1 Objectives of the Work

The objectives of the Risk Level project were that "PSA shall contribute to the establishment of a realistic and jointly agreed picture of trends in HES work which supports the efforts made by the PSA and the industry to improve the HES level within petroleum operations". This implies the following:

- Measure the impact of safety-related work in the petroleum industry, in terms of status and the trends of these actions
- Help identify areas that are critical for safety and for which priority must be given to identifying causes in order to prevent unplanned events and situations
- Improve understanding of the possible causes of accidents and unplanned situations together with their relative significance in the context of risk in order to create a reliable decision-making platform for the industry and authorities. This would enable them to direct their efforts towards preventive safety measures and emergency preparedness planning.

23.3.2   Scope of Work

The RNNP was from the initiation limited to what is considered to be 'offshore activity' and within NPD's (now PSA's) jurisdiction. Vessels associated with off-shore operations were not included in general, only when they are within the 500 m 'safety zone', because they were then considered to be 'offshore activity'. Helicopters were also from the start only considered during landing and take-off, i.e. inside the 'safety zone'. It was after a short time agreed with the Civil Aviation Authority in Norway that the helicopter transport of personnel should be included during the full transport cycle from leaving the onshore base until they returned to the base after completing their work period, and including any shuttling between installations, during the offshore work period. In addition, the inclusion of vessels outside the safety zone was considered, but an agreement with the Norwegian Maritime Directorate was not achieved.

When PSA took over the responsibility for onshore petroleum facilities in 2004, the work was extended to include these facilities, with separate indicators and a separate report. In Norway there are only eight onshore petroleum plants under petroleum safety jurisdiction. Of these eight facilities there are two refineries and three large gas plants, while the rest are relatively limited gas terminals and plants. There are also some additional facilities, such as gas fuelled power plants in association with some of these plants. Onshore facilities are not covered in this book, but many of the principles may be applied to these plants.

23.3.3   Stakeholder Interest

There was significant interest in the work from various stakeholders, workforce representatives, employers and authorities, for the reasons indicated in Sect. 23.1.1. There is a reference group for the work called 'Safety Forum' with participation from all relevant stakeholders. This reference group is considered to be the 'customer' for the work, and it accepts the results and approves the recommendations for future work.

The basic approach adopted in the Risk Level work from an early stage was that of triangulation, i.e. utilising several parallel paths to express the status and trends of HES levels. A decision was also made to use various statistical, engineering and social science methods in order to provide a broad illustration of risk levels. This applied to:

- Risk due to major hazards
- Risk due to incidents that may represent challenges for emergency response planning
- Occupational injury risk
- Occupational illness risk
- Risk perception and cultural factors.

One of the purposes of this broad approach was to establish trust by stakeholders that the work would not be biased in any direction, but present as far as possible an impartial statement of the 'state of affairs' in the offshore industry. For a company wishing to establish major hazard indicators, it may be equally important to have trust from different stakeholders in the set of indicators selected.

23.3.4   Basic Concepts and Overall Approach

The basic concepts of the Risk Level work are discussed extensively in Ref. [16]. There can be a misconception that there is a fully objective way of expressing risk levels through a set of indicators. This implies that expressing the 'true' risk level is just a matter of finding the right indicators. However, this is a misconception, as no single indicator can express all the relevant aspects of health, environment and safety. There will always be a need for parallel illustrations by invoking several approaches. We would also argue that there is no 'true' risk level, but that that discussion is outside the scope of this presentation [17].

It should be emphasised that the Risk Level approach is based on the principle that statistical indicators (such as the number of leakages) are insufficient in order to present an adequately broad basis for evaluation and assessment of safety levels and trends. However, the focus here is on statistical indicators, maybe to an extent that the basic principles as stated above may be misinterpreted. It is nevertheless stressed that triangulation and a broad basis is the fundamental approach. A brief overview of the different types of indicators is given below. More details are presented by Tharaldsen et al. [18] for safety culture and risk perception and Vinnem et al. [16] for major hazard risk.

23.3.5   Major Hazard Risk

The major hazard risk components for employees working on offshore installations are the following:

- Major hazards during stay on installations
- Major hazards associated with the helicopter transportation of personnel, for crew change purposes every two weeks and any shuttling between installations.

For the risk associated with major hazards on installations, the following types of indicators are been developed in the RNNP work:

- Indicators based on the occurrence of incidents and near-misses (i.e. precursor events)
- Indicators based on the performance of the barriers installed in order to protect against these hazards and their consequence potential.

The aforementioned indicators and barrier indicators are discussed in separate sections below. None of these indicators was readily available as data sets collated by the industry when the RNNP work started.

The same principle was applied to the indicators for risk associated with helicopter transportation. After a search for a reliable source, it was found that helicopter operators had all the required data registered, from which reliable indicators could be established, mainly for the occurrence of incidents and near-misses. Only two companies had for a long time been operating the helicopters used for all traffic to Norwegian offshore installations; thus, it was straightforward to cooperate with the relevant operators of all helicopters. Later, a third operator entered the arena.

23.3.6   Other Indicators

The emphasis in this section is on indicators for major hazards. Indicators for other hazards are also used in the Risk Level work, including:

- Indicators based on the occurrence of incidents with emergency response challenge
- Indicators based on the occurrence of occupational injuries
- Indicators based on the exposure of employees to selected hazards with occupational illness potential
- Indicators for HES culture, based on questionnaire surveys and interviews with key stakeholders representing the different parties in the industry.

The latter indicators largely correspond with the indicators proposed in a resilience engineering context [19].

23.3.7   Data Sources

23.3.7.1   Precursor Data

RNNP has collected major hazard precursor data for more than 20 years, covering the period 1996 until present.[1] The relevant major hazards for personnel on the installation have been addressed in QRA studies, which were one of the main sources when indicators were identified. Table 23.1 presents an overview of the categories of the major hazard precursor events (called 'DFUs') included in RNNP with major hazard potential. The values shown represent all production installations and mobile drilling units that have operated on the NCS in that period.

---

[1]2011 was the last year with available detailed data at the time of the manuscript preparation, 2017 is the last year with available overall data.

Table 23.1 Overview of major hazard precursor event categories ('DFUs')

DFU no.	Description	Average no/yr 2007–11
1	Non-ignited hydrocarbon leaks	12.8
2	Ignited hydrocarbon leaks	0
3	Well kicks/loss of well control	17.4
4	Fire/explosion in other areas, flammable liquids	2.8
5	Vessel on collision course	22.4
6	Drifting object	0.75
7	Collision with field-related vessel/installation/shuttle tanker	0.4
8	Structural damage to platform/stability/anchoring/positioning failure	6.8
9	Leaking from subsea production systems/pipelines/risers/flowlines/ loading buoys/loading hoses	2.0
10	Damage to subsea production equipment/pipeline systems/diving equipment caused by fishing gear	3.2

Table 23.1 shows that there are significant annual values precursors for DFU1, DFU3 and DFU5. DFU5 represents external merchant vessels on collision course, and the occurrence of such events is thus not representative for the safety management on the installation.

All incidents are reported by industry through various channels. DFU1, DFU5 and DFU6 are reported using special formats for RNNP. The remaining DFUs are also reported by industry to the authorities through other reporting schemes, from which they are extracted for use in RNNP. The volume of installations and associated manhours on the NCS, which are the basis of RNNP, are presented in Table 23.2.

It should be noted that the total number of installations is higher than that shown in Table 23.2 because of the way in which 'complexes' consisting of more than one installation are considered. A complex is considered to be one installation in RNNP irrespective of the number of bridge-linked installations. Further, the number of mobile installations is the equivalent number of mobile installation operational days divided by 365, because many mobile installations spend fractions of a year in the

Table 23.2 Overview of installations and manhours in the Norwegian sector

Installation type	Number of installations, 2011	Manhours, 2011
Fixed production installations	21	31 183 429
Floating production installations		
– with remote wells	16	
– with wells on or below installation	5	
Production complexes	10	
Normally unattended installations	17	
Mobile installations	30	13 132 836
Sum	93	39 665 547

Norwegian sector and the remaining part of the year in other national sectors. It should be noted that a typical offshore working year consists of 1460 manhours (according to agreements between the industry and the unions); over 27,000 manyears applies to production installations and just below 9,000 manyears to mobile installations.

23.3.7.2 Barrier Data

RNNP has collected barrier data for major hazards since 2002; however, data collection was not fully representative in 2002, and this year is often omitted when the data are used for statistical analysis. With one small exception, only technical barrier indicators have been covered. As for the precursor data, the values shown represent all production installations that operated on the NCS in the study period.

All barrier data are reported by industry using a special reporting format for RNNP. Table 23.3 shows an overview of the barrier elements included in RNNP for hydrocarbon-related hazards and the average number of periodic tests reported annually to RNNP.

23.3.7.3 Other Data Sources

Other data sources in addition to major hazards in the Risk Level work are the RNNP questionnaire survey, which has been conducted biannually since the first survey in early 2002. Data collection has also included serious occupational injuries on installations, following the definition of serious occupational injury in Norwegian regulations [20]. Further, noise exposure data have been supplied for each installation, for personnel with and without ear protection. Moreover, data for falling objects have been reported for each installation. These data are primarily used for the analysis of circumstances and causal factors, because frequency data are strongly influenced by different reporting practices, with respect to reporting of falling objects.

Table 23.3 Overview of major hazard barrier elements and average annual volume of tests

Barrier element	Average no of tests 2009–11
Fire detection	52,037
Gas detection	30,304
Riser ESDV	1,990
Wing/master valves	12,536
DHSV	6,078
BDV	3,531
PSV	12,884
BOP	4,030
Deluge valve	2,634
Fire pump start	7,165

23.3.8 Precursor-Based Indicators for Major Hazard Risk

The development of event based indicators in the Risk Level approach was discussed thoroughly by Vinnem et al. [16]. Several aspects were also discussed by Heide [21]. Some of the ideas and concepts are summarised here.

23.3.8.1   Individual Indicators

Figures 6.4 and 6.5 show the trend for hydrocarbon leaks with a flowrate above 0.1 kg/s, normalised against installation years, for production installations. The values are in some of the diagrams referred to as a 'relative risk indicator' because the trends are the important values rather than the absolute values.

23.3.8.2   Overall Risk Indicator

The typical picture from the individual indicators is that some will show an increase, some may show a decreasing trend, and several will usually fall within the prediction interval (middle part in the diagrams), i.e. no significant trend can be concluded. It is therefore an advantage to have an overall indicator that can balance the effects of these individual indicators, in order to identify overall development.

Although the development of the overall risk indicator was documented in some detail in Ref. [16], the main principles are summarised here. The precursor events are aggregated into the overall risk based on assigning a 'weight' to each DFU category, which represents the average expected number of fatalities, given a precursor event for each type of installation, and each severity category of the precursor events. DFU1, DFU3, DFU8 and DFU9 are subdivided into a number of sub-categories, according to severity. The weights also express barrier reliability, but only indirectly, and only as an industry average.

Finally, the values are normalised by the division of the number of manhours. Installation types are fixed production, floating production with wells on the installation or directly below, floating production with remote wells, production complexes (two or more bridge-connected installations), normally unattended installations and mobile units (for drilling and accommodation purposes).

The basis for the definition of weights is a large number of QRA studies, that cover all installation types, and provide a good basis for the definition of the weights. The implication of the weighting is that the dimension of the overall risk indicator corresponds to a FAR, although all the values are presented as relative, with the value in 2000 (which is the average of the period 1998–2000) defined as 100.

One might think that the overall indicator would always fall within the prediction interval, but this is not the case because of the large differences in weights applied to different categories, which implies that some categories dominate others.

One of the categories that has a high contribution is hydrocarbon leaks, which implies that trends regarding such leaks will influence the trends in the overall risk indicator.

One of the characteristics over more than ten years is that there have been one or two very severe precursor events per year, with a large fatality potential. Such precursor events are given an individual weight assignment, according to the specific circumstances. These events have a significant effect on the value of the overall indicator, which is why the trend has been quite stable for about ten years, with some fluctuations. This is also the reason why a rolling three-year average was found to be useful in order to reduce the effects of these variations.

It must be emphasised that the overall risk indicator is not a prediction of future risk values, but rather an indicator that presents precursor events in the past weighted by their fatality risk potential.

An illustration of the overall indicator is shown in Fig. 23.1, which presents the trend of the overall indicator for all production installations, normalised against manhours. The diagram shows a significant reduction after 2004. The overall impression is that the level was high until 2004, before an abrupt fall in 2005, and a stable level thereafter, with some variations. There has again been a reduced level since 2013, except in 2015.

Figure 23.1 shows annual values as well as three-year rolling averages in order to reduce the effect of random variations from one year to the next. Three-year rolling average values were implemented in the Risk Level report in 2005 and have continued thereafter. The latest development is that annual values as well as three-year rolling average values offer the best insight.

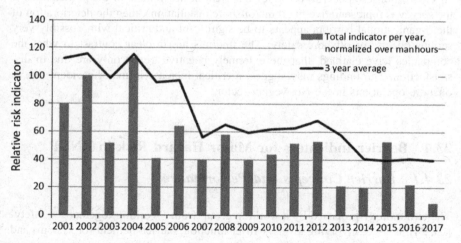

Fig. 23.1 Total risk indicator, production installations, normalized against manhours. *Source* PSA

23.3.9 Organisational Indicators on a National Level

The Risk level project includes a questionnaire survey every second year, with emphasis on HES climate, working environment conditions and perceived risk by the employees. Some of the indicators may also be interpreted as indicators of the organisational RIFs. The results have shown an exceptionally clear negative trend from 2013 through 2015 and until 2017. This applies to HES climate, perceived risk, working environment and health issues. Initially it was assumed that the negative trend mainly was due to employees who have experienced downsizing or reorganization.

A broad interview survey with offshore employees (about 80 participants on five installations) has been conducted in 2018 in order to attempt a verification of the initial assumption. The study [22] has revealed that significant organizational changes appear to have resulted from the cost cutting over the last few years relating to:

- Organizational changes and increased cost focus
- Maintenance planning
- Personnel resources and task allocation
- Psychosocial working environment and workloads
- Challenges associated with maintaining competence
- Co-operation between offshore and onshore personnel, decision-making and priority setting
- Procedures and compliance with working practices.

The responses appear to be quite consistent between most of the participants. If the survey is representative of all the offshore installations, then the deterioration of the organizational barriers appears to be significant and critical with possibly very serious effects for offshore safety. The findings are in clear contrast to what the companies have claimed, that the extremely negative trends only are due to dissatisfaction. The findings may suggest a critical increase in major accident risk in offshore operations in the Norwegian sector.

23.4 Barrier Indicators for Major Hazard Risk in RNNP

23.4.1 Barrier Concepts and Performance

The terminology proposed by a working group from 'Working together for safety' [23, 24], is used, involving levels, barrier function, barrier element (or system) and barrier influencing factor:

• Barrier function	A function planned to prevent, control, or mitigate undesired events or accidents
• Barrier element	Part of the barrier, but not sufficient alone to achieve the required overall function
• Barrier influencing factor	Factors that influence the performance of barriers

The term 'barrier' is as such not given a precise definition, but it is used in a general and imprecise sense, covering all aspects. PSA regulations require the following aspects of barrier performance to be addressed:

- Reliability/availability
- Effectiveness/capacity
- Robustness (as opposed to vulnerability).

Effectiveness and robustness are performance aspects that vary more slowly than reliability/availability; as such, they are unsuitable for regular testing, in the same manner as reliability/availability are. The characteristics of barriers and barrier management are discussed in more depth in this chapter.

PSA regulations apply to all types of barriers from an MTO perspective, namely technical barriers as well as operational barriers. Technical barrier elements are obviously technical systems that prevent incidents and accidents or limit the consequences of them. Operational barrier elements are all other barrier elements, where human operators perform the preventive or protective functions. Barrier elements require some kind of active or passive action, as also argued by Sklet [23]. With this understanding, there is no organisational barrier element, but a range of organisational factors may influence the performance of barrier elements (i.e. barrier influencing factors). Barrier elements are therefore either technical or operational, i.e. carried out by humans. The main emphasis in this section is on technical barrier elements in line with the scope of the Risk Level work. Operational barrier elements are discussed in Sect. 23.8.2.

23.4.2 Barrier Indicators in RNNP

The adoption of barrier indicators in RNNP is described in Ref. [16]. The full list of technical systems on offshore installations for which RNNP collects data was as follows at the end of 2011:

- Fire detection
- Gas detection
- Emergency shutdown valves on risers/flowlines

  - Closure tests
  - Leak tests

- Wing and master valves (X-mas tree valves)

  - Closure tests
  - Leak tests

- DHSVs
- BDVs
- PSVs
- BOP

  - Surface BOP

    Drilling BOP
    Wireline BOP
    Coiled tubing BOP
    Snubbing BOP

  - Subsea BOP

    Drilling BOP
    Wireline BOP
    Coiled tubing BOP
    Snubbing BOP

- Deluge valves
- Fire pump start.

Typical fractions of the number of failures in relation to the number of tests are shown in Figs. 23.2 and 23.3, where the distinction between the closure and leak tests of isolation valves has been disregarded. These test data are based on reports from all production operators on the NCS.

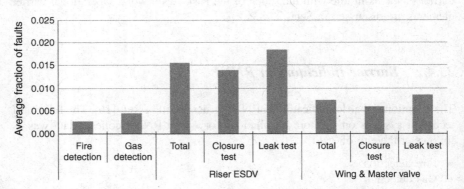

Fig. 23.2 Average fraction of failures for detection and isolation barrier elements, 2010

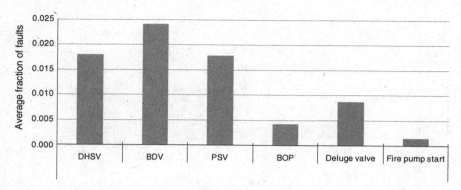

Fig. 23.3 Average fraction of failures for safety valves and fire water barrier elements, 2010

23.4.3 Availability Data for Individual Barrier Elements

The overall fraction and average fractions are presented, the former being all faults for all installations over all tests for all installations. All tests contribute equally to the average with this calculation. The average fraction is the fraction of faults over tests for each installation individually, which is then averaged across installations. With this average, an installation may have many tests or few tests, but the contribution to the average is the same in any case.

The ratio between the number of test failures and tests is an explicit expression of the on-demand unavailability for the component in question, which is dependent on the test frequency, and will also reflect those "environmental" aspects that influence the performance of the component, such as the management and human aspects of the maintenance work. However, the on-demand unavailability that may be calculated on this basis does not take into account those failures that are not detected by functional tests.

In general, the relative number of failures can be claimed to be at the same level as the industry's availability specifications for new installations, with the exception that not all potential failure sources are tested. However, some values over several years have been above the availability values specified for new installations.

Figure 23.4 shows an illustration of one specific barrier element, where all individual installations are compared in an anonymous manner. Large variations between the interval values and annual values are shown for some installations. It is further illustrated how some installations have very high unavailability fractions; some of these have had high values over several years.

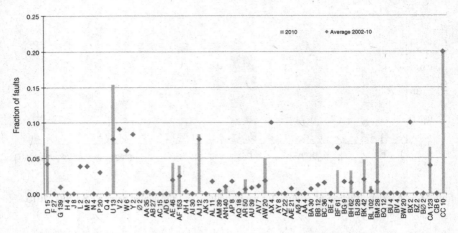

Fig. 23.4 Average fraction of failures for 2010 and average 2003–2010 for riser ESV

23.4.4   Causal Factors

The data collected are not sufficiently detailed to allow for the identification of RIFs. The circumstances of testing are not collected; however, some information on testing frequencies has from time to time been available. This has indicated that the fraction of failures and test intervals are strongly correlated, as expected according to theory.

23.5   Lessons Learned from RNNP

This section summarises some of the main lessons learned from the Risk Level work. This serves as an introduction to the more fundamental discussion of major hazard risk indicators later. It should as an opening be noted that for major hazard risk indicators developed in the mid-1990s, the following general objectives have already been stated [2]:

- The indicators should address a range of incidents, from insignificant near misses to the most severe and complex accident sequences.
- The hazard potential of an incident should be indicated, in addition to the actual outcome of the incident.
- When it comes to indicators that relate to potential major accidents, performance indicators should be capable of reflecting the importance of the incident in question. Unauthorised hot work in a safety critical area is more critical than the failure of one out of three parallel (100% capacity) fire water pumps.

- Indicators reflecting individual systems are required alongside a limited set of overall indicators that present an aggregated picture. Thus two levels of indicators are natural, with the highest level needed for the presentation of an overall indication of the safety level related to potential major accidents.

Experience with major hazard indicators over 20 years has demonstrated that these objectives should still be valid.

23.5.1 Approach to Risk Projection

One of the most important aspects of the Risk Level approach is the triangulation between several approaches to risk assessments, including technical and social science approaches. This is natural as one of the main objectives of the Risk Level work was to build trust among different stake-holders, as discussed in Sect. 23.3.3.

Another vital aspect of the Risk Level approach is that a wide spectrum of HES aspects is covered, including major hazards, occupational hazards and illness, helicopter transportation hazards, physical and psychosocial working environment aspects, attitudes and safety culture and perceived risk. Again this is crucial for building trust as mentioned above, but may be less vital for a company wishing to establish major hazard indicators.

The Risk Level approach has focused on data collection in the same manner for long periods in order to present the long term-trends for major hazard indicators. Conditions relating to major hazard prevention need to focus on long periods, because there may be many years between each occurrence of major hazards, even for a large corporation.

The Risk Level work has included a mixture of precursor-based and barrier indicators. This is a vital aspect of how to define indicators for major hazards. From the start, the focus was on precursor based indicators, while barrier indicators were introduced a few years later. This was a useful approach, because it is often easier to establish indicators based on the collection of the occurrence of accident precursors.

23.5.2 Relevance of Precursor Indicators

One of the challenges in the establishment of major hazard indicators is to define indicators that reflect protection against major hazards in a realistic manner. This often needs more than one indicator, for instance precursor based as well as barrier indicators. It can be particularly challenging to find indicators that provide a realistic prediction of major hazard risk. In the Risk Level work, this has been particularly challenging for the helicopter transportation risk, particularly in the last ten years, with the introduction of new helicopter models that have higher robustness against major accidents. The indicators established for less robust helicopters do not reflect major accident prevention equally well for the new models because of the indicators' inability to reflect the increased robustness. The consequence categories used in the company's event registration and classification are based on a 5 × 5 risk matrix and are too coarse to be able to reflect the increased robustness. The solution to this challenge has been to assess the most critical precursor events in relation to the number of remaining barriers, in order to have a consistent classification.

23.5.3   Suitability of Barrier Indicators

The barrier indicators have been successful in the Risk Level approach, in the sense that they have helped the industry focus on any negative trends or components that have too high failure fraction when tested. It may be argued that the industry should be well positioned to perform these evaluations without the help of RNNP, but the experience so far has demonstrated that the extra focus coming from the Risk Level work has been necessary. Barrier indicators have been successful in this respect.

When trends are presented for different barrier elements, it is often not possible to identify an overall trend, some barrier elements may have a positive trend, whereas others may have a negative trend or no trend at all. In the same way as the need for an overall incident indicator, there is also a need for an overall barrier indicator. This is discussed in Ref. [16].

23.5.4   Normalisation of Precursor-Based Indicators

All precursor-based indicators are presented normalised as well as not normalised. It should first be noted that the trends are virtually identical between the two sets of curves, because the volume of normalisation data does not change abruptly; changes are gradual over several years. The following normalisation parameters are used for the major hazard indicators:

- Manhours
- Number of installation years
- Number of wells drilled (for drilling associated hazards)
- Number of helicopter flight hours/person flight hours (for helicopter associated hazards).

Manhours are used as the primary parameter for normalisation. This implies that for instance the overall indicator has the dimension of FAR values, although this is not used actively in the discussions. Manhours constitute an important parameter for several hazard mechanisms. Several hazard types are correlated with activity levels, thus making manhours a natural choice. However, manhours are unsuitable for equipment malfunctions.

The number of installation years is also used extensively, although this does not discriminate between more or less complex installations. It has been suggested that production volume could be used as an indication of complexity, but this is mainly a function of reservoir characteristics, and thus not a good indicator for complexity. Production volume is therefore only established at the national level but is also available at the installation level. It has been considered that the number of leak sources (such as valves, flanges, welds, instruments, etc.) might be more suitable. Such information is available in QRA studies, but it has been found to be demanding to establish on a general basis. A case study for hydrocarbon leaks is presented in Sect. 6.3.5. The weight of processing modules has been used in one study as an approximation of complexity, as it was found that the weight of these modules and number of leak points correlated reasonably well for those installations where both sets of data were available [25].

23.5.5 Ability to Distinguish Between Companies and Installations

A further development of the methodology that may be considered in the future is to address each installation individually. The methodology is at present based on the categorisation of all installations into six broad groups (see Table 23.2), according to their overall functions and protection of personnel.

If each installation were considered individually, this would give us the opportunity to integrate fully precursor-based indicators and barrier indicators. This would in theory create an opportunity to establish overall indicator values for each installation.

It can be argued that the Risk Level project focuses on the national level and that the consideration of each installation individually is outside the scope of the project, which to some extent is correct. However, the consideration of each installation individually has two benefits:

- The representation of each installation's contribution to the national level will be more precise
- The actual results for individual installations offer information that the project may present to the operators of individual installations for their use and follow-up.

However, the main complication with such an approach is that the occurrence of major hazard precursors is a relatively rare event, typically in the order of once per year on average. This implies that there would be large fluctuations even on a three-year rolling average basis, which would severely limit the opportunity to draw conclusions for individual installations.

23.6   Precursor Events as Major Hazard Indicators

23.6.1   Proposed Approach to the Selection of Individual Indicators

23.6.1.1   PFEER Requirements

Since 1995 the operator of a UK offshore installation has been required to establish performance indicators. The number of indicators that operators have established for this purpose is variable, ranging from a few critical parameters to the total number of safety functions and actions on the platform. The requirements for performance indicators in PFEER [4] are as follows:

- PSs should relate to the purpose of the system as well as to the equipment and procedures that they describe.
- PSs should be described in terms of functionality, survivability, reliability, and availability.
- The indicators should be measurable and auditable.

These requirements are taken to imply that the number of performance indicators required to satisfy PFEER intentions is a low number, and should be limited to:447

- Systems and functions that can be shown to have a significant impact on risk level.
- Parameters that are easily observable and measurable during normal operations and emergency exercises.

It is argued on this basis that the definition of performance indicators for more than 20–30 systems (with up to several characteristics for each system) is a misunderstanding of the intentions of the PFEER regulations. It is also believed that this is detrimental to the intentions in the sense is impossible to distinguish between important and unimportant violations. This interpretation of the PFEER requirements forms the basis for the approach that is discussed in the following.

However, when it comes to selecting 'safety-critical elements' as introduced in the UK Design and Construction Regulations [26], then a considerably wider selection of systems is appropriate in accordance with the definition of a 'safety-critical element'.

23.6.1.2   Selection of Parameters

A starting point for the definition of parameters to be followed up during PFEER monitoring is the list of nodes in the event trees used in the installation QRA. This may result in the parameters as listed in Sect. 23.1.2.3.

Not all the parameters will however, have the same importance for the overall risk levels. One way to focus on those with the greatest significance is to carry out sensitivity studies for all conditional node probabilities, in the event trees in order to determine those that make the most significant contributions.

Such sensitivity studies have been carried out by many companies, sometimes extensively, but often with a rather limited scope. When comprehensive studies were carried out [27], some clear observations could be made with respect to the sensitivity of the parameters:

- Reduction in the extent of hot work hours has a clear effect on risk to personnel and material damage, the latter reduction being the most extensive.
- Reduction in the maximum overpressure from an explosion has a limited effect on the risk to personnel, but a very extensive effect on the material damage risk.
- Reduction in ESD system unavailability has some effect on the material damage risk, but not on personnel risk.

Changes in other conditional probabilities (gas detection, operator intervention, manual fire fighting, emergency shutdown) had very little, if any, effect on risk, especially risk to personnel. This could be taken to imply that the standards of these systems are adequate with respect to the protection required.

These results from the sensitivity studies are based on the change in overall risk. In a later paper [28], the results for small leaks were also examined. This study showed that additional systems and functions are important in order to limit the risk associated with small leaks. The following were shown to be the most important:

- Detection (automatic and manual) of gas leaks
- Operator intervention to isolate leaks
- Manual fire fighting
- Emergency shut-down system.

Several nodes in the event trees are associated with emergency actions and accident combatment in the case of an emergency. Training will be used to ensure proficiency in some of these actions; nevertheless, performance for some of these actions is not observable and they are unsuitable for measurement during normal operations or emergency exercises. Only those emergency actions that have observable performances were used.

The definition of a set of PSs should start with recognition that the important parameters for risk to personnel are not necessarily the same as those that are important for risk to assets. Risk to personnel should, however, be given the highest importance. Further parameters are important if one considers overall risk level as opposed to the risk associated with small (and frequent!) leaks. It is important to pay a high level of attention to the control of most frequent leaks, because this will also give operators confidence in the systems. It is also important to consider some of the aspects that subjectively are considered to be important, even if on a 'pure' quantitative basis they would not be thought of as important.

The performance indicators for the Ekofisk 2/4-T platform were presented in Sect. 23.1.2.3. The following steps are described [6] for the selection of indicators:

- Select technical indicators based on effect studies in the applicable QRA study.
- Develop indicators relating to organisational aspects based on theoretical studies combined with screening through interviews with operational personnel.
- For both the technical and organisational measures pay attention to the changes that could occur in the period of interest. Further, consider the extra work required to develop measurement systems if a parameter is not already being reviewed.

In the Ekofisk case, the indicators listed show that five out of the nine indicators are associated with organisational aspects.

The PFEER-inspired process placed less emphasis on organisational aspects, and it is interesting to consider the difference in emphasis. Three of the indicators used for the Ekofisk platform are associated with the number of personnel in different groups. This is not considered to be suitable for performance monitoring, because the number of personnel in a QRA will often be represented by average values. Apart from these, other organisational aspects are related to the use of hot work as potential ignition sources.

For the selection of performance indicators for organisational aspects, the following principles are suggested:

- First, identify all indicators that are of an organisational nature and that are included directly or indirectly in the event trees.
- Second, focus on possible indicators that relate to 'non-QRA' scenarios, with the main emphasis on the functions needed to safeguard personnel.

The following is the proposed procedure for the selection of indicators for performance monitoring:

1. Establish a 'gross list' of all possible parameters that could be defined as performance indicators. This should include those that may be inferred directly or indirectly from event trees as well as those based on more qualitative and subjective arguments by platform personnel.
2. Carry out sensitivity studies for all aspects rooted in the event trees and determine the relative sensitivities of the parameters for:

   - Overall risk
   - Risk due to small leaks
   - Risk due to escalation (by fire and explosion separately) of small leaks

3. Omit all parameters that do not influence the risk result and those that cannot be monitored in a meaningful way during normal operations
4. The remaining list should include the parameters to be followed up and the reporting of performance indicators. Monitoring intervals and reporting methods should also be identified.

23.6.1.3   Defining an Overall Indicator

The Ekofisk study [6] suggests that overall indicators could be developed but the study report is limited to the follow-up of each indicator separately. This section presents a possible approach to defining overall performance indicators. In Step 2 above, the term 'relative sensitivity' needs to be defined. Let the risk level of the installation be denoted 'R', while a specific parameter in the sensitivity study has the value 'P'. If $p = \epsilon P$ and $r = \epsilon R$, then:

- If a parameter value is changed $p_s\%$ as input to a sensitivity study,
- and if the relevant risk parameter changes by $r\%$,
- then the relative sensitivity, S_R, of parameter 'P' with respect to the risk level 'R', is calculated as follows:

$$S_R = \frac{r}{p} \tag{23.1}$$

The procedure to establish the overall performance indicators, starts with the four steps shown above, and continues as follows:

5. Classify all possible parameters as one of the following categories aa, ab, ba, bb or bc (based on the following definitions from Ref. [29]):

   (a) Probability reducing measures, with the following order of priority:

      (aa) measures that reduce the probability of a hazardous situation occurring
      (ab) measures that reduce the probability of a hazardous situation developing into an accidental event

   (b) Consequence reducing measures, with the following order of priority:

      (ba) measures relating to the design of the installation, to load-bearing structures and passive fire protection
      (bb) measures relating to safety and support systems, and active fire protection
      (bc) measures relating to contingency equipment and contingency organisation.

6. For the parameters derived in QRA, choose the highest relative sensitivity value calculated for parameters that fall in the categories 'aa' and 'ab'. For parameters in categories 'ba' and 'bb' use the highest sensitivity value for overall risk due to small leaks. The 'bc' category parameters are omitted from further consideration.
7. For the organisational factors selected, assign a relative sensitivity equal to the highest values found for the technical parameters.

8. Normalise all the chosen sensitivity values for the parameters such that they will sum to exactly one.
9. An overall indicator value is created by adding the actual observed parameter values weighted by their normalised relative sensitivity values.

Steps 1–9 are carried out separately for risk to personnel and risk to assets. The two overall indicators established should then be kept separate. With respect to Steps 5–7, the following comments can be made:

- Category 'bc' consists of the contingency aspects of an organisational nature, which in accordance with the general principles have the lowest priority. This is the reason why they are omitted so that the number of measures is reduced to the lowest number practically possible.
- The organisational factors (established in Step 1) should be given the highest sensitivity values in order to motivate platform personnel.
- The distinction between the approaches adopted for the weighting factor calculation for 'aa' and 'ab' in contrast to 'ba' and 'bb' is made in order to reflect their differences in importance.

The use of the procedure is illustrated by a case study, based on the sensitivity studies reported in Ref. [28]. The following parameters are addressed in the case study:

- Hydrocarbon gas/liquid leak
- Failure of leak detection
- Failure of operator to intervene
- Ignition internally in module or externally
- Explosion overpressure, strong explosion
- Failure of fire detection
- Failure of manual combatment of small fire
- Failure of ESD operation
- Fire water unavailability
- Escalation to neighbouring equipment
- Time to structural collapse
- Availability of ESD valves
- Availability of deluge valves.

The availability of the deluge valves is not a node in the event tree, and thus no explicitly sensitivity was available for this parameter. The unavailability of the deluge valve is on the other hand part of the basis for the fire water unavailability. The sensitivity may therefore be determined indirectly from the sensitivity of the fire water parameter.

23.6.2 Weights for Individual Indicators

When individual indicators are used to produce an overall indicator, there is a question of which weights to apply to the individual indicators. The overall indicator shown in Fig. 23.1 is based on weights established from QRA studies for six categories of indicators. These weights are assumed to represent the statistical risk picture. This also implies that the weights span several orders of magnitude. These weights are fixed values in the Risk Level project and are not dependent on the data reported for barrier performance (see Figs. 23.2 and 23.3).

A Master thesis from the University of Stavanger in 2006 compared different weighting systems from different Norwegian companies and interviews with experts [30]. One of the weighting systems is presented in Table 23.4. The basis for the weights was subjective evaluation by company risk assessment experts. One observation is that the variations between individual weights are less than one order of magnitude, which is different from the variations between the weights in the Risk Level project, which span several orders of magnitude.

Table 23.4 Weights used by one company for aggregating individual indicator values

No.	Description	Weights
PS 1	Containment	9
PS 2	Natural ventilation and HVAC	6
PS 3	Automatic gas detection	7
PS 4	Emergency shutdown	7
PS 5	Drain system	3
PS 6	Ignition source control system	8
PS 7	Automatic fire detection	5
PS 8	Blowdown and flare	7
PS 9	Fire water and firefighting systems	6
PS 10	Passive fire protection	6
PS 11	Emergency power incl. UPS and emergency lightning	3
PS 12	Process safety	6
PS 13	Alarm and communication	3
PS 14	Escape and evacuation	6
PS 15	Explosion barriers	6
PS 16	Offshore cranes	3
PS 17	Drilling and well blowout prevention	7
PS 18	Ballast and position keeping	5
PS 19	Ship collisions	5
PS 20	Structural integrity	5

23.7   Barrier Indicators for Major Hazard Risk

23.7.1   Suitability of Leading Barrier Indicators

The ideal situation would be to have leading indicators for major hazards with a significant reporting volume. The volume is essential such that there will be some variations in the indicators, which should be a useful way of attracting attention and maintaining awareness. This view is also shared by Hale [31], who argues that the indicators should be sensitive to changes and show trends.

We have shown that lagging indicators are unsuitable for major hazard focus for several reasons, not least because there will be a rather low average number of precursor events per installation per year, typically one to three, for an entire installation.

Barrier indicators have been considered to be suitable, because there is a lot of data from testing barrier elements. However, we need to examine these data somewhat more closely, in order to establish how suitable they are.

Tables 23.5 and 23.6 present overviews of the data reported from testing barriers on the two installations. Installation 1 has a significant volume of test data as well as fault data in the entire period (see Table 23.5). This implies that indicators will vary from quarter to quarter, which will provide a good way of keeping focus on major accident prevention. Installation 1 is an old installation that has several average values above the average for the industry in Norway.

Installation 2 by contrast, is a new installation that has average fractions of faults over the number of tests mainly below the industry average. Table 23.6 shows that the number of faults is low on a annual basis. For this installation the number of faults is almost as low as the number of precursor events.

Table 23.5 Barrier tests (x) and faults (y) in (x–y), Installation 1

	2004	2005	2006	2007	2008
Fire detection	1332–4	936–4	292–0	1109–1	1114–1
Gas detection	1400–4	334–3	414–2	712–12	840–2
Riser ESDV	3–0	4–2	9–2	8–0 4–0[a] 4–0[b]	12–1 6–0[a] 6–1[b]
W&M isolation valves	74–1	60–0	144–1	136–2 68–0[a] 68–2[b]	128–8 46–4[a] 82–4[b]
DHSV	0–0	106–6	167–9	176–15	151–7
BDV	0–0	22–0	222–19	71–8	162–17
PSV	0–0	241–16	288–13	235–6	304–9
BOP	0–0	0–0	194–0	355–1	368–5
Deluge	102–0	81–0	57–1	27–1	52–1
Fire pump start	312–0	232–0	155–0	155–1	156–0

[a]Closure test; [b]Leak test

Table 23.6 Barrier tests (x) and faults (y) in (x–y), Installation 2

	2003	2004	2005	2006	2007	2008
Fire detection	1065–17	1122–12	1091–6	1135–17	1135–1	1140–0
Gas detection	539–19	543–9	547–5	568–9	569–0	561–1
Riser ESDV	17–1	17–0	48–5	62–1	248–1 124–0[a] 124–1[b]	160–0 80–0[a] 80–0[b]
BDV	0–0	33–4	103–0	90–0	153–1	111–2
PSV	0–0	0–0	277–17	472–14	312–6	304–20
Deluge	58–0	45–0	32–1	48–0	32–0	55–0
Fire pump start	208–0	208–2	155–0	103–0	104–0	104–0

[a]Closure test; [b]Leak test
Reprinted from Ref. [32], with permission from Elsevier

Installation 2 had three faults during tests in each of 2007 and 2008, when well barriers and PSV were disregarded. The reasons why these barriers were disregarded was that the tests were carried out by specialist subcontractor personnel in the well barrier case and by specialist personnel in a workshop environment in the PSV case. For the reporting of barrier indicators, the use of specialist personnel is not a problem, but with respect to the general awareness among the workforce, the use of specialist subcontractor personnel implies that the experience is less well known generally.

It could be added that Installation 1 had 25 and 27 faults during testing in 2007 and 2008, respectively, even when well barriers and PSV were excluded. This implies that more data would help maintain a higher awareness. The number of faulty tests in the different categories is shown for Installations 1 and 2 in Figs. 23.5 and 23.6.

Fig. 23.5 Number of faults recorded for Installation 1

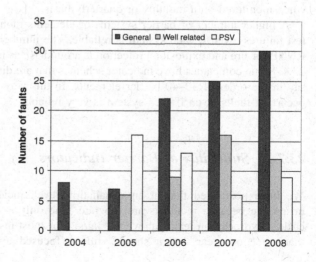

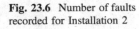

Fig. 23.6 Number of faults recorded for Installation 2

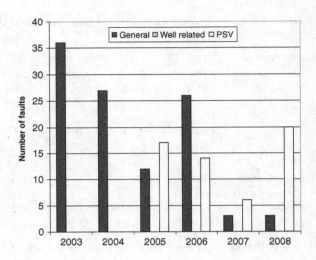

However, these arguments should not be taken as a sign that we favour installations that have a high number of faults during testing. The ideal situation is obviously to have few or none faults, from a major hazard prevention point of view.

It should be noted that the development of barriers in general in Fig. 23.6 demonstrates that Installation 2 has shown significant improvement over the period with the collection of barrier performance. It is not possible to claim that the follow-up of performance is the main reason for this, but it is likely that the follow-up has contributed to this positive development.

If we accept the assumption that a significant volume of test failures enhances motivation and awareness, then we need to have a reporting scheme that implies a certain level of test failures. Such a volume of test failures would imply that test failures are experienced regularly and thus that indicator values vary frequently when monitored on a monthly or quarterly basis.

A higher number of barrier systems or elements monitored would imply more test failures, even if each system is reliable. The number of systems monitored in RNNP for fire and explosion protection is around 10, as indicated in Figs. 23.3 and 23.4. Some companies have indicator schemes that are distinctly more voluminous, say in the order of 30–40 barrier elements. In this way there should be regularly occurring faults, even if each system is very reliable.

23.7.2   Suitability of Barrier Indicators

We have postulated that at an installation level, incident based indicators are unsuitable because incidents are too rare. The only exception is the number of critical hydrocarbon leaks (above 0.1 kg/s). For most installations this will be less than once per year, but it should still be focused upon, because of the high awareness associated with hydrocarbon leaks and their potential to cause major catastrophes (such as the Piper Alpha accident in 1988 or Texas City in 2005).

Owing to the challenges with precursor based indicators, the main effort at an installation level should be on barrier indicators. Testing safety barriers produces a large volume of data suitable for maintaining focus and awareness on major hazard risk. A thorough discussion of these opportunities is the main topic of the remainder of this section.

23.7.3 Extended Suitability of Indicators

With respect to the list of criteria for major hazard risk indicators in Sect. 23.2.2 we have so far in Sect. 23.7 discussed explicitly and implicitly all criteria except (i) and (j). These last criteria are also in need of assessment.

Seeking robustness in order to avoid manipulation is always a challenge. The precursor-based indicators in the RNNP scheme are the most severe occurrences, implying that there is limited room for manipulation, because most of these occurrences will be very well known by a large group of people. For barrier test indicators, there is somewhat more room for maneuver, especially the interpretation of failure modes in relation to recording. An example from the early phases of the RNNP's initial barrier data collection may illustrate this. The faults defined as reportable are those that are unrevealed until testing; only those faults that would prevent a safety-critical function, such as an isolation valve, closing on demand. Failure modes such as failing to be reset after closure are not safety critical and thus not reportable. However, some operators who wanted increased maintenance budget for their installation included all failure modes in order to have a high fraction of failures and then [hopefully] have a higher maintenance budget. The verification of failure mode classification is therefore crucial.

The final consideration to make is regarding the validity of the indicators. It is contended that the RNNP indicators close to the 'sharp end' are naturally valid. Precursors are closely linked to hazard outcomes and are therefore certainly valid. The barrier indicators we have used are close to the 'sharp end', in the sense that they influence strongly the accident sequence.

The weights of the categories of precursor events (see Sect. 23.3.8.2) express explicitly the degree to which individual precursor indicators will influence overall risk. Weights are also used to calculate the overall barrier indicator (see Sect. 23.3.8.2), which also express the degree to which individual barrier indicators influence overall risk. The RNNP indicators are therefore found to be valid major hazard risk indicators.

23.8   Barrier Indicators at an Installation Level

23.8.1   Technical Systems

This section discusses how technical barrier system indicators function at an individual installation level. It should be noted that RNNP implicitly assumes that such indicators are followed up on an installation and on a company basis by the industry itself. It is known that some companies have implemented such schemes, but there are very few references to published works that discuss any experience from such activity [33]. There has also been work on process safety indicators in the wake of the BP Texas City disaster.

We have used data reported for selected installations to illustrate and discuss the performance of barrier indicators for these installations. These have been selected somewhat arbitrarily and are presented in an anonymous manner. Two examples are presented in Figs. 23.7 and 23.8 for Installations 1 and 2 on the NCS. The values presented are actual values for the anonymous installations, without any adjustments.

Installation 1 is more than 20 years old. Barrier data have been reported since 2004. The fraction of failures for isolation valves on risers has been above the NCS average for all years, except 2007. For DHSVs and BDVs average values as well as values in 2008 for Installation 1 are also above the average for NCS. For the DHSV failure fraction, the trend is increasing over time, possibly because of aging.

Installation 2 is almost 15 years old, and most of its barrier elements show performance above the industry average. The installation has subsea wells, and well-related barriers are therefore not reported for this installation. The failure fractions for fire and gas detectors show values somewhat above the NCS average, and these are at stable levels. The average for Installation 2 in the period is slightly

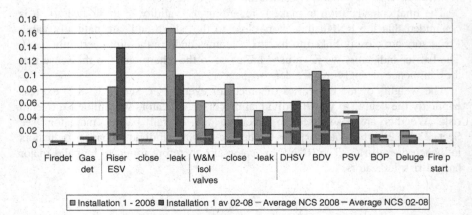

Fig. 23.7 Fraction of failures for Installation 1 and NCS average. Reprinted from Ref. [32] with permission from Elsevier

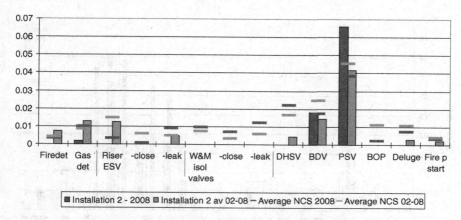

Fig. 23.8 Fraction of failures for Installation 2 and NCS average. Reprinted from Ref. [32], with permission from Elsevier

above the NCS average for riser isolation valves. The trend is falling however, and the value in 2008 is below the average for NCS.

For both Figs. 23.7 and 23.8 all the relevant barrier elements were tested in 2008, and the values shown as zero are therefore zero failures, but not zero tests. This is further illustrated in Fig. 23.9.

Figure 23.10 shows the average values for the company in relation to the Norwegian industry average. The values in 2002 have been omitted because of the uncertainty and arbitrariness of data collection for some companies in that year.

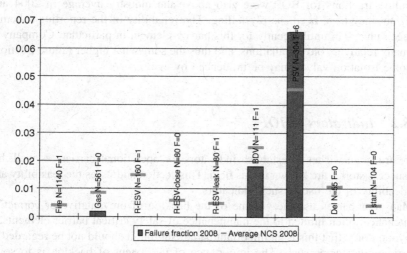

Fig. 23.9 Fraction of failures for Installation 2 and NCS in 2008 with number of test (N) and failure (T) for Installation 2. Reprinted from Ref. [32], with permission from Elsevier

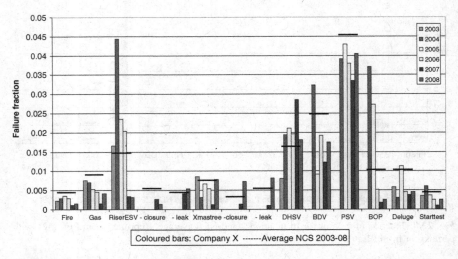

Fig. 23.10 Fraction of failures for all Company X installations and NCS in 2003–8. Reprinted from Ref. [32], with permission from Elsevier

The RNNP requirement to split the tests of isolation valves into 'Closure tests' and 'Leak tests' was not formulated until late 2006, and no companies reported this before 2007. The industry values also apply to 2007–08.

It can be observed from Fig. 23.10 that the failure fractions for riser ESDVs are above the industry average, except in 2007 and 2008. It can further be observed that failure fractions for DHSVs are above the industry average from 2004 onwards and that the trend is clearly increasing.

Failure fractions for BOP were also above the industry average in 2004 and 2005, but they have been sharply falling. The reliability of the reporting schemes has been uncertain until recently for this barrier element in particular. Company X has many relatively old installations, and thus the somewhat higher failure fractions for some isolation valves may be influenced by age.

23.8.2   Indicators—HOFs

There are often wishes to establish indicators for operational barriers as well, but few success stories are known in this field. This section addresses the feasibility and practicability of the use of such indicators.

Many companies use the volume of the backlog of preventative or corrective maintenance, often limited to the maintenance of safety critical barrier elements. It is beyond doubt that this is an important parameter, but it should not be regarded as a barrier indicator as such. The importance of the extent of backlog is to some extent similar to competence level. The absence of a backlog or high competence level are both essential prerequisites of a good safety management system, but neither of these elements is sufficient to ensure a low failure fraction or error-free operation. The level of backlog may therefore be an important indicator in addition to other indicators, but not an alternative to the failure fractions of barrier elements as presented above.

Other proposals (see for instance Ref. [14]), have included the number of operators with special education, or with special vital skills. Again we are sceptical about this as a barrier indicator. As competence level mentioned above it may be an important RIF, but not a barrier as such.

This topic is discussed based on the extensive experience available in relation to hydrocarbon leaks on offshore installations, as precursors to possible major hazards such the Piper Alpha disaster or the Texas City refinery accident. It is contended that the development of risk indicators based on technical barriers has been successful.

Based on this success, we may ask should risk indicators for organisational and operational barriers be developed in a similar manner, and if so, how should they be defined? We will address these topics based on our extensive experience with technical barrier indicators and recent experience with the risk assessment of HOFs [34, 35]. It has been shown that about 60% of the significant hydrocarbon leaks on the NCS occur because of the failure of operational barriers [25]. It is therefore important to follow up the performance of operational barriers, not only technical barrier elements. The circumstances of operational leaks fall into three categories:

- Immediate leak due to faults such as opening the wrong valve or flange during manual intervention
- Delayed leak due to latent fault introduced during manual intervention
- Overpressure due to maloperation by control room personnel.

The two first circumstances are associated with at least 90% of leaks. Overpressure cases are neglected in the discussion below. The details were presented in Sect. 6.6.3, as categories B, C and D in Fig. 6.14.

The main operational barriers in relation to the prevention of leaks are those that can prevent operational faults during intervention or reveal that a fault has occurred.

Technical barriers are special in several senses. They are subject to regular, preventive maintenance, they can be tested, and they are certainly at the sharp end in relation to influencing the development of the accident sequence. As noted above, barriers that have been included in indicator surveillance programs thus far are consequence reduction barriers, from the right-hand side of a typical Bow-tie diagram.

Organisational and operational barriers are different in several senses. These barriers are on both sides of a typical Bow-tie diagram, perhaps most importantly for the left-hand side, the barriers that may prevent the occurrence of incidents and accidents.

If we follow the focus on the sharp end, only operational barriers (i.e. human operators) are those that may influence the occurrence of incidents and accidents and the consequences thereof.

The OTS (Operational Condition Safety) project defined a number of PSs and PRs. The defined PSs are [34]:

- PS1: Work practice
- PS2: Competence
- PS3: Procedures and documentation
- PS4: Communication
- PS5: Stress and physical working environment
- PS6: Supervision and management
- PS7: Change management.

Examples of PRs for PS2 (Competence) are:

- PR2.1: Requirements for employees' competence shall be prepared and documented
- PR2.2: Requirements for employees' competence shall be complied with.

However, neither the PSs nor the PRs are suitable as indicators, mainly because there are no obvious or easy ways to measure the levels of such indicators.

By contrast, an OTS audit results in an overall classification for each PS [34], based upon interviews, questionnaire surveys and document reviews. These overall classifications or scores can be used as a basis for the definition of indicators. The challenge in this case may be that a full OTS audit is quite labour-intensive, such that there is probably some years between every full audit, and a possible indicator on this basis would change very slowly, which means it is not a good indicator.

In conclusion there are no obvious barrier indicators for human and organisational barriers. However, work is still ongoing in order to attempt to define parameters that may be suitable indicators for human and organisational barrier systems, and we may be in a position to make proposals in the future.

One important aspect of operational barriers (human operators) is that the concepts of the maintenance and testing of technical barriers can not be applied in a parallel manner. The modelling of operational barriers in Ref. [35] defines the primary influence on operational safety to be the work practice of the process and maintenance personnel. This is influenced by a number of influencing factors, according to Ref. [35]:

- Competence
- Procedures and documentation
- Communication
- Stress and physical working environment
- Supervision and management
- Change management

We could argue that the state of the organisational and operational barriers is mainly dependent on the safety culture or safety climate on the installation. The culture needs to be maintained or may need to be improved in some cases. However, this is definitely another form of 'maintenance' than the preventive maintenance of fire water pumps and gas detectors. Safety culture is not open to 'quick fix' solutions, as the maintenance of technical equipment can be. The low availability of gas detection can be improved through the replacement of detectors or more frequent preventative maintenance. Improvement in availability may have cost implications, but if funding is secured, then improvement in availability is straightforward to achieve.

Safety culture (or climate) is considerably more difficult to improve. There are many anecdotal (but not documented) examples that management has tried to change the culture but has not been in a position to succeed. Changing the safety culture requires a number of actions and focused effort over some time. This implies that the regular monitoring of the status of operational barriers in the same manner as with technical barriers is not achievable and even inadvisable. Even if a scheme could be invented, it would be a fruitless use of time and money.

Consider one possible root cause focused upon by some authors, namely 'silent deviations' [36]. Such deviations occur when it is customary (i.e. accepted by supervisors and management) on the installation that procedures are not followed. This is probably done each day and may have positive effects on production; skipping elements of a procedure often leads to less time on manual tasks, and possibly more production. It is likely that shortcuts may be taken over long periods without leading to severe consequences, possibly a near miss occasionally. Even these will probably not result in a change in work practices in an organisation characterised by a negative safety culture. We would expect that if a work practice were mapped, it would come out as negative for organisations where silent deviations are frequent. 'Silent deviations' can also be found in the detailed study of organisational culture in Statoil (now Equinor) after a serious near miss in 2010 [37].

A monitoring effort such as OTS [34] is intended to be redone about once every five years. A questionnaire survey has been developed as part of the OTS survey, which could be used more frequently, say once per year. Work practice and deviations are natural parts of this questionnaire. Many oil companies have culture surveys each year, and this is an opportunity to combine the general survey with a major hazard-specific survey and conduct an annual survey. The results could be aggregated into a few overall scores, which could serve as the indicators of operational RIFs.

One example of such factors is presented in Ref. [25], based on a correlation study between hydrocarbon leaks, barriers, the safety climate questionnaire and noise data. This study analysed the data structured in the following factors:

- Values in the organisation
- Personal attitudes
- Organisation

- Management/supervision
- Planning
- Corrections
- Use of procedures
- Competence
- Time used on each task.

This approach would be in accordance with the view expressed by Rosness et al. [38] who propose the following in order to monitor major accident risk: "Monitor the structural and cultural preconditions for organisational redundancy". The structural dimensions are defined as "Possibility of direct observation, overlapping competence, tasks or responsibility". The cultural dimensions are defined as "Capability and willingness to exchange information, provide feedback, and reconsider decisions made by oneself and colleagues".

The structural and cultural distinction may also be valuable in order to describe the difference between major accidents and occupational accidents. The structural components related to observations, competence and tasks of importance for the avoidance of leaks during process system interventions are significantly different from observations, competence and tasks relating to the avoidance of occupational accidents.

The structural components for major hazards, typically the avoidance of leaks during process system interventions are related to work practices involved in the maintenance of process systems, competence in maintenance work and responsibilities for checking and verifications related to the integrity of process systems. The structural components of occupational accidents are related to how work with hazardous equipment and work on elevated platforms is carried out as well as the use of protective equipment and avoidance of exposure to hazards.

In conclusion, indicators for selected RIFs are recommended in the human and organisational sphere, but not indicators for operational barrier elements as such.

23.9   Proposed Major Hazard Indicators for Companies

The preceding sections documented the indicators for major hazards in the Risk Level initiative by PSA. One of the objectives of that work was to inspire individual companies to develop their own scheme of major hazard indicators.

Performance indicators for personnel risk have traditionally focused on occupational accidents, and less on major accidents. Major accidents are rare events even at a national scale, and certainly for an operator with relatively few installations. Even major hazard precursors (such as hydrocarbon leaks of a critical size) are relatively seldom occurrences for an operator with few installations.

When companies have focused on indicators for occupational hazards, this presumably reflects the assumption that there is a strong correlation between occupational hazards and major hazards. This was found to be one of the problems in the Texas City refinery accident [7]. The investigation found that this invalid assumption was one of the factors leading to conditions that would allow such an accident to occur.

One company has been a forerunner in this regard in the Norwegian offshore petroleum operations, and it has established major hazard indicators for internal company follow-up. Statoil (now Equinor [33]) has established several indicators and has developed an information system in order to focus the awareness of management on such aspects. The proposals presented in this section are to some extent inspired by the follow-up of indicators in Statoil. Total E&P Norway also took some early initiatives.

It is essential that companies develop their internal major hazard indicator schemes, for individual installations as well as companywide operations, in order not to end up in the same situation as in the Texas City refinery case, where only occupational hazards were monitored. The background from RNNP provides a starting point in order to propose such schemes. This section discusses how such schemes may be outlined, based on experience with schemes for major hazard indicators for some companies.

It should be noted that different sets of indicators might be relevant for a companywide presentation as opposed to a presentation for an individual installation (or plant). This is addressed in the subsequent discussion.

Barrier indicators focus on the performance of the maintenance of safety-critical equipment, which is an area where PSA has expressed that some improvement should be achieved by the petroleum industry.

Precursor-based indicators should not be disregarded completely, because they provide valuable input about how organisations have performed in the past. It is also a way to show the organisation the result of unwanted occurrences. It is sometimes a slogan in the industry: "What is measured will be focused on". The implication of this is that both types of indicators should be included, in order to focus on barrier performance as well as on major hazard precursors.

The overall recommendation is to use both types of indicators, if possible. The emphasis should be on barrier indicators, supplemented by precursor based indicators. Both types of indicators are discussed in the following.

23.9.1   Precursor Based Indicators

A company with some few installations on the NCS registered only five precursor events in RNNP during a five-year period. This provides a good illustration of the challenge with precursor based major hazard indicators for small or moderate size companies on the NCS. Five occurrences are considered far too few events in order to provide reliable indicators of major hazard risk trend based on precursor events.

In the actual case it could also be considered including data from the company's other national sectors in order to extend the data sources. If the combined number of installations is sufficiently large, this may be a solution. On the NCS, only two companies are considered to have a sufficiently high number of installations so that precursor-based indicators would be meaningful at a company level.

The only precursor-based indicator that may be considered in spite of the low number of occurrences is the number of 'significant' (above 0.1 kg/s initial leak rate) leaks of hydrocarbons. Figures 6.4 and 6.5 suggest that for an average installation, the incidence rate is likely to be one reportable leak every few years. This is too infrequent to allow trends to be presented, but this indicator should nevertheless be focused upon, because of the high interest in the industry. It should also be noted that some installations have a few reportable leaks per year, i.e. more than 10 times the average incidence rate. This implies that trend analyses may be meaningful in these cases.

The importance of high focus on such leaks must be emphasised. The offshore industry in the UK had a campaign to reduce the number of such leaks in the latter 1990s. When the campaign ended, the number of leaks started to increase again. The Norwegian offshore industry has had a similar campaign from 2003, which was planned to be terminated in 2008. However, the target was achieved in 2007, and the focus was reduced. The level increased immediately, and stayed at 50% plus in relation to the 2007 value in the period 2008–2010, before a new campaign was launched in the first half of 2011.

The authorities in Norway have made a clear split between leaks with an initial leak rate below 0.1 kg/s and leaks above this limit. Only the latter leak sizes are considered to have the potential to cause major accidents. In the UK however, the authorities also record the number of leaks below this limit. As expected, the number of leaks below 0.1 kg/s is considerably higher than those above this threshold, on some installations more than 10 leaks per year.

Although it might be assumed that the number of leaks below 0.1 kg/s is a good indicator of more serious leaks, it was explained in Sect. 23.1.2 why this is not the case.

23.9.2   Barrier Indicators

Leading indicators for major hazard risk are in practice performance data for barrier elements or systems. In theory this could apply to physical (technical) as well as non-physical (human) barrier elements. The companies that have developed such indicators have focused on technical barrier elements. The data in Figs. 23.2 and 23.3 are thus limited to technical barrier elements.

There is usually a sufficient amount of data in order to allow for the follow-up of individual installations; this is one of the main advantages of barrier indicators. This implies that barrier indicators can be illustrated at an installation level as well as at a company level.

It is important for the use of barrier indicators that the reporting of safety critical failures of barriers is achieved and maintained at a high quality level in order to avoid under- as well as over-reporting. When companies started to report barrier failures on the NCS, over-reporting was actually more of a problem compared with under-reporting. If an emergency shutdown valve does not close on demand, this is a safety critical failure. However, if the valve cannot be reopened, this may be critical for the production assurance, but is not a safety critical failure. Only the first type of failure should be reported in a barrier data reporting scheme for safety critical faults.

The barriers in RNNP are limited to technical barriers, with response times from muster drills being the only exception. Other non-technical barriers that may be used for the follow-up of PRs, that other operators have used [16] are:

- Performance of man overboard pick-up in exercises
- Mobilisation of emergency teams.

23.9.3 Proposal—Barrier Indicators

For an individual company, the following leading indicators may be used for each installation (plant) as well as for the company as such:

- Topside barrier elements

  - As a minimum: Barrier elements adopted by RNNP (see Sect. 23.4.1)
  - Possibly additional barrier elements

- Marine barrier elements

  - Only relevant for floating installations
  - As a minimum: Barrier elements adopted by RNNP (i.e. ballast valves, anchor winch brakes, closure of water tight doors)
  - Possibly additional barrier elements.

The primary use of such indicators should be to follow up trends over long periods. Some companies in Norway have however cooperated in order to establish target values, against which observed values can be compared. Such target values may be essential in order to define the status of individual barrier elements. As long as the same barrier elements as on the NCS are used, reference to average values for the NCS can be used in order to define status (see further discussion below).

Barrier indicators may be followed up separately for each installation as well as for the company overall. Some operators have established a system for the periodic reporting and follow-up of the performance of major hazard barriers, using what is called a barrier performance panel.

Such a system can be used in order to focus the attention of management and operational personnel on the follow-up of performance. Figures 23.11 and 23.12 present example barrier panels for Company X and Installation Y.

Fig. 23.11 Barrier panel for Company X. Reprinted from Ref. [32], with permission from Elsevier

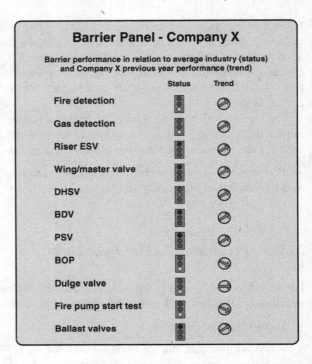

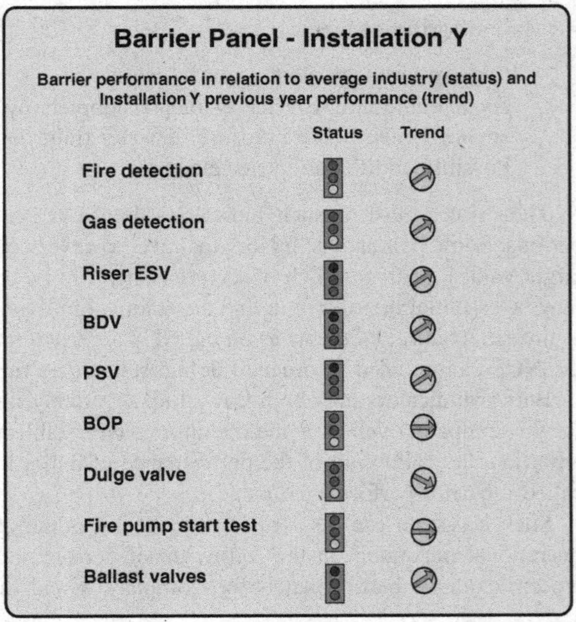

Fig. 23.12 Barrier panel for Installation Y. Reprinted from Ref. [32], with permission from Elsevier

It is suggested that a barrier panel be updated each three months or six months, with a rolling 12-month average. It should be noted that the average for the NCS is updated annually.

The barrier panel shows the status for each barrier element in relation to the incidence rate of faults during testing as well as the trend based on the past 12 months in relation to the previous 12-month period. The status could be related to the industry average, say for the past three years.

For the example shown in Fig. 23.11 the following observations can be made for the overall barrier performance of Company X:

- Status: 5 of 11 are in the red zone, 1 is in the yellow zone
- All red zone, yellow zone have increasing trend
- 2 green zones have increasing trend.

For the example shown in Fig. 23.12 the following observations can be made for the overall barrier performance of Installation Y:

- Status: 4 of 9 are in the red zone
- All red zone and 2 green zones have increasing trend
- Only one green zone has falling trend.

It should be noted that there has been significant improvement of the computer software for presentation of barrier status during the last few years. One recent example is shown in Figs. 23.14, 23.15 and 23.16. Another example is presented in Ref. [39].

The data in the diagram are taken from an installation in the Norwegian sector. The installation here has shown a significant improvement in the period covered by the diagram.

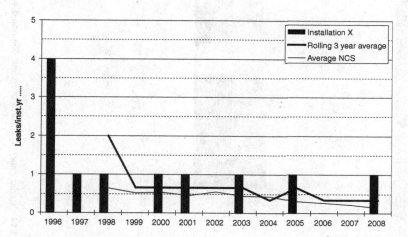

Fig. 23.13 Number of leaks >0.1 kg/s for installation X and average NCS, 1996–2008. Reprinted from Ref. [32], with permission from Elsevier

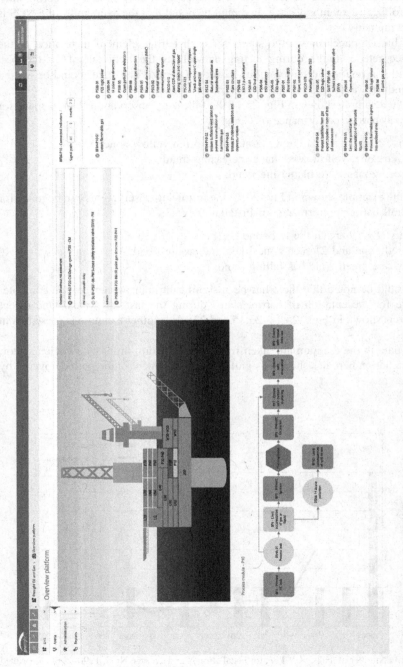

Fig. 23.14 Barrier panel—platform overview. Reprinted with permission from Presight

23.9.4 Proposal—Precursor-Based Indicators

As already mentioned the incidence rate of hydrocarbon leaks for offshore and onshore petroleum installations should be stated as a lagging indicator. Figure 23.13 presents a suggestion for the illustration of such a lagging indicator. It would be natural to compare a rolling three-year average for an installation with the average for the industry as a whole.

Figure 23.14 shows the overview of the barrier panel with three sections placed side by side. The first section is the graphical outline of the installations, the barrier functions and the status of the barrier functions. This is shown separately also in Fig. 23.15, which is the left part of the screen with overview of the barrier function in the process module P10 (see also www.presight.com).

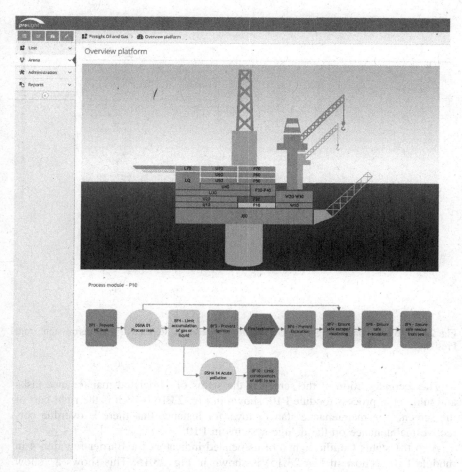

Fig. 23.15 Barrier panel—left side of platform overview. Reprinted with permission from Presight

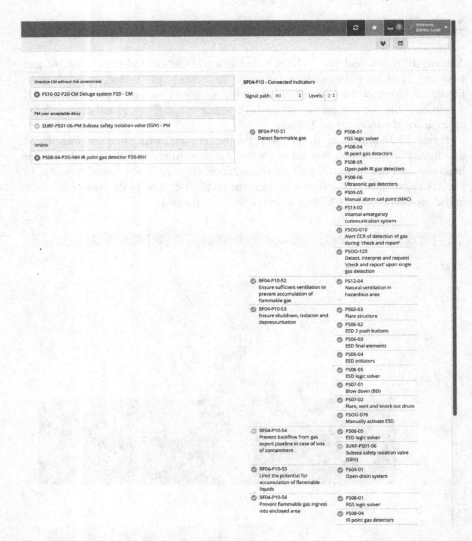

Fig. 23.16 Barrier panel—right side of platform overview. Reprinted with permission from Presight

The second section of the screen is the status of associated maintenance tasks and inhibits in process module P10, shown in Fig. 23.16, which is the right part of the screen. The maintenance status shows for instance that there is overdue corrective maintenance on the deluge systems in P10.

Also the status ('traffic light') of associated indicators for Barrier function 4 in module P10 (yellow in Fig. 23.15) is shown in Fig. 23.16. This shows a yellow ('warning') traffic light for the barrier element 'subsea safety isolation valve'.

References

1. Safety Science (2009) Special issue, vol 47, no 4
2. Vinnem JE (2000) Risk monitoring for major hazards, SPE paper 61283. In: Presented at the SPE international conference on health, safety and environment in oil and gas exploration and production, Stavanger, 26–28 June 2000
3. Øien K, Sklet S (1999) Risk control during operation of offshore petroleum installations. In: Proceedings of ESREL=99, Munich, October, 1999. Balkema, pp 1297–1302
4. HSE (1995) Prevention of fire and explosion, and emergency response regulations. HMSO, London
5. Total E&P Norway (1998) Health, safety and environment. Annual report for 1997. Total E&P, Stavanger
6. Øien K, Sklet S, Nielsen L (1997) Risk level indicators for surveillance of changes in risk level. In: European safety and reliability conference 1997. Lisboa, Portugal
7. Baker J (2007) The report of the BP U.S. refineries independent safety review panel, 'Baker report', January 2007
8. Kjellén U (2009) The safety measurement problem revisited. Saf Sci 4(47):486–489
9. Ale B (2009) More thinking about process safety indicators. Saf Sci 47(4):470–471
10. Zwetsloot GIJM (2009) Prospects and limitations of process safety performance indicators. Saf Sci 47(4):495–497
11. Kvitrud A, Ersdal G, Leonardsen RL (2001) On the risk of structural failure on Norwegian offshore installations. In: Proceedings of ISOPE 2001, 11th international offshore and polar engineering conference, Stavanger, Norway
12. Kjellén U (2000) Prevention of accidents through experience feedback. Taylor & Francis, London and NY
13. Rockwell TH (1959) Safety performance measurement. J Ind Eng 10:12–16
14. Øien K (2001) A framework for the establishment of organizational risk indicators. Reliab Eng Syst Saf 74(2):147–167
15. Aven T (2003) Foundations of risk analysis. Wiley, New York
16. Vinnem JE et al (2006) Major hazard risk indicator for monitoring of trends in the Norwegian offshore petroleum sector. Reliab Eng Syst Saf 91(7):778–791
17. Aven T, Vinnem JE (2007) Risk management, with applications from the offshore oil and gas industry. Springer
18. Tharaldsen JE, Olsen E, Rundmo T (2008) A longitudinal study of safety climate on the Norwegian continental shelf. Saf Sci 46(3):427–439
19. Huber et al (2008) Learning from organisational incidents: resilience engineering for high-risk process environments. Process Saf Prog 28(1):90–95
20. PSA (2011) Guidelines to regulations relating yo material and information in the petroleum activities (the information duty regulations). Petroleum Safety Authority, 1 January 2011
21. Heide B (2009) Monitoring major hazard risk for industrial sectors. PhD thesis, University of Stavanger
22. PSA (2019) Trends in risk level on the Norwegian continental shelf, main report, (in Norwegian only, English summary report). Petroleum Safety Authority, Stavanger, 11 April 2019
23. Sklet S (2006) Safety barriers; definition, classification and performance. J Loss Prev Process Ind 19:494–506
24. Vinnem JE, Aven T, Sørum M, Øien K (2003) Structured approach to risk indicators for major hazards. In: Presented at the ESREL 2003 conference. Maastrict, 16–18 June 2003
25. Vinnem JE, Hestad JA, Kvaløy JT, Skogdalen JE (2010) Analysis of root causes of major hazard precursors in the offshore petroleum industry. Reliab Eng Syst Saf 95(11):1142–1153
26. HSE (1996) The offshore installations and wells (design and construction, etc.) regulations. HMSO, London

27. Vinnem JE, Pedersen JI, Rosenthal P (1996) Efficient risk management: use of computerized QRA model for safety improvements to an existing installation. In: 3rd international conference on health, safety and environment in oil and gas exploration and production, New Orleans, USA. SPE paper 35775

28. Vinnem JE (1997) On the sensitivity of offshore QRA studies. In: European safety and reliability conference, 1997, Lisboa, Portugal

29. NPD (1992) Regulations relating to explosion and fire protection of installations in the petroleum activities. Norwegian Petroleum Directorate, Stavanger, 1992, February

30. Syvertsen R (2006) Weight of importance for barriers in the offshore oil and gas industry. MSc thesis, University of Stavanger

31. Hale A (2009) Why safety performance indicators. Saf Sci 47(4):479–480

32. Vinnem JE (2010) Risk indicators for major hazards on offshore installations. Saf Sci 48 (6):18p

33. Thomassen O, Sørum M (2002) Mapping and monitoring the safety level, SPE 73923. In: Presented at SPE international conference on health, safety and environment in oil and gas exploration and production, Kuala Lumpur, 20–22 March 2002

34. Vinnem JE et al (2007) Operational safety condition—concept development. In: Paper presented at ESREL 2007, Stavanger, 25–27 June 2007

35. Vinnem JE (2008) On causes and dependencies of errors in human and organizational barriers against major accidents. In: Presented at ESREL2008, Valencia, 22–25 September 2008

36. Høivik D, Moen BE, Mearns K, Haukelid K (2009) An explorative study of health, safety and environment culture in a Norwegian petroleum company. Saf Sci 47(7):992–1001

37. Austnes-Underhaug R et al (2011) Learning from incidents in Statoil (in Norwegian only). IRIS report 2011/156, 21 September 2011. Available from www.statoil.com

38. Rosness R, Guttormsen G, Steiro T, Tinmannsvik RK, Herrera IA (2004) Organisational accidents and resilient organisations: five perspectives. SINTEF report STF38 A 04403

39. Hansen HN, Rekdal O (2015) Barrier management for Goliat (in Norwegian only). ESRA Norway, 25 March 2015

Chapter 24
Barrier Management for Major Hazard Risk

24.1 Background

Barrier management is one of the buzzwords in Norwegian offshore risk management society, following the publication of a philosophy document by PSA late in 2011, namely 'Principles for barrier management in the petroleum industry' [1]. This document, often referred to as the PSA barrier memorandum, has later been updated twice, in 2013 [2], and in 2017 [3]. This philosophy document is largely the basis of this chapter.

On one hand, PSA argue that the barrier memorandum brings nothing new; it is well covered by the current regulations. On the other hand, it argues that barrier management has been weak in many companies. So what is PSA trying to achieve?

PSA has probably identified a number of shortfalls in the implementation of the regulatory requirements by companies, which has been trying to find a solution to for several years.

One common complaint is that there is limited connection between the risk management work done in the planning and engineering phases and in the operations phase. More specifically, QRA studies from the engineering phase are used only to a limited extent during the day-to-day operations of the installations, much less than what seems to be reasonable and relevant.

It is also observed that the emphasis on the analysis of barrier performance is weak even in the engineering phase, to the extent that specific requirements for technical barrier elements often cannot be established based on QRA studies. This is true for most barrier functions, but even more so for barrier functions that are supposed to prevent a loss of stability and/or buoyancy for floating installations.

A further complaint is that QRA studies with too much focus on FAR or AIR values (and tolerance criteria) are unsuitable for establishing barrier system performance requirements (PRs). Often it is claimed by specialists supplying QRA

J.-E. Vinnem and W. Røed, *Offshore Risk Assessment Vol. 2*, Springer Series in Reliability Engineering, https://doi.org/10.1007/978-1-4471-7448-6_24

studies that improvement in barrier performance would have no or limited effect on overall FAR values. This is probably true for some barrier systems, but may also be a limitation of the usefulness of QRA studies because the models for escalation or accident sequence development are too coarse (see also Chap. 15 in relation to the choice of details to be included in event trees). When these models are coarse, the ability to reflect differences in barrier system performance may be limited. It is therefore often argued that QRA studies do not reflect what is important in order to provide a suitable basis for the effective management of barrier systems and elements. At this stage it should be added that not only QRA studies are used as inputs to the establishment of barrier strategies and specific performance standards (PSs).

It is argued that focus on barrier management throughout all life cycle phases may provide the 'missing link' to solve many, if not all, of these challenges simultaneously.

The applicability of the barrier management approach to various hazards and risks may serve as a final introductory comment. The regulatory requirements are general and seem to be applicable for all hazards and risks within the HES field. The authors' recommendation is nevertheless that the primary focus should be on major hazards, which is the framework for the present chapter.

In the last version of the barrier memorandum, it is argued that there is a need to distinguish between 'safe and robust solutions' and 'barriers'. The motivation has been to clarify what barriers are and what they are not. Barriers are measures whose function is to offer protection in failure, hazard and accident situations. Their function is provided by barrier elements which may be technical, organisational or operational. Technical elements might, for example, be sensors that measure the pressure in a well, while organisational and operational elements might be mud loggers and drillers who monitor, detect and implement measures [3].

In many cases a barrier function is obtained by a combination of organisational, operational and technical barrier elements. One way of ensuring that all three categories are addressed is by asking the question; 'Who does what with which equipment?'. Then 'who' addresses the organisational element, 'what' addresses the operational element and 'with what equipment' addresses the technical element. While previous work on barrier management has paid attention primarily to technical barrier elements, many accident investigation reports show that operational and organisational elements are at least as important to ensure that barrier functions are obtained.

As mentioned above, barrier management is relevant not only for the operations phase, but also for the preceding planning and engineering phases. This is emphasized in the latest version of the barrier memorandum by referring to continuous improvement (plan, do, check, act) in all relevant phases.

Compared with previous versions of the memorandum, the latest version provides examples from various specialty fields, such as the security field.

24.2 Regulatory Requirements

The Norwegian regulations for offshore facilities and operations contain many requirements that can be related both directly and indirectly to barriers. The integrated and consistent management of barriers requires understanding and dealing with key relationships in the regulations. Section 5 of the management regulations [4], specifies that (see also Sect. 1.5):

Barriers shall be established that at all times can

- identify conditions that can lead to failures, hazard and accident situations,
- reduce the possibility of failures, hazard and accident situations occurring and developing,
- limit possible harm and inconveniences.

Where more than one barrier is necessary, there should be sufficient independence between barriers.

It is also required that the operator or the party responsible for the operation of an offshore or onshore facility should stipulate the strategies and principles that form the basis for the design, use and maintenance of barriers, so that the barriers' function is safeguarded throughout the offshore or onshore facility's life.

Personnel should be aware of the barriers that have been established and the functions that they are intended to fulfil, as well as the performance requirements that have been defined in respect of the concrete technical, operational or organisational barrier elements necessary for the individual barrier to be effective.

Finally, it is required that personnel should be aware of which barriers and barrier elements are not functioning or have been impaired, and necessary measures should be implemented to remedy or compensate for missing or impaired barriers.

These requirements to barriers are quite strict, and they refer to strategies, principles and PRs. Barrier management is not explicitly referred to, but the process during all life-cycle phases is clearly described. This process can be referred to as 'barrier management'. The process is discussed in Sects. 24.3 and 24.4.

The ISO13702 standard for the control and mitigation of fires and explosions [5], has long required two strategies: a fire and explosion strategy (FES) and an emergency response strategy. The above standard is also referred to by PSA in the guidelines to its regulations. The requirement for a barrier strategy can be regarded as a generalisation of these two strategies. Further, the FES starts with a description of the hazards and risk picture, including the characterisation of potential consequences. This also needs to be done when describing a barrier strategy.

Health and Safety Executive [UK] has issued a 'Fire and explosion strategy (Issue 1)' that presents the strategy development carried out by HSE's Offshore Division to address fire and explosion issues on offshore installations. It provides an overview of the current state of knowledge with regard to fire and explosion hazards, their prevention, control and mitigation. It identifies current areas of uncertainty and describes the strategy areas where areas of significant uncertainty require clarification. This document focuses on individual strategies and functional requirements, rather than on the management process.

24.3 Barrier Concepts

The main concepts regarding barriers are briefly introduced in Sect. 2.5.2, and are elaborated on in this section. PSA has expressed the following: *Barrier management implies coordinated activities for establishing and maintaining barriers so that they fulfil their functions at all times* [3].

Barrier management thereby includes the processes, systems, solutions and measures that must be in place to ensure the necessary risk reduction through the implementation and follow-up of barriers.

In order to manage risk in an acceptable manner, the necessary barrier functions and elements must be identified, as shown in Fig. 24.1.

These have to be based on the specific context and risk picture. Further, the necessary PRs must be established, so that barrier functions can be implemented as intended, and sufficient maintenance and follow-up activities must be initiated to ensure that the performance of the barriers is maintained over time. The requirements for effective barrier management apply throughout the life cycle of the offshore facility or land-based plant, including in the execution of every activity and operation. This means that, even after the design and construction phase, many conditions need to be monitored and continuously followed up. Apart from normal operational and maintenance activities, systems/routines must be in place to ensure efficient communication, expertise, monitoring of results, changes in context and change management.

Barrier strategies and PRs are developed and described in the planning phase, and actively followed up in the operations phase through the monitoring, evaluation and implementation of further improvements in a typical control loop (or quality loop) fashion.

The implementation of the barrier management model throughout the operations phase is strongly dependent on the barrier management in the planning phase, which therefore becomes extremely crucial. The following text therefore starts with a discussion of the implementation in the planning phase.

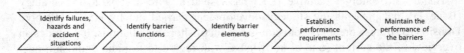

Fig. 24.1 Key points in barrier management [3]

24.4   Barrier Management in Life Cycle Phases

24.4.1   Planning Phases

PSA's model for barrier management in the planning phases is presented in Fig. 24.2. The model is based on the ISO31000 model for risk assessment and management, as presented in Chap. 3. The upper part of the model is consistent with the risk assessment and risk management models in Chap. 3, whereas the lowest box is specific for barrier management. This involves establishing a specific barrier strategy and specific performance requirements.

24.4.2   Establishing Barrier Strategy

As emphasized in the barrier memorandum [3] a (specific) barrier strategy is a plan for how barrier functions, on the basis of a risk picture, are to be implemented in order to reduce risk. This highlights that the barrier strategy should be based on the risk picture for the installation.

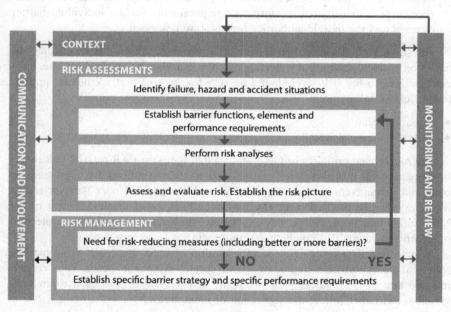

Fig. 24.2 PSA's model for barrier management in the planning phase *Source* Ref. [3]

This section provides an overview of the barrier functions that together define the barrier strategy. These barrier functions must, according to PSA requirements, be applicable for each area. The following principles should form the basis of a barrier strategy [3]:

- be designed so that it helps the parties involved gain a common understanding of the basis of the requirements for the different barriers, including

  - which phases, operations and activities the strategy has been established for
  - which failure, hazard and accident situations may occur in the phases, operations and activities the strategy has been established for
  - which barrier functions are required to handle these situations
  - where additional information is to be found about the performance requirements which specifically apply for the individual barrier.

- be sufficiently fine-grained for the individual plant (e.g. area, system, equipment) and phases, operations and activities
- be kept updated at all times
- identify which roles/tasks the different barrier functions have
- identify important assumptions which are significant for the individual barrier function and the individual barrier element
- identify the relationship between strategy and performance requirements established for the individual barrier. The strategy should provide information about where the different performance requirements for the individual barrier element and the individual barrier function are described.

The starting point for the barrier strategy is the risk picture; this will usually be based on a QRA study, but other approaches can also be applicable. The main contributions to the risk level should be stated as part of the risk picture presentation, for the following:

- Possible hazards
- Different areas on installations.

The much-cited criticism of QRA studies that focus on FAR or AIR values has been mentioned. It is therefore essential that risk contributions are expressed for other risk parameters in addition to FAR or AIR values. Under the Norwegian regulatory regime this will naturally be the frequencies of the impairment of the Main Safety Functions; escape ways, safe areas, prevention of escalation, protection of significant rooms [for combatment] and main load carrying capacity.

A large installation may consist of a handful of main areas, but these may be split into many subareas and rooms with special protection needs. It is recommended that barrier strategies do not apply to each subsea and each room individually, but rather follow the separation of main areas.

24.4.3 Overview of Barrier Functions and Systems

The following outlines typical barrier functions for an integrated fixed structure drilling, production and accommodation installation:

- BF1: Prevent hydrocarbon leaks
- BF2: Limit size of hydrocarbon leak
- BF3: Prevent ignition of hydrocarbons
- BF4: Prevent escalations due to fire or explosion
- BF5: Prevent and limit non-hydrocarbon fires
- BF6: Prevent hydrocarbon release from well operations
- BF7: Prevent and limit damage to the external environment
- BF8: Prevent impact from ship on collision course
- BF9: Prevent the loss of structural integrity from environmental loads
- BF10: Prevent the loss of stability or station-keeping
- BF11: Prevent fatalities during escape, evacuation and rescue
- BF12: Protection of safety critical rooms
- BF13: Protection of main and load bearing structures
- BF14: Prevent and limit fire from helicopter accidents.

24.4.4 Structure of Barrier Functions

Figure 24.3 presents an overview of the structure of barrier functions, and the dependencies between them, with respect to the following:

- Prevention of fatalities
- Prevention and limitation of damage to the external environment
- Protection of safety critical rooms
- Protection of main and load bearing structures
- Prevention and limitation of fire from helicopter accidents.

The coding in Fig. 24.3 is used in order to distinguish between the connections between the barrier functions and the protection of life (BF1 and BF14), environment (BF7), safety critical rooms (BF12), main and load bearing structure (BF13) and helideck and passengers (BF14).

The hazards and barrier functions may also be coupled, in order to describe how different barrier functions apply to various hazards, as presented in Fig. 24.4.

The structure of hazards shown in Fig. 24.4 is considered to be a general structure of hazard categories applicable for major hazards:

HAZ01: Hydrocarbon leak/fire in process area
HAZ02: Hydrocarbon leak/fire in wellhead area
HAZ03: Hydrocarbon leak from pipeline/fire on sea
HAZ04: Fire in gas turbine/generator

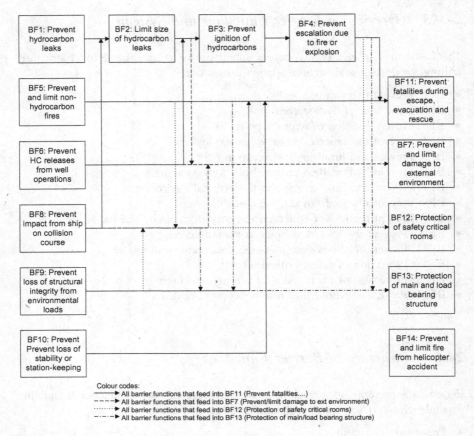

Colour codes:
→ All barrier functions that feed into BF11 (Prevent fatalities....)
---→ All barrier functions that feed into BF7 (Prevent/limit damage to ext environment)
······→ All barrier functions that feed into BF12 (Protection of safety critical rooms)
—·—·→ All barrier functions that feed into BF13 (Protection of main/load bearing structure)

Fig. 24.3 Structure and dependencies of typical barrier functions

HAZ05: Fire in utility area
HAZ06: Fire in accommodation
HAZ07: Collision risk
HAZ08: Structural damage due to environmental load
HAZ09: Ballast and station keeping failure
HAZ10: Acute oil spill
HAZ11: Fire on helideck/helicopter accident.

The diagram in Fig. 24.4 shows for instance that BF1, BF2, BF3, BF4 and BF11 are involved in protecting personnel from hazards associated with hydrocarbon leaks. Further BF8, BF2, BF3, BF4 and BF11 are involved in protecting personnel from hazards associated with impact from commercial vessel on a collision course. Similarly, the barrier functions BF6, BF3, BF4 and BF11 are involved in protecting personnel from hazards associated with releases from well operations. B5 and B11 are involved in protecting personnel from hazards associated with non-hydrocarbon

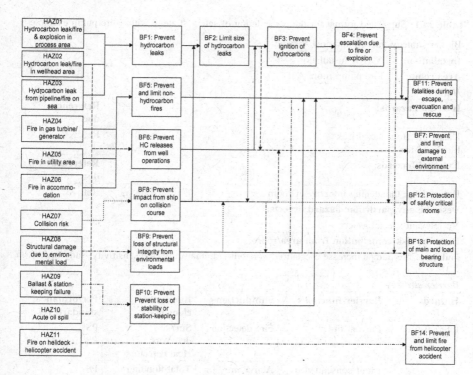

Fig. 24.4 Structure of hazards and typical barrier functions

fires. B9 and B11 are involved in protecting personnel from hazards associated with structural threats from environmental loads.

Similar combinations are included in Fig. 24.4 for the prevention and limitation of damage to the external environment (BF1, BF2, BF6 and BF8). The protection of safety critical rooms is a Main Safety Function, and the following barrier functions are involved: BF1, BF2, BF3, BF4, BF5, BF6, BF8, BF9 and BF12. The protection of main and load bearing structures is also a Main Safety Function, with the following involved barrier functions: BF1, BF2, BF3, BF4, BF5, BF6, BF8, BF9 and BF13.

Barrier strategies should express the link between hazards, areas/rooms and barrier functions and elements, as well as present these links in a manner that can also be used by installations personnel during operations. It is therefore essential that it provides a good overview. A possible format for the presentation of safety strategy is shown in Table 24.1, with some illustrations filled in for a fire pump room.

The top of the sheet is used for installation and area/room identification. The second part is the overview of hazards and risk contributions. The overview of hazards is split into hazards inside and outside the area/room. It is further important to focus on any particular hazard aspects that are specific to the installation in

Table 24.1 Suggested format for the presentation of safety strategy with a fire pump example

Barrier strategy, hazard and area specific				
Installation:	XXXX installation			
Area/room:	Fire pump room A			
Hazards				
Internal hazards:				**Reference doc.**
Diesel fire from diesel leak Electrical fire				Ref. X1
External hazards:				**Reference doc.**
External fire threatening integrity of room				Ref. X2
Describe any particular hazard aspects:				
No relevant special aspects				
Describe risk contribution from area/room:				
Failure of fire pumps increase escalation probability dramatically (see sensitivity study in Ref. X3)				

Barrier strategy				
Hazards	**Barrier functions**	**Subfunctions**	**Barrier elements**	**Performance standard ref**
Internal fire	Prevent fire	Fire detection	Smoke detectors Heat detectors	PS… PS…
	Limit consequence of fire	Active fire protection	Total flooding system in room	PS…
External fire	Prevent exposure of room to fire	Fire detection outside	Smoke detectors Heat detectors	PS… PS…
	Limit consequence of external fire	Passive fire protection	Fire walls around room	PS…
		Active fire protection	Total flooding system in room	PS…

question compared with other similar offshore installations. This is important because such special hazards may call for additional barrier functions or elements. If no such aspects exist, then an industry standard solution with respect to barrier functions will be relevant.

The final element under hazards and risks is a brief description of the influence on risk levels (in QRA studies or similar), in order to identify those hazards that have high criticality. The bottom part of the sheet is the actual presentation of the strategy for each hazard, with barrier functions, subfunctions, barrier elements and reference to PSs.

24.4.5 Establishing PRs

Barrier functions, barrier systems and barrier elements need to be structured. This can be carried out through a barrier diagram as shown in Fig. 24.5, which is an example presentation for BF3: Prevent the ignition of hydrocarbons. The PRs are mainly stated for barrier elements, but sometimes also for barrier systems and barrier functions or subfunctions. Some requirements with respect to PSs:

- The connection between hazards and requirements should be explicit
- There has to be a connection between strategies and barrier functions, systems and elements
- The role of the maintenance of barrier elements needs to be explicitly stated
- PRs and their rationales need to be known on the installation.

The starting point for definition of PSs is hazard identification and risk assessment (see Fig. 24.2). It has been recommended by PSA that more emphasis than usual be placed on hazard identification. It is also essential that risk analysis studies are sufficiently detailed, in order to provide the basis for detailed PRs.

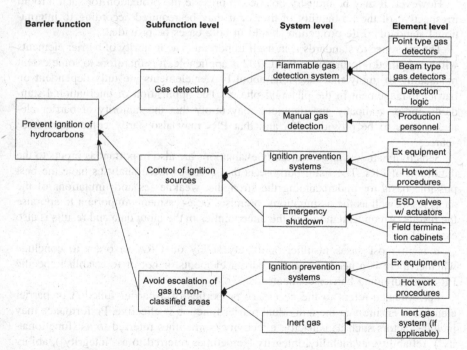

Fig. 24.5 Barrier diagram for barrier functions, systems and elements

There are usually quite a high number of assumptions and premises made, during the execution of risk analysis studies, especially in detailed studies. These may need to be made visible and known, if they have implications for barrier performance, and therefore need to be stated as PRs.

It is important that risks are established, presented and evaluated in terms of what it actually will be used for. The risk picture is required to determine the required barrier functions and to ensure that barrier elements have the required properties. Risk assessment and risk tolerance criteria must be adapted to the specific issues and purpose of the risk assessment process. It is insufficient to evaluate the results against the risk tolerance criteria for personnel safety and/or the loss of main safety functions for several of the PRs.

Consider the following illustration: The results of risk analyses may in certain circumstances indicate that some barrier elements are not required, as they do not contribute positively to the quantitative results generated in a traditional QRA. This could be the fire protection of certain rooms. This should not be interpreted to imply that the room for that reason does not need to withstand fires, but it means that the risk analysis in this case is not a suitable tool to provide input to the PSs that should apply.

However, it may be industry practice to provide fire protection for such a room irrespective of the probability of fire, or it may be required according to international standards. Fire protection should in both cases be provided.

The reference to standards is in itself important. The majority of barrier elements will be required if for instance ISO13702 is applicable. It is therefore to some extent misleading to imply that decisions about barrier elements are only dependent on barrier management in the planning phases. The application of international standards and/or company specifications may imply that the majority of barrier elements have to be provided for, and that PRs may also partly come from these sources.

Sensitivity studies and uncertainty evaluations are also important as inputs to the definition of PRs. The main purpose is that users of these analyses have the best possible basis for understanding the strengths, weaknesses and limitations of the analyses as well as the assumptions, premises or assessments important to appraise the results. An understanding of the uncertainties in the input data and results is also essential.

It is in most cases insufficient to rely solely on QRA in order to conclude whether there is a need for certain barrier elements or not or to establish specific PRs for individual barrier elements.

Performance refers to the necessary properties of a barrier function or barrier element to ensure that the individual barrier function is effective. Performance may include factors such as capacity, effectivity (sometimes referred to as 'functionality'), reliability, availability, integrity (sometimes referred to as 'integrity'), ability to resist loads and robustness (sometimes referred to as 'vulnerability').

24.5 Barrier Management in Operations Phase

Some main barrier management activities in the operations phase will be related to the follow-up of strategies and PRs in order to ensure that barriers are in accordance with what was specified during the planning phase. Indicators are therefore essential in this context (see Chap. 22). The PSA barrier memorandum [3], has an extensive description of the follow-up of monitoring and review in the operations phase.

It was argued in the introduction to this chapter that the connection between planning phases and the operations phase has long been a challenge. If the barrier elements and the PSs are actively followed-up during the operations phase, this may create a close connection between the planning phases and operations phase, bridging the gap often found in audits. The maintenance and inspection of barrier elements will be essential activities in the operations phase in order to meet the performance requirements for technical barrier elements.

24.6 Challenges for Implementation

This chapter has suggested some tentative solutions, but several aspects are unclear at the moment. These are discussed below.

Figure 24.4 presents a proposed structure of barrier functions in relation to a standardised set of hazard categories. This structure is supposed to be general. The details within the categories will obviously vary according to the installation type, concept and process media. However, there are other ways to structure these hazards, not the least a more detailed structure. The overriding goal here has been to keep these hazard categories at a low level in order to enable easy communication.

The structure of barrier functions, barrier elements and PRs is also a challenge. Some companies have chosen to structure strategies and requirements for barrier systems, in stead of barrier functions. The following has been used by some companies [6]:

PS1: Containment
PS2: Natural ventilation and HVAC
PS3: Gas detection system
PS4: Emergency shutdown
PS5: Open drain
PS6: Ignition source control
PS7: Fire detection system
PS8: Blowdown and flare
PS9: Active firefighting
PS10: Passive fire protection
PS11: Emergency power and lighting

PS12: Process safety
PS13: PA, alarm and emergency communication
PS14: Escape and evacuation
PS15: Explosion barriers
PS16: Offshore deck cranes
PS17: Drilling and well intervention
PS18: Ballast system and positioning
PS19: Ship collision barriers
PS20: Structural integrity.

The list above focuses on barrier systems. However, the most important over-view needed is the relation between the barrier systems and the barrier functions: which systems support each function, and what is the overall performance for each of the barrier functions?

The process as outlined above (see Fig. 24.2) should take the context into account. This will imply that several contexts will have to be considered, such as normal production, starting-up shut-down for maintenance, etc. This is in parallel with what traditionally been included in HAZOP studies, for the process safety systems. It is recommended that a limited number of relatively general contexts are considered, without going into extreme details.

The main challenge with barrier management is in any case how to distinguish between what is special for the concept and its context and what is in accordance with common industry practices. The philosophy document gives the impression that the establishment of barrier strategies would always be tailored to the concept and its context.

It is strongly argued that with respect to barrier functions and barrier systems, the vast majority (probably at least 80%) on an installation are systems and require-ments according to standard industry practices. Consider as an illustration a fire pump room. The hazards inside the fire pump room (and outside) are well known and independent of the installation in question, and the solutions and requirements will also be according to standard practice. It would be strongly counterproductive to argue that strategies and requirements need to be tailor-made for the actual installation. Such arguments will only foster disrespect for the value of tailor-made solutions where they are appropriate. It will therefore be crucial to make the dis-tinction between what needs to be tailored and what can follow industry practices.

If this distinction is not taken seriously, it may be impossible to prevent the establishment of barrier strategies and requirements becoming a 'Xerox engineer-ing' task, i.e. that barrier strategies and requirements are copied from the previous installation with a minimum of thought and consideration. This is probably worse than anything.

The last aspect is a challenge that PSA has also been fully aware of, namely applying the philosophies to organisational and operational barriers in addition to technical barriers. Although this has been focused on specifically in the latest version of the barrier memorandum [3], it is still a challenge to fully implement the ideas in real applications.

References

1. PSA (2011) Principles for barrier management in the petroleum industry (in Norwegian only), 1st edn. PSA. Accessed 6 Dec 2011
2. PSA (2013) Principles for barrier management in the petroleum industry, 2nd edn. PSA. Accessed 29 Jan 2013
3. PSA (2017) Principles for barrier management in the petroleum industry, 3rd edn. PSA. Accessed 15 Mar 2017
4. PSA (2017) Regulations relating to management and the duty to provide information in the petroleum activities and at certain onshore facilities (the management regulations). Accessed 18 Dec 2017
5. ISO (1999) Petroleum and natural gas industries—control and mitigation of fires and explosions on offshore production installations—requirements and guidelines, ISO13702:1999
6. Vinnem JE, Ravdal BA (2006) Analysis of barrier indicators, presented at ESREL2006, Estoril, 18–22 September

Appendix A
Overview of Software

A.1 Introduction

Quite extensive software tools have become available over the past 20 years. A brief overview over some of the main tools which are oriented towards offshore/oil and gas applications are presented in this appendix. These tools have been categorised into the following:

- QRA software
- QRA software tools for scenario and probability analysis
- QRA software tools for consequence analysis
- Risk management software
- Qualitative risk assessment software
- Reporting and analysis of incidents and accidents.

Brief summaries are presented as an overview, followed by brief sections presenting some of the main characteristics of these products. These summaries have been prepared by the software vendors. Only those products are detailed where a response was received from the vendors. The descriptions are structured as follows:

- Name and purpose of software
- Scope of software
- License conditions, pricing etc.

It should be stressed that there is a large amount of general software tools for CFD from many different suppliers. These have not been included in the presentations that follow throughout this appendix. Some of these may have quite valid applications during estimation of loads from fire or explosion, or for gas dispersion or oil slick movements. Because there are so many software tools available in this category, it becomes impossible to give an overview of all relevant tools. None of these are therefore included.

© Springer-Verlag London Ltd., part of Springer Nature 2020   419
J.-E. Vinnem and W. Røed, *Offshore Risk Assessment Vol. 2*, Springer Series
in Reliability Engineering, https://doi.org/10.1007/978-1-4471-7448-6

Table A.1 Overview of software for quantitative risk analysis

Software	Purpose	Contact
ASAP®	3D geometrical description and analysis of a fixed set of event trees	Lilleaker Consulting, Oslo, Norway
COSAC®	Risk assessment tool for early project phases of a field development for concept evaluation and screening. The tool can also be used as an aid for HAZID of offshore platforms	Lloyd's Register
Safeti Offshore®	Comprehensive Offshore Quantitative Risk Assessment software	DNV GL Limited, London, UK
RiskSpectrum® PSA	Fault tree and event tree software	Lloyd's Register
RISK®	Comprehensive offshore quantitative risk assessment tool	ESR Technology, UK
SAFETI®	Comprehensive QRA tool for Flammable, Explosive and Toxic Impact	DNV GL Limited, London, UK
Shepherd®	Risk Management Software Tool	Gexcon AS

Table A.2 Overview of QRA software for scenario and probability risk analysis

Software	Purpose	Contact
BlowFAM®	Evaluation of blowout risk during specific well operations	Lloyd's Register
Egress®	Mustering and evacuation simulation for evacuation/rescue modelling	ESR Technology, UK
LEAK®	Calculation of the frequency of leaks at an installation	DNV GL Limited, London, UK
R-DAT Plus®	Bayesian data analysis	Prediction Technologies, MD, USA
US Offshore Energy Model™	A fully probabilistic risk model that quantifies prospective risk from hurricanes in the Gulf of Mexico	Corelogic, USA
BlockSim®	System analysis with RBDs or Fault Trees	HBM Prenscia
RENO®	Probabilistic Event and Risk Analysis	HBM Prenscia
RiskSpectrum® HRA	Human reliability analysis tool to evaluate and quantify the probability of human errors.	Lloyd's Register
CORAM	Cumulative Risk assessment tool	ESR Technologies
RISKCURVES®	Quantitative Risk Analysis	Gexcon AS

Software tools that are only directed at onshore usage are not included in the reviews, neither are tools for production/transport regularity analysis.

All software tools that are mentioned in the following are commercially available from the vendor as listed. However, there is no implicit recommendations given (Tables A.1, A.2, A.3, A.4, A.5 and A.6).

Table A.3 Overview of QRA software for consequence analysis

Software	Purpose	Contact
FLACS®	Explosion simulation	Gexcon, Bergen, Norway
KAMELEON FireEx-KFX®	Fire and gas dispersion simulation	DNV GL AS, Høvik, Norway
OLGA®	Transient multiphase flow simulator for systems comprising flow lines, risers and process equipment.	Schlumberger
PHAST®	Windows-based toolkit for determination of consequences of accidental releases of hazardous material	DNV GL Limited, London, UK
FRED®	Consequence modelling tool for fire, release, explosion and dispersion	Gexcon AS
USFOS/FAHTS	Structural fire and explosion analysis	USFOS AS, Norway
PIPENET™	Pressure loss and flowrate model for offloading, firewater and ventilation systems	Sunrise System Ltd, UK
VessFire®	Heat transfer, depressurisation and stress modelling from fires	Petrell AS, Norway
HYENA®	Fire sprinkler/hydrant analysis	ACADS-BSG, Australia
ANSYS Fluent	Physical modeling of flow, turbulence, heat transfer, and reactions for industrial applications	ANSYS, USA
ANSYS CFX	Fluid dynamics program to solve wide-ranging fluid flow problems	ANSYS, USA
GASP	Modelling of pool spreading and evaporation from liquid spills on water or land	ESR Technology
DRIFT	Toxic and flammable gas dispersion modelling	ESR Technology
EFFECTS	Consequence analysis and modelling of physical effects and consequences of release	Gexcon AS

Table A.4 Overview of software for qualitative risk analysis

Software	Purpose	Contact
HAZOP Manager	HAZOP Tool	Lihou Technical & Software Services, UK

Table A.5 Overview of software for accident/incident analysis

Software	Purpose	Contact
CGE IncidentXP	Incident analysis software	CGE Risk Management Solutions, Netherlands
DNV BSCAT^{TM}	Modern risk-based safety management approaches to systematic root cause incident investigation	CGE Risk Management Solutions, Netherlands
BFA	Modern risk-based safety management approaches to systematic root cause incident investigation	CGE Risk Management Solutions, Netherlands
Stichting Tripod Foundation— Tripod Beta	Modern risk-based safety management approaches to systematic root cause incident investigation	CGE Risk Management Solutions, Netherlands
RCA	Modern risk-based safety management approaches to systematic root cause incident investigation	CGE Risk Management Solutions, Netherlands
TOP-SET RCA	Modern risk-based safety management approaches to systematic root cause incident investigation	CGE Risk Management Solutions, Netherlands

Table A.6 Overview of risk management software

Software	Purpose	Contact
Synergi RBI$^{®}$	Onshore and offshore risk based inspection tool	DNV GL Limited, London, UK
BowTieXP$^{®}$	Management of major risks to people, the environment, assets and reputation by means of Abow-tie@ graphical interface diagram	CGE Risk Management Solutions, Netherlands
THESIS	Management of major risks to people, the environment, assets and reputation by means of "bow-tie" graphical interface diagram	ABS Consulting, Warrington, UK
BowTie Pro^{TM}	Qualitative, semi-quantitative and illustrative risk analysis and risk management	BowTie Pro, Aberdeen, UK

(continued)

Table A.6 (continued)

Software	Purpose	Contact
Presight® Operations & Barrier Safety KPI Management	Reduce major accident risk with operations and context focused KPI management for barrier and process safety improvements and reduce down-time events. Convert complex multi-systems information into indicators, supporting risk awareness and safer operations decisions	Presight Solutions AS, Norway
Synergi Life	Complete business solution for risk and QHSE management, managing all non-conformances, incidents, risk, risk analyses, audits, assessments and improvement suggestions	DNV GL Limited, London, UK
Wellmaster RMS	Reliability Management System	Exprosoft AS

A.2 Electronic Contacts

The following is a listing of electronic contacts to the software providers:

- ESR Technology, UK—www.esrtechnology.com
- Lilleaker Consulting AS, Norway—www.lilleaker.com
- ANSYS, US—www.ansys.com
- DNV GL Limited, London, UK—www.dnvgl.com
- CoreLogic, USA—www.corelogic.com
- Prediction Technologies Inc—www.prediction-technology.com
- ABS Consulting, UK—www.absconsulting.com
- BowTie Pro, UK—www.bowtiepro.com
- Lloyd's Register—www.lr.org
- Gexcon, Bergen, Norway—www.gexcon.com
- Schlumberger—www.software.slb.com
- CGE Risk Management Solutions, Netherlands—www.cgerisk.com.

- Lihou Technical & Software Services, UK—www.lihoutech.com
- Sunrise System Ltd, UK—www.sunrise-sys.com
- Petrell AS, Norway—www.petrell.no
- ACADS-BSG, Australia—www.members.ozemail.com.au
- HBM Prenscia—www.hbmprenscia.com
- Presight Solutions AS, Norway—www.presight.com
- ExproSoft AS, Trondheim, Norway—www.exprosoft.com
- USFOS AS, Norway—www.usfos.no.

A.3 Quantitative Risk Analysis Software

A.3.1 ASAP®

The following is a brief description of this software, the function, vendor, pricing and main features.

• Function	3D geometrical description and analysis of a fixed set of event trees
• Vendor	Lilleaker Consulting a.s., Sandvika-Norway
• Pricing	Single User license cost is NOK 252.000 annually Multiuser license cost is minimum NOK 350.000 annually Prices as per 2018

The Advanced Safety Analysis Package (ASAP) is a software tool that estimates the risk to personnel and loss of main safety functions related to flammable and toxic leaks, fires and explosions on oil and gas facilities both offshore and onshore. The software analyses a chain of events following a leak by taking into account ventilation conditions, process conditions, safety barriers, escape possibilities, layout and manning distributions.

The development of the risk analysis tool Advanced Safety Analysis Package (ASAP) started in 1988. The program specifications were written by the risk analysts in Aker Engineering's Safety department and programmed by professional software developers. Substantial development has been invested in the program continuously over the last 30 years. The increase in computing capacity has enabled more complex models and large number of scenarios to be handled in one analysis. The program has also been tailored to provide risk measures and models required by the Norwegian facilities regulations and NORSOK. In the latter years the major changes in the software have been to implement the transient cloud, detection and ignition model (required) in the Z-013 appendix G as well as contour plots. New ASAP versions have been issued approximately once every other year, more frequent in the latter years. Program patches are implemented more frequently with minor bug fixes, user interface changes, changes to improve performance and/or minor functionality changes. Accessibility to the analyses in ASAP and/or from the Oracle database is not affected by implementation of patches and the analyses are therefore fully available in the same version. In 2000 Lilleaker Consulting was established and the year after the ownership of ASAP was bought by Lilleaker. Today the development of ASAP is administrated by YouTrack, which is an agile software project management tool. The development of the software is based on iterative and incremental development where requirements and solutions evolve through collaboration between the model developers (risk analysts in Lilleaker), software developers (Veloxit), program testers (risk analysts in Lilleaker) and users. Each new release of ASAP (versions) is validated before a new version of ASAP is officially released. In the last part of 1990 an ORACLE database was implemented to store all input data and results. This facilitated the reporting and testing of results, as the analysts is free to «slice and dice» the data to meet client requirements or examine particular areas of interest for the specific QRA.

A.3.2 COSAC®

The following is a brief description of this software, the function, vendor, pricing and main features.

• Function	Risk assessment tool for early project phases of a field development for concept evaluation and screening
• Vendor	Lloyd's Register
• Pricing	Available on request

COSAC® is a computerised tool for efficient risk assessment in the early project phases of a field development.

COSAC® analysis and results are tailor-made for concept evaluation and screening. Its aim is to increase the safety of new offshore developments by utilising 20 years of experience gained from risk analyses. Some of the main features of COSAC® include reducing uncertainty, improving the quality and efficiency of early phase safety evaluations. COSAC® provides a safety score for every risk factor associated with an offshore field development concept. A low score indicates safety concerns and/or lack of documentation of important safety issues. Therefore, a low safety score in COSAC® puts these issues in focus. In addition the user is provided with information on how to resolve the problems identified by COSAC®.

COSAC® has a unique explosion risk prediction model which calculates explosion loads for a variety of module configurations. The predictions are based on FLACS® simulations for comparable geometries.

A.3.3 Safeti Offshore®

The following is a brief description of this software, the function, vendor, pricing and main features.

• Function	Offshore risk analysis
• Vendor	DNV GL Limited, London, UK
• Pricing	Available upon request

Safeti Offshore® is the successor to Neptune and OHRAT; it was released in 2012 and has been continuously developed since then. Safeti Offshore® is a comprehensive software tool for designing, calculating and providing full traceability of a quantitative risk assessment. The system architecture has been designed to provide a methodological approach to risk analyses using best practice techniques. Safeti Offshore® contains models for calculation of discharge, dispersion, pool formation and evaporation, flammable and toxic effects and impact. Also contained with Safeti Offshore® are models designed for the specific needs of offshore installations, such as inventory calculation from isolatable sections, safety systems for isolation and blowdown, fire and blast wall success and failure, event escalation, smoke generation, evacuation, endangerment of muster areas, collision with ships, in module dispersion, ignition and explosion and many more. Safeti Offshore® offers flexible modelling of hazards and risks through a wide range of analytic capabilities including consequence modelling, leak frequency calculations, sensitivity and what-if analysis. Safeti Offshore® operates under MS-Windows®.

A.3.4 RiskSpectrum® PSA

The following is a brief description of this software, the function, vendor, pricing and main features.

• Function	Fault tree and event tree software
• Vendor	Lloyd's Register
• Pricing	On request

RiskSpectrum® PSA offers Fault Tree and Event Tree modelling interface, analysis, reliability data management and quality assurance all in one application. The separate MCS and BDD Engine—RSAT—is used for the conversion of Fault Trees, Sequences and Consequences to MCS or BDD to enable quantification of e.g. CDF and LERF.

RiskSpectrum® PSA is the complete linked Fault Tree and Event Tree tool. Its basic blocks of functionality are:

- Fault Tree editor
- Event Tree editor
- Analysis Tool (MCS generator)

All data are stored in a relational database, which makes it easy to browse, find relations and update—data is never repeated but only stored in one place. RiskSpectrum® PSA has also a system for storing information about each records status—edit, review and approve status.

A.3.5 RISK®

The following is a brief description of this software, the function, vendor, pricing and main features.

• Function	Comprehensive offshore quantitative risk assessment tool
• Vendor	ESR Technology, UK
• Pricing	Not available

RISK is a flexible and dynamic QRA model developed on EXCEL®. It enables users to clearly identify the key stages of the risk assessment process and follow individual major hazard events from their initiation, through accident development, to the contribution they make to accident scenarios, Temporary Refuge Impairment, SECE Impairment, Individual Risk and Potential Loss of Life. Key features of RISK are:

- Developed using industry standard spreadsheet software package (EXCEL®);
- Is user friendly and can be interrogated by engineers without the need for formal training;
- Is easy to tailor to meet specific project requirements;
- Is transparent and focuses on key scenarios at an appropriate level of detail;
- Inbuilt validated consequence models, including time-varying releases, gas dispersion and fire modelling.

A.3.6 SAFETI®

The following is a brief description of this software, the function, vendor, pricing and main features.

• Function	Comprehensive QRA software tool for flammable, explosive and toxic impact
• Vendor	DNV GL Limited, London, UK
• Pricing	Available on request

SAFETI® (Software for the Assessment of Flammable, Explosive and Toxic Impact) is the most comprehensive and widely used onshore QRA package available. It is a Windows® based system that provides a user friendly, industry standard method for quantifying major chemical risks. It enables analysis of the likelihood and severity of major hazards and makes use of the PHAST® models to predict the consequence of major releases. By combining these with their frequencies and taking account of population location and density, along with ignition source location for flammable and explosive effects, a number of presentations of 'risk' are possible. These include risk contours, F/N curves, risk transects and risk ranking at specific points. Additionally, effect exceedance results including flammable radiation and explosion overpressure provide complete support for occupied building analysis. SAFETI® performs 3D explosion calculations using the Multi-Energy and Baker Strehlow Tang explosion methodologies.

A.3.7 Shepherd®

The following is a brief description of this software, the function, vendor, pricing and main features.

• Function	Shepherd is a risk management software tool, tailored to do quantitative risk analysis for onshore facilities and operations. It allows fast and reliable prediction of the risk related to incidents such as releases of flammable or toxic fluid, fires and explosions It can be combined with the consequence modelling tool FRED to allow for rapid consequence modelling Typical outputs are risk and frequency contours for a range of consequences and types of risk (toxic risk, flammable risk, total individual risk). It produces fN curves and can be used to calculate Individual Risk per Annum (IRPA) for various workgroups
• Vendor	Gexcon AS, Bergen, Norway
• Pricing	Available on request through sales@gexcon.com

Shepherd has been continuously developed and validated by Shell since the 1990s, and has been extensively used by oil, gas and petrochemical operating companies, engineering contractors, insurers and regulators throughout the world. Since January 2018, Gexcon AS has been the exclusive commercial distributor of the Shepherd tool.

Typical use is in QRA for onshore Oil & Gas installations, such as upstream exploration and production, transportation and refining of petrochemical products.

A.4 QRA Tools for Scenario and Probability Analysis

A.4.1 BlowFAM®

The following is a brief description of this software, the function, vendor, pricing and main features.

• Function	Evaluation of blowout risk during specific well operations through assessment of approximately 300 elements, which influence the probability of a blowout
• Vendor	Lloyd's Register
• Pricing	Available on request

BlowFAM® is a PC-tool for evaluation of blowout risk during specific well operations. BlowFAM® has been developed in close cooperation with drilling/well intervention professionals in the participating companies. In addition, drilling specialists from several contractor companies have contributed.

The BlowFAM® model has identified approximately 300 elements, that influence the probability of a blowout. Many of these are applicable for the whole well life while others are only relevant for a specific well phase, e.g. drilling of the well. These elements are rated in regard to their importance to the risk. Main risk contributors for a specific development can be identified and cost-efficient risk reducing measures may be implemented.

The BlowFAM® model is also a valuable tool for communicating risk elements to the drilling professionals involved in the well operations.

A.4.2 EGRESS®

The following is a brief description of this software, the function, vendor, pricing and main features.

• Function	Mustering and evacuation simulation for evacuation/rescue modelling
• Vendor	ESR Technology, UK
• Pricing	Not available

The Egress code allows the movement of large numbers of personnel, such as when mustering on an installation, to be simulated. The platform layout is modelled as a matrix of interconnecting cells. The code covers both the physical movement and behavioural decision-making of personnel. The output is graphical and the movement watched as a real-time graphical representation. It was developed as part of a joint industry project in the UK between ESR Technology, industry, and the Health and Safety Executive.

The code has been used both offshore and onshore for the oil and gas and other industries to provide assessments of the movement of people during incidents.

A.4.3 LEAK®

The following is a brief description of this software, the function, vendor, pricing and main features.

• Function	Calculation of the frequency of leaks at an installation
• Vendor	DNV GL Limited, London, UK
• Pricing	Available on request

LEAK® is a software tool which calculates the frequency of leaks at an installation, typically an oil platform. Each installation is broken down into a number of areas which are themselves split into a number of segments each containing a list of equipment groups. Each equipment group is built up of base elements such as valves, flanges, pipes, etc. LEAK® will calculate the leak frequency for the installation, area, segment or equipment group based on built-in historical leak frequency data. The total frequency for each user defined category is reported together with each contributor. The model used expresses the frequency of a leak being larger than a certain size as a continuous function of the equivalent hole size diameter. The historical data used in the calculations is read from a database, enabling the most up-to-date data to be included.

A.4.4 R-DAT Plus®

The following is a brief description of this software, the function, vendor, pricing and main features.

• Function	Bayesian data analysis
• Vendor	Prediction Technologies, MD, USA
• Pricing	Not available

http://www.prediction-technologies.com/products/r-dat.brochure.pdf R-DAT Plus® is a full-featured Bayesian data analysis package for risk and reliability analysts. It is designed for users who need to perform system specific analyses, but who also have a need to develop generic prior distributions based on industry data.

R-DAT Plus® provides the user with a powerful, yet simple and flexible environment for storing and organising many types of reliability data and related information. A hierarchical structure enables the user to develop functional or structural or any other type of breakdown, at any level of detail. The elements of this hierarchy act as folders containing the reliability data and the results of Bayesian analyses performed on the data sets.

With R-DAT Plus® the user may specify the prior distribution in many different ways depending on the type and level of information available. These include a wide variety of parametric distributions (e.g. lognormal, beta and loguniform) using any of a number of input options such as lower and upper bounds, mean and variance, or the distribution parameters. Furthermore, R-DAT Plus® enables the user to develop generic distributions based on industry data (counts of failures in other applications) as well as expert estimates. The resulting distributions will represent the plant-to-plant variability of failure rate of a given class of components or initiating events, and can be used in a plant-specific analysis in order to perform the Two-Stage Bayesian procedure.

A.4.5 US Offshore Energy Model™

The following is a brief description of this software, the function, vendor, pricing and main features.

US Offshore Energy Model™ is a fully probabilistic risk model that quantifies

• Function	A fully probabilistic risk model that quantifies prospective risk from hurricanes in the Gulf of Mexico
• Vendor	CoreLogic, US
• Pricing	Available on request

prospective risk from hurricanes in the Gulf of Mexico. The model is part of EQECAT's global multi-peril catastrophe modeling platform, RQE™ (Risk Quantification & Engineering). Since the initial release of the model in 2007, the model has been enhanced and updated to incorporate improvements to vulnerability functions, wave modelling, exposure definition, data processing functionality, and the application of complex insurance policy structures.

The model analyzes risk in the Bureau of Ocean Energy Management (BOEM) planning regions, as well as in US lease waters. Meanwhile, onshore oil and gas delivery points and processing facilities are modeled for wind and storm surge damage (and for effects on shut-in oil and gas production).

The model handles the risks derived from the following hazards:

- Wind: Gust and sustained wind speeds, as defined by the probabilistic event set.
- Waves: A separate wave module calculates wave heights from complex hazard elements, including key hurricane wind parameters, water depth, sea-floor slope, wind duration and ocean fetch.

- Landslides and sub-sea currents: These perils threaten sub-surface equipment and pipelines. Hazard is derived from hurricane and wave parameters, sea-floor slope and water depth. Regions of mudslide hazard to pipelines are defined.

A.4.6 BlockSim®

The following is a brief description of this software, the function, vendor, pricing and main features.

• Function	System analysis with reliability block diagrams (RBDs) or Fault Trees
• Vendor	HBM Prenscia

ReliaSoft BlockSim provides a comprehensive platform for system reliability, availability, maintainability and related analyses. The software offers a sophisticated graphical interface that allows you to model the simplest or most complex systems and processes using reliability block diagrams (RBDs) or fault tree analysis (FTA), or a combination of both approaches. BlockSim supports an extensive array of RBD configurations and FTA gates and events, including advanced capabilities to model complex configurations, load sharing, standby redundancy, phases and duty cycles. Using exact computations and/or discrete event simulation, BlockSim facilitates a wide variety of analyses for both repairable and non-repairable systems. This includes:

- System Reliability Analysis
- Identification of Critical Components
- Optimum Reliability Allocation
- System Maintainability Analysis
- System Availability Analysis
- Throughput Calculation

A.4.7 RENO®

The following is a brief description of this software, the function, vendor, pricing and main features.

• Function	Probabilistic Event and Risk Analysis
• Vendor	HBM Prenscia

ReliaSoft RENO is a powerful and user-friendly platform for building and running complex analyses for any probabilistic or deterministic scenario using an intuitive flowchart modeling approach and simulation. Flowchart models can be created for complex reliability analyses, risk and safety analyses, decision making or maintenance planning. Users can also build models for other applications, such as optimizing your stock portfolio or testing your blackjack strategy.

RENO can be used for a wide variety of applications including, but not limited to:

- Risk/Safety Analysis
- Complex Reliability Modelling
- Decision Making
- Maintenance Planning
- Optimization
- Operational Research
- Financial Analysis

A.4.8 RiskSpectrum® HRA

The following is a brief description of this software, the function, vendor, pricing and main features.

• Function	Human Reliability Analysis (HRA) tool
• Vendor	Lloyd's Register
• Pricing	Available on request

RiskSpectrum® HRA is a human reliability analysis tool to evaluate and quantify the probability of human errors.

RiskSpectrum® HRA helps the user standardize HRA analysis process: by going through necessary steps to generate human error probabilities for human failure events, and to document important assumptions, conditions, inputs and results in the HRA process at the same time. With the tool the user can consistently conduct HRA and produce results of high quality, good traceability and documentation.

RiskSpectrum® HRA can help the HRA analysis meet the requirements of the ASME PSA Standard and NRC HRA Good Practices. RiskSpectrum® HRA includes a number of commonly-used HRA methods, such as THERP, ASEP, HCR/ORE, SPAR-H and HEART, etc.

For each human failure event, multiple methods can be used in quantification. Different results from different methods could therefore be compared. Conservative screening value can be defined and assigned to the selected human failure events.

A.4.9 CORAM

The following is a brief description of this software, the function, vendor, pricing and main features.

• Function	Cumulative Risk assessment tool
• Vendor	ESR Technology, UK
• Pricing	Not available

The purpose of the Cumulative Operational Risk Assessment Model (CORAM) is to provide a means to assess the changes in risk that could arise for example if one or more of the intended safeguards are not fully functional or subject to an outage, and hence determine if the risks are still tolerable or can be reduced.

CORAM is available in EXCEL spreadsheet format or as a dynamic web-based platform that employs the latest web technologies to bring risk information instantaneously to the user.

A.4.10 RISKCURVES®

The following is a brief description of this software, the function, vendor, pricing and main features.

• Function	RISKCURVES is an affordable and easy-to-use software tool that allows the user to do extensive Quantitative Risk Analysis for (petro-) chemical installations and sites. It offers the calculation of individual risk, societal risk and consequence risk. The software has all consequence modelling built in for scenarios that result from the (accidental) release of toxic and/or flammable gases, liquefied gases and liquids. Additionally, consequence calculations from external sources can be included, making it a very flexible solution to do QRA's
• Vendor	Gexcon AS, Bergen, Norway
• Pricing	Available on request through sales@gexcon.com

The RISKCURVES software was originally developed by TNO in The Netherlands and is based on the TNO "coloured books": the "Purple Book", the "Yellow Book" and the "Green Book". The software has been available since the 90's and is one of the leading tools world-wide for quantitative risk analysis (QRA) using integral modelling. End of 2018 TNO and Gexcon AS started a joint venture to further develop and market the software.

Typical use is in onshore (petro-)chemical installations.

A.5 QRA Tools for Consequence Analysis

A.5.1 FLACS®

The following is a brief description of this software, the function, vendor, pricing and main features.

• Function	3D CFD software tool for analysing gas and air flows in industrial environments as well as in the atmosphere. Major application areas include ventilation, dispersion, pool/jet-fires and gas explosions and subsequent blast effects
• Vendor	Gexcon AS, Bergen, Norway
• Pricing	Available on request through sales@gexcon.com

The development of FLACS® has been carried out continuously since 1980 with the co-operation, support, direction and funding of ten international oil and gas companies as well as legislative bodies of three countries. Application specific validation, wide applicability and efficiency when using FLACS® has been given high priority in the development work.

FLACS® has been the leading tool for explosion consequence prediction on off-shore and onshore petrochemical installations for more than two decades. It is also used in many other industries including nuclear, mining, marine and auto-motive. Every year FLACS® is used in safety studies and risk assessments on more than one hundred offshore and onshore facilities worldwide. In addition, FLACS® is approved for LNG Vapor Dispersion Modelling under U.S. Federal Regulations (49 CFR 193.2059).

The full FLACS-Standard version can handle explosion, dispersion, ventilation, mitigation and pool modelling. Separate/additional modules of FLACS include:

- FLACS-Dispersion
- FLACS-GasEx
- FLACS-Hydrogen
- FLACS-DustEx (for dust explosions)
- FLACS-Fire (for pool and jet-fires)
- FLACS-Risk

Typical applications of FLACS® include:

- Quantitative risk assessments
- Accident/incident investigations
- Identify/evaluate worst-case
- Safety evaluation of modifications
- Explosion venting of coupled and non-standard vessels
- Room/module layout optimization
- Predicting the effect of mitigation
- Blast waves and control rooms

- Drag loads on piping
- Exhaust pipe explosions
- Toxic gas dispersion
- LNG dispersion studies
- Gas detector optimization
- Planning and QA of experiments
- Assist certification processes
- Jet and pool-fires
- Dust explosions
- Tunnel fire and explosion safety
- Blast propagation from condensed explosives

A.5.2 KAMELEON FireEx-KFX®

The following is a brief description of this software, the function, vendor, pricing and main features.

• Function	CFD-based tool for prediction of gas dispersion and fire characteristics and response in complex geometries, as well as fire mitigation and extinguishment analysis and design
• Vendor	DNV GL AS, Hovik, Norway
• Pricing	Short and long term leasing contracts, licenses ranging from academic to commercial licenses

Kameleon FireEx-KFX® is an advanced simulator dedicated to gas dispersion and fire simulation with the following main characteristics:

- Three-dimensional transient finite-volume CFD code.
- Includes CAD import capabilities (PDS, PDMS, IGES, Flacs macro, others).
- Interfaced with the finite-element structure response codes Fahts/Usfos for dynamic structural response analysis.
- Includes detailed Lagrangian models for fire mitigation by water systems, for instance water mist systems, water curtains, deluge, sprinklers.
- Includes efficient and user—friendly pre- and post-processor capabilities, including options for animation of simulation results and "moving cameras" through simulations
- Originally developed by ComputIT/NTNU/SINTEF with the partners Statoil (now Equinor, N), Total (F), ENI-group (now Vår Energi, N), Hydro (now Equinor), ConocoPhillips (N), Gaz de France (now Neptune, N), Ruhrgas (D) and Sandia National Laboratories (USA).
- Extensively validated against small and large scale experimental data.
- Used for a large number of industrial analyses worldwide for more than 20 years.

Industrial analyses performed by KFX® can typically be:

- Simulation of all kind of fires; pool fires, jet fires, spray fires, flares, fire in enclosures, in complex geometries, in open space, in still air or in windy conditions. This includes detailed calculation of temperatures, radiation, smoke, visibility, concentrations of species, toxic gases, noise etc.
- Fire impact on structures and process equipment
- Optimization of passive fire protection
- Fire temperature, radiation and smoke impact on humans
- Evaluation of escape routes
- Simulation and evaluation of fire mitigation by water systems; sprinklers, deluge, mist, curtains
- Flare simulations; radiation, noise (not standard KFX® version), detailed tip simulations, ignition, startup
- Dispersion of gas
- Calculation of explosive cloud sizes
- Gas and fire detection systems
- Simulation and evaluation of LNG spills, including dispersion of LNG vapour and LNG fires
- Combustion in incinerators, furnaces, engines, burners and other combustion devices.
- Reduction of emissions; CO, NO_x, others
- HVAC (ventilation simulations)
- Turbulent flow analysis with respect to safe helicopter operation
- Fluid flow and combustion in general
- 3D visualization, animations, contour plot in real CAD geometry.

A.5.3 OLGA®

The following is a brief description of this software, the function, vendor, pricing and main features.

• Function	Transient multiphase flow simulator covering both hydraulic and thermal phenomena. Simulation of wells, flow lines, risers and process equipment separately or as an integrated system. OLGA® is also used extensively for blowout simulation and relief well design
• Vendor	Schlumberger
• Pricing	Available as rental, permanent licence or as cloud application

OLGA® is a simulator for transient multi-phase flow phenomena. OLGA® can model general networks of wells, flowlines, risers as well as process equipment. OLGA® is also used extensively for blowout simulation, relief well design and drilling related scenarios. The simulator covers the full spectrum from single phase oil, water or gas though the multiphase flow region. OLGA stability to predict release behaviour from condensate pipeline reflecting the impact of the topography, is of significant importance in feasibility and risk analysis of offshore developments. As the use of multiphase transportation of hydrocarbons has become the most economic and preferred solutions for field developments, OLGA® continues developed to extend its application and use. It is also connected to real time data for production monitoring as a digital twin.

OLGA® was first developed in 1980 and has been continuously improved to accurately capture and predict pressure gradients, liquid hold-up, flow regimes and flow rates. With the latest High Definition modelling (HD), further steps in accuracy are being made. OLGA® is verified and validated against more than 10,000 experiments at the two and three-phase-flow test loop operated by SINTEF in Trondheim and IFE at Kjeller. Additionally, OLGA® is verified by the largest database of multiphase wells and pipeline data in multiclient project, OLGA® Verification and Improvement Project (OVIP).

A.5.4 PHAST®

The following is a brief description of this software, the function, vendor, pricing and main features.

• Function	Windows-based toolkit for determination of consequences of accidental releases of hazardous material
• Vendor	DNV GL Limited, London, UK
• Pricing	Available on request

PHAST® (Process Hazard Analysis Software Tools) is a Windows-based toolkit, which determines the consequences of accidental releases of hazardous material. It examines the progress of a potential incident from initial release, through formation of a cloud, with or without a pool, to its dispersion. The program uses DNV GL's unique Unified Dispersion Model (UDM) to apply the appropriate entrainment and dispersion models as the conditions change and to integrate the relevant individual models such that the transition from one behaviour pattern to another is smooth, continuous and automatic. The discharge, pool formation and vapour and gas dispersion results are used to automatically calculate toxic dose, probit and lethality, flammable effects including pool fire, fireball, jet fire, flash fire and Multi Energy and Baker Strehlow Tang explosions. Access to specific calculations at any step is also possible for detailed analysis of individual phenomena. It is applicable to all stages of design and operation across a range of process and chemical industry sectors and may be used to identify situations which present potential hazards to life, property or the environment.

A.5.5 FRED®

The following is a brief description of this software, the function, vendor, pricing and main features.

• Function	The FRED software gathers in one tool state-of-the-art Fire, Release, Explosion and Dispersion models that predict consequences of accidental and design releases of products from process, storage, transport and distribution operations Typical uses include the modelling of release rates, fires, explosions, and gas dispersion. It also includes specialized models, including for flare modelling, congested and confined explosions, compartment fires, and tank overfills
• Vendor	Gexcon AS, Bergen, Norway
• Pricing	Available on request through sales@gexcon.com

The FRED software has been continuously developed and validated by Shell since the 1980s, and has been extensively used by oil, gas and petrochemical operating companies, engineering contractors, insurers and regulators throughout the world. Since January 2018, Gexcon AS has been the exclusive commercial distributor of the FRED tool.

Typical use is in onshore and offshore Oil & Gas installations, such as upstream exploration and production, transportation and refining of petrochemical products, often combined with the QRA software Shepherd.

A.5.6 USFOS®/FAHTS®

The following is a brief description of this software, the function, vendor, pricing and main features.

• Function	Structural fire and explosion analysis
• Vendor	USFOS AS, Norway
• Pricing	Available on request

USFOS® is a computer program for collapse analyses and accidental load analyses of fixed offshore structures, intact or damaged. The program simulates the collapse process of space frame structures, from the initial yielding, through to the formation of a complete collapse mechanism and on to the final toppling of the structure.

The USFOS® program has been in commercial use since 1985 by oil companies and engineering consultants all over the world. The program has been extensively used in areas such as inspection planning, lifetime extension and integrity assessment of ageing structures, and in fire protection assessment for new designs. It is verified through participation in extensive benchmark activities both in Europe and

USA, through comparison with experiments and through extensive scientific publication.

The particular characteristic features of USFOS® include:

- The program traces the entire collapse and post collapse behaviour of the structure, including global unloading, member unloading and redistribution of forces.
- The program requires only one finite element per physical member of the structure.
- The program employs efficient solution algorithms (SPARSE technology), performing complete collapse analyses or time-domain dynamic analyses in short time.
- Robust incremental/iterative solution procedures are implemented, with automatic step scaling and verification of numerical solution accuracy.
- The program comes with an extremely powerful and versatile graphical post processor with full 3D graphics and image plots.

FAHTS® is a specialized tool for framed structures. The technology is based on non-linear finite element technique with a special handling of boundary conditions (such as insulation etc.).

FAHTS® has an interface to Kameleon Fire Ex® (KFX), which ensures effective and accurate transfer of data to the structure. The software is used to prepare temperature data for structural response analysis with USFOS®.

FAHTS® can be used in the following applications:

- Evaluation of need for thermal insulation (passive fire protection of steel and aluminium structures
- Simulation of pipelines and pressurized vessels exposed to fire
- Optimization of Passive Fire Protection, PFP
- Estimation of structural integrity during fire (together with USFOS®)
- Evaluation of effect from deluge

A.5.7 PIPENET™

The following is a brief description of this software, the function, vendor, pricing and main features.

• Function	Steady state and dynamic fluid flow analysis in pipe and duct networks
• Vendor	Sunrise System Limited, Cambridge, UK
• Pricing	Available on request

PIPENET® is the global leader in the field of flow analysis of pipe and duct networks in the steady state as well as dynamic state. The software can be used for hydrocarbon fluids, water, gases and steam. It is widely used and specified around the globe by the largest companies in the oil and gas, power, petrochemical and shipbuilding industries.

The following three modules of PIPENET are used extensively:

- PIPENET Spray/Sprinkler Module sets the global standard for the design of fire protection systems, especially in the oil and gas, petrochemical and power industries—deluge, ringmain, sprinkler, foam solution and foam concentrate systems complying with the NFPA and other rules.
- PIPENET Transient Module is ideal for analysing dynamic flow events like pressure surge, water hammer, steam hammer, modelling control loops and calculating hydraulic transient forces for pipe stress analysis and other safety related applications.
- PIPENET Standard Module is a perfect tool for solving general steady-state flow problems with liquids, gases or steam—in pipe and duct networks—cooling water systems, steam distribution systems, HVAC systems.

Typical applications include:

- Design of fire protection systems for offshore platforms, FPSOs, onshore terminals, refineries, power stations and ships.
- Pipelines, loading/unloading systems, water injection systems, crude oil transfer lines, main steam lines in power stations, reheat lines, cooling water systems and fuel oil lines.
- Steam distribution systems, utility systems, ventilation systems and water distribution systems.

A.5.8 VessFire®

The following is a brief description of this software, the function, vendor, pricing and main features.

• Function	Heat transfer, depressurisation and stress modelling from fires
• Vendor	Petrell AS, Norway
• Pricing	Available on request

VessFire® is a simulation program for time-dependent non-linear analysis of thermo-mechanical response during blow-down of process segments and process equipment exposed and unexposed to fire. VessFire® solves the problem of heat transfer, conduction, thermodynamics of object contents and stress using a coupled approach. VessFire® also includes the strain-based approach for calculation of time to rupture. VessFire® is based on a coupled solution of problems using a combined numerical and analytical approach to simulate:

- Heat transfer from the fire onto the vessel, flow line, heat exchanger and/or pipe work surface.
- Heat transfer through the fire protective coating, thermal insulation or protective shield.
- Heat conduction through the object shell.
- Heat transfer from the inner object surface to the object contents.
- Thermodynamics of the object contents.
- Variation of pressure in the object due to depressurization counter-acted by the increase of the pressure due to evaporation, boiling and expansion of object contents.
- Stress in the object shell.
- Temperature in the depressurization pipe work for material selection.
- Time to object failure.
- Each component is calculated separately having its own temperature and content mixture.

A.5.9 HYENA®

The following is a brief description of this software, the function, vendor, pricing and main features.

• Function	Fire sprinkler/hydrant analysis
• Vendor	ACADS-BSG
• Pricing	Available on request

HYENA® can be used to analyze automatic fire sprinkler systems with a simple end, side or center fed configuration or more complicated looped and gridded systems. It may also be used to analyze fire hydrant and hose reel installations or combined sprinkler, hydrant and/or hose reel systems or any other systems where the discharges can be represented by a k factor and minimum flow. With a given sized network the program performs a complete hydraulic analysis determining the water flow in, and pressure drop though, each pipe in the entered network taking account of all fittings entered by the user.

Main features of HYENA® include:

- The program is capable of analysing looped and gridded systems as well as the more conventional tree configurations.
- The program can be used to carry out a sprinkler system analysis in accordance with NFPA, NZ4541, AS2118 or SSPC52 (Singapore) or GB50084 (China) or to carry out an analysis of hydrant systems with or without hoses or hose reels
- The piping system can have up to 10 input points and these can be modelled as a fixed pressure, a town mains water supply or a pump

- The program can work in a wide range of units including Metric and British or US and uses the Hazen-Williams formula for the hydraulic analysis.
- The program can also analyse mist systems in accordance with NFPA750 or A54587 using the Darcy Weisbach formula. This also allows the analysis of systems using fluids other than water.
- The program operates under WINDOWS® and all input data is via a series of screens with numerous features including drop down lists, selection lists, various sort options, etc; to facilitate easy data input
- All nodes that are not nominated as discharges, or input points are automatically assigned as reference nodes to save input, the user only having to assign elevations.
- The program can handle a range of pipe materials including flexible and HDPE pipes as well as any user defined pipe materials.
- Balancing devices can be included (orifice plates, pressure regulating valves etc.).
- The program can determine the required input pressure or the required flow and pressure of a booster pump or determine the flow from all discharges for a given input pressure.

A.5.10 ANSYS Fluent®

The following is a brief description of this software, the function, vendor, pricing and main features.

• Function	Physical modeling of flow, turbulence, heat transfer, and reactions for industrial applications
• Vendor	ANSYS, USA
• Pricing	Available on request

ANSYS Fluent® software contains the broad physical modeling capabilities needed to model flow, turbulence, heat transfer, and reactions for industrial applications ranging from air flow over an aircraft wing to combustion in a furnace, from bubble columns to oil platforms, from blood flow to semiconductor manufacturing, and from clean room design to wastewater treatment plants.

User-defined functions allow the implementation of new user models and the extensive customization of existing ones. The interactive solver setup, solution and post-processing capabilities of ANSYS Fluent® make it easy to pause a calculation, examine results with integrated post-processing, change any setting, and then continue the calculation within a single application.

Case and data files can be read into ANSYS CFD-Post® for further analysis with advanced post-processing tools and side-by-side comparison of different cases.

A.5.11 ANSYS CFX®

The following is a brief description of this software, the function, vendor, pricing and main features.

• Function	Fluid dynamics program to solve wide-ranging fluid flow problems
• Vendor	ANSYS, USA
• Pricing	Available on request

ANSYS CFX® software is a fluid dynamics program that has been applied to solve wide-ranging fluid flow problems for over 20 years. The highly parallelized solver is the foundation for an abundant choice of physical models to capture virtually any type of phenomena related to fluid flow. The solver and its many physical models are wrapped in a modern, intuitive, and flexible GUI and user environment, with extensive capabilities for customization and automation using session files, scripting and a powerful expression language.

Modeling Capabilities of ANSYS CFX® include the following:

- Laminar and turbulent
- Steady-state and transient
- Incompressible to fully compressible (subsonic, transonic, supersonic)
- Ideal and real gases
- Newtonian and non-Newtonian fluids
- Heat transfer
- Radiation
- Rotating and stationary
- Eulerian multiphase
- Free surfaces (VoF)
- Lagrangian particle tracking
- Chemical reactions and combustion
- Mesh motion and remeshing
- Immersed solids
- Fluid structure interaction

A.5.12 GASP

The following is a brief description of this software, the function, vendor, pricing and main features.

• Function	Modelling of pool spreading and evaporation from liquid spills on land or water
• Vendor	ESR Technology, UK
• Pricing	Not available

GASP (Gas Accumulation over Spreading Pools) is a pool spread and vaporisation model that has been developed in association with the UK Health and Safety Executive. The GASP model includes:

- a unified model for wind driven vaporisation (evaporation) and boiling;
- spreading over smooth and rough land;
- spreading over deep water;
- interface with gas dispersion model DRIFT.

A.5.13 DRIFT

The following is a brief description of this software, the function, vendor, pricing and main features.

• Function	Toxic and flammable gas dispersion modelling
• Vendor	ESR Technology, UK
• Pricing	Not available

DRIFT (Dispersion of Releases Involving Flammables or Toxics) is an advanced gas dispersion model that has been developed in association with the UK Health and Safety Executive. DRIFT Version 3 incorporates a considerable number of modelling enhancements over previous versions of DRIFT. As well as modelling heavy gas dispersion, DRIFT now also models:

- buoyant lift-off and rise;
- momentum jets;
- finite-duration and time-varying releases;
- thermodynamics of multi-component mixtures.

A.5.14 EFFECTS®

The following is a brief description of this software, the function, vendor, pricing and main features.

• Function	EFFECTS is an affordable and easy-to-use software tool that allows the user to calculate the physical effects and consequences of the (accidental) release of toxic and/or flammable gases, liquefied gases and liquids. Typical scenarios include various types of fires, explosions, dispersion and releases
• Vendor	Gexcon AS, Bergen, Norway
• Pricing	Available on request through sales@gexcon.com

The EFFECTS software was originally developed by TNO in The Netherlands and is based on the TNO "coloured books": the "Yellow Book" and the "Green Book". The software has been available since the 90's and is one of the leading tools world-wide for consequence analysis using integral modelling. End of 2018 TNO and Gexcon AS started a joint venture to further develop and market the software.

Typical use is in onshore (petro-)chemical installations, often combined with the QRA software RISKCURVES.

A.6 Qualitative Risk Assessment Software

A.6.1 HAZOP Manager®

The following is a brief description of this software, the function, vendor, pricing and main features.

• Function	Recording and Reporting of Safety Reviews
• Vendor	Lihou Technical & Software Services, UK
• Pricing	Available on request

HAZOP Manager® is a comprehensive Personal Computer program for the management of Hazard and Operability Studies (HAZOPs) and other similar safety-related reviews (e.g. PHA, HAZID, FMEA, SIL, etc.) It is extensively used to conduct more efficient and effective studies.

HAZOP Manager® incorporates features and facilities that:

- Serve as a framework within which preparation for the review can be structured.
- Ease the task of recording the meeting minutes, and help to maintain the team's focus of attention and interest.
- Give speedy access to material useful to the study team, such as previously identified problems, failure rate data and other such historical information.
- Allow professionally formatted reports to be produced with the minimum of effort.
- Permit additional management information to be extracted from the study records.
- Provide a comprehensive and easy to use system for effective action follow-up and close-out, without the significant administrative burden that this usually entails.

A.7 Reporting and Analysis of Incidents and Accidents

A.7.1 CGE IncidentXP

In the upcoming section, five different methods will be highlighted that are present in the incident analysis software called IncidentXP. The advantages and differences of each will be presented. In addition to the methods, each method comes with the Timeline tool as well as the Incident Manager.

The Incident manager helps guide users through the steps of incident investigation. It provides a structured approach consisting of 6 steps in total: Details, Investigation, Analysis, Recommendations, Attachments and Report. The user is free to go back and forth between these 6 steps as he or she sees fit.

The Timeline tool allows users to enter all of their evidence and events in a simple clear diagram. The data can be reordered when necessary and highlighted to indicate extra importance. In short, the Timeline is used to support the fact finding part of an incident analysis.

A.7.2 DNV BSCAT™

The following is a brief description of this software, the function, vendor, pricing and main features.

• Function	Modern risk-based safety management approaches to systematic root cause incident investigation
• Vendor	CGE Risk Management Solutions, Netherlands
• Pricing	Available on request

The BSCAT™ method refers to a method that links modern risk-based safety management approaches to systematic root cause incident investigation. The "B" refers to barrier-based as each barrier identified in bowtie risk assessments is tested for why it failed.

SCAT (Systematic Cause Analysis Technique) is a well-established root cause analysis approach which incorporates the DNV loss causation model. The model is a sequence of dominos establishing the hierarchy of accident progression from the type of event, to the immediate cause back to fundamental root causes and system failures, and hence to necessary actions for improvement.

In short, BSCAT™ is the barrier based extension to DNV's SCAT method. The SCAT model was developed to help incident investigators apply the DNV loss causation model to actual events. This is done by means of the SCAT chart. The chart was created to build-out an event using standardized event descriptions that can fit the whole range of incidents and near misses. Due to using standardized categories, this assists investigators to assess events in a systematic manner and making these suitable for aggregation, leading to more insight into the weak areas of the safety management system and the underpinning risk assessment. The barrier-based accident investigation still applies the SCAT model but now it is applied to each barrier separately, not to the incident as a whole.

The BSCAT™ software allows investigators to reuse and link existing risk assessment information (bowties) and to do full integration of incident analysis and risk analysis. If applicable bowtie diagrams are available for use during the investigation, the analysis can bring events and barriers from the bowtie directly into the BSCAT™ analysis. This means that every incident investigation refers back to the risk assessment.

By reusing the bowtie risk analysis and/or describing the barriers in the incident analyses, more value is extracted from the incident analyses. Incidents highlight weaknesses in barriers and these in turn may indicate the risk assessment is too optimistic and that specified risk targets are not being met. This information is beyond that normally identified in incident investigations.

This entire process makes it possible to gauge barrier effectiveness and availability based on real operations and linked to ongoing incident analyses. This process is potentially the richest source of barrier performance information available to a facility.

In general, BSCAT™ is a well proven root cause analysis technique which uses standardized immediate and basic cause categories, and this allows incident analyses to aggregate for trends, leading to more insight into the weak areas of the facility safety management system. It has been updated to link it directly to facility risk assessments and thereby combines root cause analysis and risk assessment into a single tool.

A.7.3 BFA

The following is a brief description of this software, the function, vendor, pricing and main features.

• Function	Modern risk-based safety management approaches to systematic root cause incident investigation
• Vendor	CGE Risk Management Solutions, Netherlands
• Pricing	Available on request

The Barrier Failure Analysis (BFA) method is a method developed by CGE that tries to find the root cause of each barrier failure. It is a simplified form of the BSCAT method that does not use any of the lists that the BSCAT method uses.

The model uses a sequence of events with barriers in-between each event to show the progression of the incident. Each barrier's state can be assigned to display exactly how it failed: failed, missing, inadequate, unreliable, or, in the case of a near miss, effective. This can then be used to stimulate the thought processes of the causation process. Once all these steps are complete, areas of improvement can be found.

The BFA software allows investigators to reuse and link existing risk assessment information (bowties) and to do full integration of incident analysis and risk analysis. If applicable bowtie diagrams are available for use during the investigation, the analysis can bring events and barriers from the bowtie directly into the BFA analysis. This means that every incident investigation refers back to the risk assessment.

By reusing the bowtie risk analysis and/or describing the barriers in the incident analyses, more value is extracted from the incident analyses. Incidents highlight weaknesses in barriers and these in turn may indicate the risk assessment is too optimistic and that specified risk targets are not being met. This information is beyond that normally identified in incident investigations.

This entire process makes it possible to gauge barrier effectiveness and availability based on real operations and linked to ongoing incident analyses. This process is potentially the richest source of barrier performance information available to a facility.

In general, BFA is a root cause analysis technique which allows incident analyses to aggregate for trends, leading to more insight into the weak areas of the facility safety management system.

A.7.4 Stichting Tripod Foundation—Tripod Beta

The following is a brief description of this software, the function, vendor, pricing and main features.

• Function	Modern risk-based safety management approaches to systematic root cause incident investigation
• Vendor	CGE Risk Management Solutions, Netherlands
• Pricing	Available on request

Developed in the early 1990s, Tripod Beta is an incident investigation methodology designed in line with the human behavior model. It was explicitly created to help accident investigators model incidents in a way that allows them to understand the influencing environment and uncover the root organizational deficiencies that allowed that incident to happen. It differentiates itself from the BSCAT and BFA methods described above through the way it sets up the causal chain of events. Rather than just going from event to event, Tripod Beta uses a tripod, or trio structure. The trios consist of an Agent, an Object and an Event. The Agent is a force that acts on the Object resulting in an Event. There are barriers on the lines between Agent and Event as well as on the line between and Object and Event.

Sometimes it is possible to have multiple Agents or Objects resulting in the same Event.

Tripod Beta is a well-established root cause analysis approach which has been developed by famous researchers such as James Reason and Jop Groeneweg. On each barrier, the investigator is encouraged to analyse the causation path of the barrier failure, for which the Stichting Tripod Foundation has developed clear guidelines. Once the underlying causes have been found, actions can be assigned either on the system level, the underlying cause, or on immediate level, the barrier. Additionally, the Tripod Beta method can help find trends in barrier failures through the basic risk factors (BRF). Basic risk factors have been defined by looking at numerous incidents and distilling the most frequently occurring type of incidents. These basic risk factors assist investigators to assess events in a systematic manner and allow better aggregation of data, leading to more insight into the weak areas of the safety management system and the underpinning risk assessment.

The Tripod Beta software allows investigators to reuse and link existing risk assessment information (bowties) and to do full integration of incident analysis and risk analysis. If applicable bowtie diagrams are available for use during the investigation, the analysis can bring events and barriers from the bowtie directly into the Tripod Beta analysis. This means that every incident investigation refers back to the risk assessment.

By reusing the bowtie risk analysis and/or describing the barriers in the incident analyses, more value is extracted from the incident analyses. Incidents highlight weaknesses in barriers and these in turn may indicate the risk assessment is too optimistic and that specified risk targets are not being met. This information is beyond that normally identified in incident investigations.

This entire process makes it possible to gauge barrier effectiveness and availability based on real operations and linked to ongoing incident analyses. This process is potentially the richest source of barrier performance information available to a facility.

In general, Tripod Beta is a well-established root cause analysis technique which uses clear guidelines to assist investigators into finding the right causation pathways. Using the basic risk factors it also this allows incident analyses to aggregate for trends, leading to more insight into the weak areas of the facility safety management system. It has been updated to link it directly to facility risk assessments and thereby combines root cause analysis and risk assessment into a single tool.

A.7.5 RCA

The following is a brief description of this software, the function, vendor, pricing and main features.

• Function	Modern risk-based safety management approaches to systematic root cause incident investigation
• Vendor	CGE Risk Management Solutions, Netherlands
• Pricing	Available on request

Root Cause Analysis (RCA) is a simple and straightforward incident analysis technique. It starts with an incident and drills down into the chain of events that led to that incident until the root causes are identified. This method is widely used throughout the world, and the idea of drilling down to the root cause is also present in all of our other incident analysis methods.

However, a traditional root cause analysis has the potential to turn into a jumble of elements. The RCA software tries to improve on this in two ways. First, subtle categorizations have been added that help the viewer immediately spot where the real problem areas are. These categorizations are optional. Second, it is possible to cut up a large diagram into smaller pieces and link them together. Separating the main diagram from sub-diagrams avoids a situation where the diagram becomes so large you lose overview.

The main difference between RCA and the other incident analysis methods is that RCA is not barrier based. Everything in RCA is an event, including those things that would be considered barriers or barrier failures in BSCAT, Tripod or BFA. However, the aggregation of incident data through the bowtie method is considered so valuable, that an artificial connection can be created in the software. Using the added categorizations, it is possible to signal that an RCA event is a failed barrier, which can then be linked to existing barriers in bowtie diagrams.

In general, RCA is a well proven root cause analysis technique. It has been updated to link it directly to facility risk assessments and thereby combines root cause analysis and risk assessment into a single tool.

A.7.6 TOP-SET RCA

The following is a brief description of this software, the function, vendor, pricing and main features.

• Function	Modern risk-based safety management approaches to systematic root cause incident investigation
• Vendor	CGE Risk Management Solutions, Netherlands
• Pricing	Available on request

TOP-SET® was developed in 1989 to provide an incident investigation methodology that takes users from the start of the process of investigation right through to the end. The Kelvin TOP-SET RCA method is a more structured approach of the regular RCA method. In the software four additional layers are added to the diagram, guiding analyses. The layers are: 'What happened', 'Consequences', 'Intermediate causes', 'Underlying causes' and 'Root causes'.

The TOP-SET acronym stands for Technology, Organization, People, Similar Events, Environment and Time. These are used in the Timeline feature of the software.

The main difference between TOP-SET® RCA and the other incident analysis methods is that TOP-SET® RCA is not barrier based. Everything in TOP-SET® RCA is an event, including those things that would be considered barriers or barrier failures in BSCAT, Tripod or BFA. However, the aggregation of incident data through the bowtie method is considered so valuable, that an artificial connection can be created in the software. Using the added categorizations, it is possible to signal that an TOP-SET® RCA event is a failed barrier, which can then be linked to existing barriers in bowtie diagrams.

Note: The Kelvin TOP-SET RCA method is a complete incident investigation method that walks you through the very start of an incident to the analysis of it. For instance, Kelvin TOP-SET makes use of reference lists to ensure better quality incident investigations. The software of CGE, however, helps mostly with the analysis part of incident investigation. Therefore, it is advised to contact Kelvin TOP-SET directly to gain a better understanding of their full methodology.

A.8 Risk Management Software

A.8.1 Synergi RBI®

The following is a brief description of this software, the function, vendor, pricing and main features.

• Function	Onshore and Offshore risk based inspection tool
• Vendor	DNV GL Limited, London, UK
• Pricing	Available upon request

Synergi RBI® are software tools that use DNV GL's Risk Based Inspection (RBI) techniques to help users optimise their inspection management programme. DNV GL's RBI technique, as described in DNV GL's RP-G 101 "Recommended Practice for Risk-Based Inspection of Topsides Static Mechanical Equipment", is used to calculate the risk due to corrosion, erosion and cracking for pressure equipment in a marine environment. DNV GL's onshore RBI method fully comprises API 580 and 581 and goes further, providing complete support for a best practice approach to risk-based inspection for onshore facilities.

Synergi RBI® is designed to help users sustain high productivity and reliability of their offshore platforms by minimising lost production and downtime through effective inspection and maintenance. It assists in the management of safety and equipment integrity to user specified levels.

It helps users to achieve these objectives systematically and efficiently. It allows you to quantify the risk for process and utility systems and equipment on topsides and FPSOs. Risk can be defined in terms of potential loss of life, cost, or both. A cost-effective inspection programme can then be devised based on the greatest risk reduction per cost of inspection.

A.8.2 BowTieXP®

The following is a brief description of this software, the function, vendor, pricing and main features.

• Function	Management of major risks to people, the environment, assets and reputation by means of 'bow-tie' graphical interface diagram
• Vendor	CGE Risk Management Solutions, Netherlands
• Pricing	Available on request

The Bowtie method is a risk evaluation method that can be used to analyze and demonstrate causal relationships in high risk scenarios. The method takes its name from the shape of the diagram that you create, which looks like a men's bowtie. A Bowtie diagram does two things. First of all, a Bowtie gives a visual summary of all plausible accident scenarios that could exist around a certain Hazard. Secondly, by identifying control measures, the Bowtie displays what a company does to control those scenarios.

However, this is just the beginning. Once the control measures are identified, the Bowtie method takes it one step further and identifies the ways in which control measures fail. These factors or conditions are called Escalation factors. There are possible control measures for Escalation factors as well, which is why there is also a special type of control called an Escalation factor control, which has an indirect but crucial effect on the main Hazard. By visualizing the interaction between Controls and their Escalation factors one can see how the overall system weakens when Controls have Escalation factors.

Besides the basic Bowtie diagram, management systems should also be considered and integrated with the Bowtie to give an overview of which activities keep a Control working and who is responsible for a Control. Integrating the management system in a Bowtie demonstrates how Hazards are managed by a company. The Bowtie can also be used effectively to assure that Hazards are managed to an acceptable level (ALARP).

By combining the strengths of several safety techniques and the contribution of human and organizational factors, Bowtie diagrams facilitate workforce understanding of Hazard management and their own role in it. It is a method that can be understood by all layers of the organization due to its highly visual and intuitive nature, while it also provides new insights to the HSE professional.

A.8.3 THESIS

The following is a brief description of this software, the function, vendor, pricing and main features.

• Function	Management of major risks to people, the environment, assets and reputation by means of 'bow-tie' graphical interface diagram
• Vendor	ABS Consulting Ltd., Warrington, UK
• Pricing	Please visit www.absgroup.com/thesis and contact us for the most up-to-date pricing information

The THESIS BowTie™ (THESIS) risk management software application delivers simplified, integrated risk management solutions for the entire business portfolio. Enhanced visuals in THESIS make the core components of the management process more readily understandable at all levels across an organization.

Supported by ABS Consulting Ltd., a subsidiary of ABS Group of Companies, Inc. (ABS Group), THESIS helps clients analyze and manage a range of hazards and risks facing a global enterprise. Through a rich graphical interface, the software displays the relationship between hazards, controls, risk reduction measures and business activities. While communicating critical procedures and individual responsibilities to employees, the software demonstrates compliance clearly across all management levels to external stakeholders such as regulators, principal investors and the public.

To address global corporate governance requirements, a web-based version of THESIS was created to supplement the more traditional standalone version of the software.

Areas where THESIS can be used:

- High level hazard identification (hazard register) and risk assessment
- Detailed HAZID—derivation of threats, consequences and controls
- GAP analysis
- Management of controls (safety critical elements and soft/non hardware types)
- Derivation of tasks and procedures
- Document management
- Shortfall and Remedial Action tracking
- Focusing on personnel critical tasks
- As a complement to the Safety Case, HSE Case or HSEIA
- Incident investigation
- Illustrates the status and management of risk within a business to senior management, the workforce and regulators

- Enterprise Risk Management
- As the Safety/Live Risk module

In order to define a standardized Bowtie methodology for process industries, the Center for Chemical Process Safety (CCPS), together with the UK Energy Institute (EI), developed Bow Ties in Risk Management: A Concept Book for Process Safety (Wiley, 2018). ABS Consulting Ltd. provided guidance to CCPS and EI while developing the concept book and has updated the THESIS software to align with this industry standard text.

A.8.4 BowTie Pro™

The following is a brief description of this software, the function, vendor, pricing and main features.

• Function	Qualitative, semi-quantitative and illustrative risk analysis and risk management
• Vendor	BowTie Pro, Aberdeen, UK
• Pricing	Available on request

BowTie Pro™ is a tool to facilitate the creation of risk assessments utilizing the latest Microsoft.NET technology. The bowtie diagram is a powerful visual presentation of the risk assessment process that can be readily understood by the non-specialist.

BowTie Pro™ can have up to 6 diagrams open at once allowing the copy and paste between diagrams and tailor each diagram by reordering and changing the display properties. BowTie Pro™ also allows a great deal of detailed information to be recorded against the controls such as tasks, task assignment, documents, hyperlinks, verification method etc. There is also the facility to link the controls to incident investigation packages and record BRFs etc.

The analysis features of BowTie Pro™ include:

- Risk Profiling allows an interactive version of the hazard register to be analysed and modified on the screen.
- Risk Profiling Matrix allows the user to see how the risks have been assessed and how many times each item has been used.
- Critical Task Listings allows the visibility, filtering and use of tasks in an easy to use screen across a BowTie Pro™ file.
- The People Matrix displays responsibilities for items across a BowTie Pro™ file.
- Deficiency Analysis allows a range of functions to analyse any deficiencies identified when creating a diagram.
- The Layers of Protection allow a numerical analysis of each strand of a BowTie Pro™ diagram.

- The Quality Check module ensures that the data is relevant based on various criteria
- The Document Matrix shows where a reference document is used within a BowTie Pro™ file.
- The Permitted Operations and Permitted boundary operations modules allow the creation of a matrix based on various criteria similar. This allows the creation of the Manual of Permitted Operation (MOPO).
- File Searching searches for all instances of a word within the current BowTie Pro™ file. These can easily be edited by double clicking on the item. There is also Report Searching where you search for the text within a report.

A.8.5 Presight® Operations & Barrier Safety KPI Management

The following is a brief description of this software, the function, vendor, pricing and main features.

• Function	Provide digitalisation solutions for increased safety and efficiency. Solution provides operational reporting and barrier monitoring. Visualize status of barriers and performance including technical and non-technical (Human factor) in a dashboard environment. Status is used by management to take informed decision with regards to daily operations. Presight utilize data for customers data sources to compile and present current status
• Vendor	Presight Solutions AS, Norway (www.presight.com)
• Pricing	Available on request

The strength in Presight Performance Monitoring is the ability to extract data from any underlying source system. Presight PM aggregates information regarding risk, strategy, operations, quality, projects, processes, HR and finance in a single interface to provide the right business context to make the right decisions. Presight PM presents the information easily and transparently in various interactive dashboards for different decision makers. The information is presented in various formats such as tables, graphs, images, figures and lists in a visual interface. Quick navigation to relevant underlying data helps locate the source of potential problems. The result is easier and more automated access to the important information.

The software Presight Barrier Monitor (BM) is built to compile large amounts of data from all relevant source systems. The data is then structured, aggregated and visualized in an easy and understandable way. The aim of the Presight BM is to provide companies with leading barrier performance indicators for crew and management in decision making situations.

Presight Operations Reporting automates production reporting by consolidating manual data entry, automatically collect data where possible and deliver accurate and timely reports. Presight OR will provide many advantages such as saving man-hours to fill in the data, reduce inaccuracy, combine all information into a single database, have an on-line dashboard, prepare customized reports etc. Presight OR combines data regarding well status, daily and weekly drill, flight and marine traffic, bit/coring, 24 h operation, HSEQ, misc. stock, consumables, weight onboard etc. Presight OR provides the ability to register data by handheld devices, to deliver useful operations and end-user focused indicators. Drill-down to underlying indicators, bow-tie visualization or source system to identify the specific safety critical equipment, process, organization or human factor performance failures for follow up and action between off- and onshore units.

Early warning when status today and trend is not enough. The Energy Institute suggests that "*most well-run organizations can state how many accidents occurred over a certain period in the past, but the "real challenge" is to assess the likelihood of an incident happening tomorrow*." Prevent the escalation of major accident risk scenarios through use of Presight advanced control and forecast indicators in decision making situations. Implement automatic barrier fail and early warnings notifications to responsible person or role, off- and onshore—through web, email or mobile devices—for corrective and preventive actions. Track deviations and override decisions against individual indicators. Easily compare different indicators against period for performance analysis and experience transfer opportunities. Identify underlying causes and focus areas for operations safety performance improvement initiatives.

Enterprise strength KPI administration features. Presight off-the-shelf approach to development includes administration features to meet regional and global operations and regulatory standards. Presight is language and measure of unit independent, SharePoint ready, data integrity status, log of all changes, powerful search and multi-indicator edit and copy features for global and multi-asset operations.

A.8.6 Synergi Life®

The following is a brief description of this software, the function, vendor, pricing and main features.

• Function	Complete business solution for risk and QHSE management, managing all non-conformances, incidents, risk, risk analyses, audits, assessments and improvement suggestions
• Vendor	DNV GL Limited, London, UK
• Pricing	Available on request

The Synergi Life® software (previously named Synergi) is a complete business solution for risk and QHSE management, managing all non-conformances, incidents, risk, risk analyses, audits, assessments and improvement suggestions.

The Synergi Life® software covers every workflow process, such as reporting, processing, analysing, corrective actions, communication, experience transfer, trending and KPI monitoring.

Synergi Life® is a module based HSE and Risk Management solution developed with a full set of optional modules for the various business needs relevant to our clients. These modules can be used as stand-alone solutions, or in combination to fit the exact needs and focus of each individual client and user. It is the combinations of several modules that contribute to a total risk and QHSE Management solution. Modules in the Synergi Life software package include:

- Synergi Life® Incident Management software
- Synergi Life® Quality Management software
- Synergi Life® Audit Management software
- Synergi Life® Activity Management software
- Synergi Life® Risk Management software
- Synergi Life® Environmental Management software
- Synergi Life® Improvement Management software
- Synergi Life® Anonymous Incident Management software
- Synergi Life® Inspection and BBS Management software
- Synergi Life® Deviation Management software
- Synergi Life® Hospital Infection Management software
- Synergi Life® Adverse Drug Reaction Management software (ADR)

A.8.7 WellMaster RMS

The following is a brief description of this software, the function, vendor, pricing and main features.

• Function	Reliability Management System
• Vendor	ExproSoft AS
• Pricing	Contact vendor for price

WellMaster RMS is a well and subsea equipment reliability database and analysis solution used by oil and gas operators worldwide. It provides equipment reliability analysis and RAM simulation capabilities in an easy to use graphical user interface. The solution is utilized through the full life cycle, from designing better wells and selecting better equipment, to risk assessment, well integrity analysis and remaining useful life assessments.

Appendix B
Overview of Fatalities in Norwegian Sector

An overview of all fatalities in the Norwegian sector of North Sea and Norwegian Sea is presented. The focus is on a statistical overview, split in occupational and major accidents, diving accidents and helicopter accidents.

B.1 Introduction

B.1.1 Background

The research programme 'Safety Offshore' (SPS) established the first overview of all fatalities in the Norwegian offshore industry. But this effort stopped when the programme ended after a five year period from 1978 until 1983. The author was able to obtain a printout from the SPS project team, which was the starting point of work conducted in Safetec Nordic AS, partially funded by the Norwegian Research Council, until the author left the company. The records were given to the author from Safetec Nordic AS, and the work has been continued by the author since 1993, financed by Preventor AS. This work has been the source of overviews presented annually by PSA in the Risk Level project.

The first well was spudded in the Norwegian sector in July 1966 using the Ocean Traveller mobile drilling unit. The first serious accident occurred on 6. November 1966, when the supply vessel Smith Lloyd 8 rammed into Ocean Traveler, puncturing 2 columns. Over 50 persons jumped overboard, whereas five persons remained onboard and managed to stabilize the installation before it capsized. All personnel in the sea were rescued by the supply vessel, and no fatalities occurred.

© Springer-Verlag London Ltd., part of Springer Nature 2020
J.-E. Vinnem and W. Røed, *Offshore Risk Assessment Vol. 2*, Springer Series
in Reliability Engineering, https://doi.org/10.1007/978-1-4471-7448-6

The first fatal accident occurred in 1967, this was a diving accident. The first accident on a mobile installation occurred in 1969, when the drilling manager was killed during testing of a manometer onboard Glomar Grand Isle. The first fatality on a production installation occurred in 1974, in a crane accident on Ekofisk B installation, when the crane fell overboard.

B.1.2 Limitations

The fatalities included in the tables in this appendix are limited mainly to what falls under the petroleum law, with some additions. This implies that the following are included in the statistics:

- Fatalities on offshore production installations
- Fatalities on offshore mobile installations when operating at an offshore field
- Fatalities on attending vessels when operating within the safety zone of an offshore installation
- Fatalities on attending vessels when operating in association with offshore installations
- Fatalities on crane and pipe laying vessels when operating in association with offshore installations
- Helicopter fatalities during all phases of helicopter transportation to/from shore.

The implications of what is listed above are that the following are excluded in the statistics:

- Fatalities on production installations during construction inshore or yard
- Fatalities on production installations during inshore decommissioning activities
- Fatalities on mobile installations during transit between offshore fields or to/from shore
- Fatalities on supply vessels when en route from shore to offshore installations
- Diving accidents during inshore training and testing.

B.2 Production Installations

Table B.1 presents the overview of fatalities on production installations since the first fatality in 1974 (see above), until 31.12.2018. 1974 was the first year with manhours logged on production installations. The fatalities are split in occupational accidents and major accidents.

It should be noted that the three occupational fatalities in 1991 occurred when a helicopter was employed on Ekofisk to replace a flare tip. The rotor hit the structure causing a crash of the helicopter in the sea, and the three persons onboard perished. This accident is not associated with transport of personnel by helicopter, and does

Table B.1 Fatalities on production installations, Norwegian sector, 1974–2018

Year	Fatalities in occupational accidents	Fatalities in major accidents	Year	Fatalities in occupational accidents	Fatalities in major accidents
1974	2		1997	0	
1975	2	3	1998	0	
1976	2		1999	1	
1977	2		2000	1	
1978	1	5	2001	0	
1979	0		2002	1	
1980	0	123	2003	0	
1981	0		2004	0	
1982	0		2005	0	
1983	0		2006	0	
1984	1		2007	0	
1985	1		2008	0	
1986	0		2009	1	
1987	0		2010	0	
1988	0		2011	0	
1989	1		2012	0	
1990	1		2013	0	
1991	3		2014	0	
1992	0		2015	0	
1993	2		2016	0	
1994	1		2017	0	
1995	1		2018	0	
1996	0				

not belong in the section with helicopter fatalities during helicopter transportation of personnel. It has therefore been included with the occupational accidents.

It should further be noted that the Alexander Kielland accident has been classified as production installation, although the Alexander Kielland installation was a mobile installation, a flotel, which was connected to a fixed installation.

There are two accidents after year 2000, on Gyda in 2002 (falling object) and Oseberg B (fall to lower level) in 2009.

B.3 Mobile Installations

Table B.2 presents the overview of fatalities on mobile installations since the first fatality in 1969 (see above), until 31.12.2018. The fatalities are split in occupational accidents and major accidents.

Table B.2 Fatalities on mobile installations, Norwegian sector, 1969–2018

Year	Fatalities in occupational accidents	Fatalities in major accidents	Year	Fatalities in occupational accidents	Fatalities in major accidents
1969	1		1994	0	
1970	1		1995	0	
1971	2		1996	0	
1972	0		1997	0	
1973	0		1998	0	
1974	1		1999	0	
1975	0		2000	0	
1976	0	6	2001	0	
1977	0		2002	1	
1978	0		2003	0	
1979	0		2004	0	
1980	0		2005	0	
1981	0		2006	0	
1982	1		2007	0	
1983	2		2008	0	
1984	0		2009	0	
1985	1	1	2010	0	
1986	0		2011	0	
1987	0		2012	0	
1988	0		2013	0	
1989	2		2014	0	
1990	1		2015	1	
1991	0		2016	0	
1992	0		2017	1	
1993	2		2018	0	

It should be noted that the six fatalities in a major accident in 1976 (grounding of Deep Sea Driller) occurred when the unit was towed to shore, just outside Fedje (north of Bergen). This accident should not be included if the criteria listed in Sect. B.1 are strictly adhered to, but this accident has for a long time been counted as an offshore accident.

There are three fatal accidents after year 2000, on Byford Dolphin (falling object in the derrick) in 2002, on COSL Innovator (wave impact in quarters) in 2015 as well as on Maersk Interceptor (dropped object) in 2017.

B.4 Vessels

Table B.3 presents the overview of fatalities on attending and pipe laying vessels since the first fatality in 1972 (see above), until 31.12.2012. The fatalities are split on attending vessels and pipe laying vessels separately. Attending vessels include supply vessels, standby vessels as well as anchor handling and tug vessels.

There were quite a number of fatal accidents on attending vessels in the period from 1994 until 2001. As a result of these eight fatalities, the authorities pushed the industry to focus on risk reduction, and it can be seen to have paid off, with no fatalities on attending vessels after 2001.

There are three accidents after year 2000, on Viking Queen in 2001 (hit by steel wire during anchor handling operations) and two on Saipem 7000 (falling object and fall overboard) in 2003 and 2007.

Table B.3 Fatalities on attending vessels and pipe laying vessels, Norwegian sector, 1972–2018

Year	Occupational fatalities on attending vessels	Occupational fatalities on pipe laying vessels	Year	Occupational fatalities on attending vessels	Occupational fatalities on pipe laying vessels
1972	2		1996	2	1
1973	1		1997	0	
1974	1		1998	0	
1975	1		1999	1	
1976	0		2000	1	
1977	5		2001	1	
1978	2		2002	0	
1979	1		2003	1	
1980	0		2004	0	
1981	0	1	2005	0	
1982	0		2006	0	
1983	0		2007	1	
1984	0		2008	0	
1985	0		2009	0	
1986	0		2010	0	
1987	0		2011	0	
1988	0		2012	0	
1989	0		2013	0	
1990	0		2014	0	
1991	1		2015	0	
1992	0		2016	0	
1993	0		2017	0	
1994	2		2018	0	
1995	1				

B.5 Shuttle Tankers

There has only been one fatal accident on shuttle tankers during the period from start of operations. The fatality occurred on the shuttle tanker Polytraveller on 1st August 1980, during off-loading on the Statfjord field. The mooring line failed, thus causing the loading hose to rupture and spill crude oil. The spill was ignited and the tanker captain who was positioned in the stern of the vessels was fatally injured by the fire.

No other fatalities have occurred during off-loading or during transit to onshore terminals and refineries.

B.6 Diving Accidents

Table B.4 presents the overview of diving fatalities in the Norwegian offshore operations since the first fatality in 1967 (see above), until 31.12.2012. No diving fatalities have occurred after 1987, but the volume of manned diving has been considerably reduced from the volume in the 1970s and 1980s. Use of Remote Operated Vehicle (ROV) has replaced the use of divers to a large extent.

It should be noted that fatalities during diving training inshore and similar activities have not been included. There are no diving accidents after 1987.

Table B.4 Fatalities during diving accidents, Norwegian sector, 1967–1987

Year	Fatalities in diving accidents	Year	Fatalities in diving accidents
1967	1	1978	0
1968	0	1979	0
1969	0	1980	0
1970	0	1981	0
1971	2	1982	0
1972	0	1983	5
1973	0	1984	0
1974	3	1985	0
1975	2	1986	0
1976	0	1987	1
1977	0		

B.7 Helicopter Accidents

Table B.5 presents the overview of fatalities during helicopter transport of personnel between offshore installations and heliports onshore in the Norwegian sector the first fatality in 1973, until 31.12.2018. there has been one accident after 1997, the Turøy accident in 2016, with 13 fatalities.

The fatal accident in 1997 occurred while the Norne FPSO was in the commissioning phase, prior to start-up of production operations. Personnel were being shuttled on a daily basis between shore and the FPSO on location, due to limited accommodation capacity onboard. Shuttling of personnel on such a scale was criticized after the accident, and has not been practiced since then. There has been a significant focus from the authorities to limit as far as possible the use of shuttling.

Table B.5 Fatalities during helicopter transportation of personnel to/from shore, Norwegian sector, 1967–2018

Year	Fatalities in helicopter accidents	Year	Fatalities in helicopter accidents
1973	4	1996	0
1974	0	1997	12
1975	0	1998	0
1976	0	1999	0
1977	12	2000	0
1978	18	2001	0
1979	0	2002	0
1980	0	2003	0
1981	0	2004	0
1982	0	2005	0
1983	0	2006	0
1984	0	2007	0
1985	0	2008	0
1986	0	2009	0
1987	0	2010	0
1988	0	2011	0
1989	0	2012	0
1990	0	2013	0
1991	0	2014	0
1992	0	2015	0
1993	0	2016	13
1994	0	2017	0
1995	0	2018	0

Appendix C
Network Resources

An overview of network resources that may be used for offshore risk assessment is presented, including data sources, investigation organisations and some recent investigation reports.

C.1 Data Sources

An overview of data sources is presented in Sect. 14.9. The following is an overview of the network accessible resources:

- All types of events

  - Oil and gas producers—https://www.iogp.org/
  - RNNP—http://www.ptil.no/rnnp

- HC leaks

  - RNNP—http://www.ptil.no/rnnp
  - HCR database—https://www.hse.gov.uk/hcr3/
  - IRF—https://irfoffshoresafety.com/country-performance/

- Structural & marine accidents

  - WOAD—https://www.dnvgl.com/services/world-offshore-accident-database-woad-1747 (login required)
  - IMCA—https://www.imca-int.com/ (login required)

- Reliability of safety systems etc.

  - OREDA—http://www.oreda.com/
  - PDS Forum—https://www.sintef.no/PDS

C.2 Investigation Bodies

The following is an overview of investigation boards and organisations for offshore and helicopter transportation accidents on a worldwide scale:

- Air Accidents Investigation Branch—https://www.gov.uk/government/organisations/air-accidents-investigation-branch
- Transportation Safety Board (Canada)—http://www.tsb.gc.ca
- NTSB—National Transportation Safety Board (US)—https://www.ntsb.gov
- US Chemical Safety Board (CSB)—https://www.csb.gov
- Bureau of Safety and Environmental Enforcement—https://www.bsee.gov
- PSA—Petroleum Safety Authority (Norway)—http://www.ptil.no
- Accident Investigation Board Norway (Statens havarikommisjon for transport)—https://www.aibn.no
- Health & Safety Laboratory—https://www.hsl.gov.uk
- National Offshore Petroleum Safety and Environmental Management Authority (NOPSEMA, Australia)—https://www.nopsema.gov.au
- Canada-Nova Scotia Offshore Petroleum Board (CNSOPB, Canada)—https://www.cnsopb.ns.ca
- Danish Maritime Authority, Division for Investigation of Marine Accidents (Denmark)—https://www.dma.dk
- National agency of oil, natural gas and biofuels (ANP, Brazil), Incident investigation reports—
  http://www.anp.gov.br/exploracao-e-producao-de-oleo-e-gas/seguranca-operacional-e-meio-ambiente/comunicacao-e-investigacao-de-incidentes/fpso-cidade-de-sao-mateus/368-exploracao-e-producao-de-oleo-e-gas/seguranca-operacional-e-meio-ambiente/comunicacao-de-incidentes/relatorios-de-investigacao-de-incidentes

C.3 Investigation Reports—Available Online

The following is an overview of some recent investigation reports for offshore and helicopter transportation accidents on a worldwide scale:

- The Public Inquiry into the Piper Alpha Disaster (Cullen report), 1988—http://www.hse.gov.uk/offshore/piper-alpha-public-inquiry-volume1.pdf. http://www.hse.gov.uk/offshore/piper-alpha-public-inquiry-volume2.pdf
- Roncador P-36 accident (Petrobras), 2001—http://www.anp.gov.br/?dw=23387
- Texas City fire and explosion accident (BP), 2005—http://www.csb.gov/assets/document/baker_panel_report1.pdf
- Cougar helicopter crash (New Foundland), 2009—http://www.tsb.gc.ca/eng/rapports-reports/aviation/2009/a09a0016/a09a0016.pdf

- Helicopter ditching in North Sea (UK), 2009—http://www.aaib.gov.uk/cms_resources.cfm?file=/AAR%201-2011%20Eurocopter%20EC225%20LP%20Super%20Puma,%20G-REDU%2010-11.pdf
- Helicopter crash in North Sea (UK), 2009—http://www.aaib.gov.uk/cms_resources.cfm?file=/2-2011%20G-REDL.pdf
- Macondo burning blowout (BP), 2010—http://www.oilspillcommission.gov/chief-counsels-report   http://www.bp.com/liveassets/bp_internet/globalbp/globalbp_uk_english/gom_response/STAGING/local_assets/downloads_pdfs/Deepwater_Horizon_Accident_Investigation_Report.pdf   http://www.oilspillcommission.gov/final-report   http://ccrm.berkeley.edu/pdfs_papers/bea_pdfs/DHSG_ThirdProgressReportFinal.pdf   http://www.deepwater.com/_filelib/FileCabinet/pdfs/00_TRANSOCEAN_Vol_1.pdf   http://www.deepwater.com/_filelib/FileCabinet/pdfs/12_TRANSOCEAN_Vol_2.pdf
- Montara blowout (PTTEP, Australasia), 2009—
  Volume 1
  https://www.google.com/url?q=https://www.nopsema.gov.au/assets/Safety-resources/REPORT-Montara-Investigation-Volume-One.pdf&sa=U&ved=0ahUKEwjh0arD2ajfAhUBpCwKHVupCmQQFggLMAI&client=internal-uds-cse&cx=012550091042421336782:mgkeaimjxqc&usg=AOvVaw3oo59QDIs6cXU3Yv15JzQb
  Volume 2
  https://www.google.com/url?q=https://www.nopsema.gov.au/assets/Safety-resources/REPORT-Montara-Investigation-Volume-Two.pdf&sa=U&ved=0ahUKEwjh0arD2ajfAhUBpCwKHVupCmQQFggWMAY&client=internal-uds-cse&cx=012550091042421336782:mgkeaimjxqc&usg=AOvVaw2Ycj3ifxDiSCMPbbJW35dO
  Volume 3
  https://www.google.com/url?q=https://www.nopsema.gov.au/assets/Safety-resources/REPORT-Montara-Investigation-Volume-Three.pdf&sa=U&ved=0ahUKEwjh0arD2ajfAhUBpCwKHVupCmQQFggOMAM&client=internal-uds-cse&cx=012550091042421336782:mgkeaimjxqc&usg=AOvVaw1JXxy4mum7tZ-0ockIKXrC
- São Mateus explosion (Petrobras, Brazil), 2015—(Brazilian version)
  http://www.anp.gov.br/images/EXPLORACAO_E_PRODUCAO_DE_OLEO_E_GAS/Seguranca_Operacional/Relat_incidentes/Sao_Mateus/relatorio_investigacao_11-02-15.pdf
- Helicopter accident Turøy 29.4.2016 (N), 2016—https://www.aibn.no/Aviation/Published-reports/2018-04?pid=SHT-Report-ReportFile&attach=1

C.4 Investigation Reports—Online Sites

The following is an overview of some sites that regularly publish new investigation reports for offshore incidents and accidents:

- PSA conducted incident and accident investigations—http://www.ptil.no/investigations/category893.html
- Norwegian Maritime Directorate previous investigations—https://www.sdir.no/en/shipping/accidents-and-safety/safety-investigations-and-reports/older-investigations/
- Marine Accident Investigators' International Forum—https://maiif.org/links/members-investigation-reports/
- Helicopter Safety Study (HSS3)—https://norskoljeoggass.no/contentassets/0a1817b142dd4effbc64b7b550266aa6/100610sintefa15753helicoptersafetystudy3hss-3mainreport-100610071828-phpapp02.pdf

C.5 Investigation Reports—No Longer Available Online

It is a disadvantage that some investigation reports for well known accidents in the offshore petroleum sector no longer are available online. Some reports were never available online in the first instance but have been published in the public domain. The following is an overview of some investigation reports for offshore petroleum accidents (& one onshore) on a worldwide scale that are no longer available online (Ocean Ranger report was never there in the first place):

English reports:

- Usumacinta blowout (MX), 2007
- Ocean Ranger accident (CA), 1982

Please contact the authors to obtain access to any of these reports.
Norwegian reports:

- Alexander L. Kielland accident (N), 1980, NOU1981:11—https://www.nb.no/items/URN:NBN:no-nb_digibok_2007062804027
- West Vanguard accident (N), 1985, NOU1986:16—https://www.regjeringen.no/globalassets/upload/kilde/odn/tmp/2002/0034/ddd/pdfv/154614-nou1986-16.pdf.

Glossary

The following definitions are coordinated with NORSOK Z-013 (which reflects ISO terminology (ISO/IEC Guide 73:2002 and ISO31000:2018) where relevant, except that 'risk tolerance criteria' replaces 'risk acceptance criteria', in accordance with what is used internationally.

Accidental Event (AE) Event or a chain of events that may cause loss of life or damage to health assets or the environment.

Accidental Effect The result of an accidental event expressed as heat flux, impact force or energy, acceleration, etc. which is the basis for the safety evaluation.

Acute release The abrupt or sudden release in the form of a discharge emission or exposure, usually due to incidents or accidents.

Area exposed by the accidental event (AEAE) Area(s) on the facility (or its surroundings) exposed by the accidental event.

Area risk Risk personnel located in an area is exposed to during a defined period of time.

As Low as Reasonably Practicable (ALARP) ALARP expresses that the risk shall be reduced to a level that is as low as reasonably practicable.

Average individual risk (AIR) Risk an average individual is exposed to during a defined period of time.

Barrier element Technical operational and organisational measures or solutions involved in the realisation of a barrier function.

Barrier function The task or role of a barrier.

Barrier Performance (or risk) influencing factor Factors identified as having significance for barrier functions and the ability of barrier elements to function as intended.

Barrier system System designed and implemented to perform one or more barrier function.

BLEVE Boiling Liquid Expanding Vapour Explosion is defined as rupture of a hydrocarbon containing vessel due to being heated by fire loads.

Causal analysis The process of determining potential combinations of circumstances leading to a top event.

Consequence Outcome of an event.

Consequence evaluation Assessment of physical effects due to accidents such as fire and explosion loads.

Contingency planning Planning provision of facilities training and drilling for the handling of emergency conditions, including the actual institution of emergency actions.

Control (of hazards) Limiting the extent and/or duration of a hazardous event to prevent escalation.

Cost/benefit evaluation Quantitative assessment and comparison of costs and benefits. In the present context often related to safety measures or environmental protection measures where the benefits are reduced safety or environmental hazard.

Chronic release The continuous or ongoing release in the form of a discharge emission or exposure.

Defined situations of hazard and accident (DSHA) Selection of hazardous and accidental events that will be used for the dimensioning of the emergency preparedness for the activity.

Design Accidental Event Accidental events that serve as the basis for layout dimensioning and use of installations and the activity at large, in order to meet the defined risk tolerance criteria.

Design Accidental Load (DeAL) Chosen accidental load that is to be used as the basis for design.

Dimensioning Accidental Event (DAE) Accidental events that serve as the basis for layout dimensioning and use of installations and the activity at large.

Dimensioning accidental load (DiAL) Most severe accidental load that the function or system shall be able to withstand during a required period of time in order to meet the defined risk tolerance criteria.

Emergency Preparedness Technical operational and organisational measures, including necessary equipment that are planned to be used under the management of the emergency organisation in case hazardous or accidental situations occur, in order to protect human and environmental resources and assets.

Emergency preparedness analysis (EPA) Analysis which includes establishment of DSHA accident including major DAEs, establishment of emergency response strategies and performance requirements for emergency preparedness and identification of emergency preparedness measure, including environmental emergency and response measures.

Emergency preparedness assessment Overall process of performing a emergency preparedness assessment including: establishment of the context performance of the EPA, identification and evaluation of measures and solutions and to recommend strategies and final performance requirements, and to assure that the communication and consultations and monitoring and review activities, performed prior to, during and after the analysis has been executed, are suitable and appropriate with respect to achieving the goals for the assessment.

Emergency preparedness organisation Organisation which is planned established and trained in order to handle occurrences of hazardous or accidental situations.

Emergency preparedness philosophy Overall guidelines and principles for establishment of emergency response based on the operator vision goals, values and principles.

Emergency response Action taken by personnel on or off the installation, to control or mitigate a hazardous event or initiate and execute abandonment.

Emergency response strategy Specific description of emergency response actions for each DSHA.

Environment Surroundings in which an organization operates including air, water, land, natural resources, flora, fauna, humans and their interrelation.

Environmental impact Any change to the environment whether adverse or beneficial, wholly or partially resulting from an organization's activities, products or services.

Environmental resource Includes a stock or a habitat defined as:
Stock: A group of individuals of a stock present in a defined geographical area in a defined period of time.
Alternatively: The sum of individuals within a species which are reproductively isolated within a defined geographical area.
Habitat: A limited area where several species are present and interact. Example: a beach.

Environment Safety Safety relating to protection of the environment from accidental spills which may cause damage.

Escalation Escalation has occurred when the area exposed by the accidental event (AEAE) covers more than one fire area or more than one main area.

Escalation factor Conditions that lead to increased risk due to loss of control mitigation or recovery capabilities.

Escape Actions by personnel on board surface installations (as well as those by divers) taken to avoid the area of accident origin and accident consequences to reach an area where they may remain in shelter.

Escape way Routes of specially designated gangways from the platform leading from hazardous areas to muster areas, lifeboat stations, or shelter area.

Escape route Route from an intermittently manned or permanently manned area of a facility leading to safe area(s).

Establishment of emergency preparedness Systematic process which involves selection and planning of suitable emergency preparedness measures on the basis of risk and emergency preparedness analysis.

Essential safety system System which has a major role in the control and mitigation of accidents and in any subsequent EER activities.

Evacuation Planned method of leaving the facility in an emergency.

Event tree analysis Inductive analysis in order to determine alternative potential scenarios arising from a particular hazardous event. It may be used quantitatively to determine the probability or frequency of different consequences arising from the hazardous event.

Explosion load Time dependent pressure or drag forces generated by violent combustion of a flammable atmosphere.

External escalation When the area exposed by the accidental event (AEAE) covers more than one main area external escalation has occurred.

Facility Offshore or onshore petroleum installation facility or plant for production of oil and gas.

Fault Tree Analysis Deductive quantitative analysis technique in order to identify the causes of failures and accidents and quantify the probability of these.

Fire area Area separated from other areas on the facility either by physical barriers (fire/blast partition) or distance, which will prevent a dimensioning fire to escalate.

Functional requirements to safety and emergency preparedness Verifiable requirements to the effectiveness of safety and emergency preparedness measures which shall ensure that safety objectives risk tolerance criteria, authority minimum requirements, and established norms are satisfied during design and operation.

Group individual risk (GIR) Average IR for a defined group.

Hazard Potential source of harm.

Hazardous event Incident which occurs when a hazard is realized.

Immediate vicinity of the scene of accident Main area(s) where an accidental event (AE) has its origin.

Individual risk (IR) Risk an individual is exposed to during a defined period of time.

Inherently safer design In inherently safer design the following concepts are used to reduce risk:
- reduction, e.g. reducing the hazardous inventories or the frequency or duration of exposure
- substitution, e.g. substituting hazardous materials with less hazardous ones (but recognizing that there could be some trade-offs here between plant safety and the wider product and lifecycle issues)
- attenuation, e.g. using the hazardous materials or processes in a way that limits their hazard potential, such as segregating the process plant into smaller sections using ESD valves, processing at lower temperature or pressure
- simplifications, e.g. making the plant and process simpler to design, build and operate, hence less prone to equipment, control and human failure.

Internal control All administrative measures which are implemented to ensure that the work is in accordance with all requirements and specifications.

Internal escalation When the area exposed by the accidental event (AEAE) covers more than one fire area within the same main area internal escalation has occurred.

Main area Defined part of the facility with a specific functionality and/or level of risk.

Main load bearing structures Structure which when it loses its main load carrying capacity, may result in a collapse or loss of either the main structure of the installation or the main support frames for the deck.

Main safety function Most important safety functions that need to be intact in order to ensure the safety for personnel and/or to limit pollution.

Major accident Acute occurrence of an event such as a major emission fire, or explosion, which immediately or delayed, leads to serious consequences to human health and/or fatalities and/or environmental damage and/or larger economical losses.

Material Damage Safety Safety of the installation its structure, and equipment relating to accidental consequences in terms of production delay and reconstruction of equipment and structures.

Mitigation Limitation of any negative consequence of a particular event.

MOB-boat Man Over Board Boat.

Muster Station A place where personnel may gather in a Safe Haven prior to evacuation or abandonment from emergency situations.

Muster area Area on the platform where the personnel may be sheltered from accidental conditions until they embark into the lifeboats.

Normalisation The normalisation phase starts when the development of a situation of hazard or accident has stopped.

Occupational Accidents Accidents relating to hazards that are associated with the work places (falls slips, crushing etc.), thus other hazards than hydrocarbon gas or oil under pressure. These accidents are normally related to a single individual.

Performance requirements for safety and emergency preparedness Requirements to the performance of safety and emergency preparedness measures which ensure that safety objectives RAC, authority minimum requirements and established norms are satisfied during design and operation.

Personnel Safety Safety for all personnel involved in the operation of a field.

Probability Extent to which an event is likely to occur.

Recovery time Time from an accidental event causing environmental damage occurs until the biological features have recovered to a pre-spill state or to a new stable state taking into consideration natural ecological variations and are providing ecosystem services comparable to the pre-spill services.

Reliability Analysis Analysis of causes and conditions of failure inspection, maintenance and repair, and the quantitative assessment of up-times and down-times.

Residual Accidental Event Accidental event which the installation is not designed against therefore it will be part of the risk level for the installation.

Residual risk Risk remaining after risk treatment.

Risk PSA (2015): Consequences of the activities with associated uncertaintyTraditionally: Combination of the probability of occurrence of harm and the severity of that harm.

Risk acceptance Decision to accept risk.

Risk analysis Structured use of available information to identify hazards and to describe risk.

Risk assessment Overall process of performing hazard identification risk analysis and risk evaluation.

Risk avoidance Decision not to become involved in or action to withdraw from, a situation that involves risk.

Risk control Actions implementing risk management decisions.

Risk evaluation Judgement on the basis of risk analysis, sometimes involving RAC, of whether the risk is tolerable or not

Risk identification Process to seek for list and characterise elements of risk.

Risk management Coordinated activities to direct and control an organisation with regard to risk.

Risk management system Set of elements of an organisation's management system concerned with managing risk.

Risk perception Way in which a stakeholder views a risk based on a set of values or concerns.

Risk picture Synthesis of the risk assessment with the intention to provide useful and understandable information to relevant decision makers.

Risk reduction Actions taken to reduce the probability negative consequences, or both, associated with a hazard.

Risk tolerance criteria (RAC) Criteria that are used to express a risk level that is considered as the upper limit for the activity in question to be tolerable.

Risk transfer Sharing with another party the potential burden of loss or benefit of gain. (Contracts and insurance are two examples).

Risk treatment Process of selection and implementation of measures to modify risk.

Rooms of significance to combating accidental events CCR and other equivalent room(s) that are essential for safe shutdown blowdown and emergency response.

Safe area(s) Area(s) which depending on each specific defined situation of hazard and accident (DSHA), are defined as safe until the personnel are evacuated or the situation is normalized.

Safety barrier A measure intended to identify conditions that may lead to failure hazard and accident situations, prevent an actual sequence of events occurring or developing, influence a sequence of events in a deliberate way, or limit damage and/or loss.

Safety function Measures which reduce the probability of a situation of hazard and accident occurring or which limit the consequences of an accident.

Safety Goals Concrete targets against which the operations of installations at the field are measured with respect to safety. These targets shall contribute to avoidance of accidents or resistance against accidental consequences.

Safety objective Objective for the safety of personnel environment and assets towards which the management of the activity will be aimed.

Serious Accidents See major accidents.

Shelter Area An area on the platform where the crew will remain safe for a specific period of time in an emergency situation.

Stakeholder Any individual group or organization that can affect, be affected by, or perceive itself to be affected by, a hazard.

System Common expression for installation(s) plant(s), system(s), activity/activities, operation(s) and/or phase(s) subjected to the risk and/or emergency preparedness assessment.

System basis Inputs (regarding the system subjected to assessment) used as basis for the assessment.

System boundaries System boundaries defines what shall and what shall not be subjected to the assessment.

Working Accidents Accidents relating to other hazards than hydrocarbon gas or oil under pressure (falls crushing etc.) normally related to a single individual.

Worst case consequence The worst possible consequences to health environment and safety resulting from a hazardous event. For this to occur, all critical defences in place must have failed.

Index

Printed in the United States
By Bookmasters